Abiotic Stress and Biotechnology

The Editor

Prof. T. Pullaiah obtained his M.Sc. (1973) and Ph.D. (1976) degrees in Botany from Andhra University. He was a Post Doctoral Fellow at Moscow State University, Russia during 1976-78. He traveled widely in Europe and USA and visited Universities and Botanical Gardens in about 17 countries. Professor Pullaiah joined Sri Krishnadevaraya University as Lecturer in 1979 and became Professor in 1993. He held several positions in the University, which include Dean, Faculty of Life Sciences, Head of the Department of Botany, Chairman, BOS in Botany, Head of the Department of Sericulture, Co-ordinator and Chairman, BOS in Biotechnology, Vice Principal and Principal, S.K. University College. He retired from active service on 31st May 2011. He was selected by UGC as UGC-BSR Faculty Fellow and is working in Sri Krishnadevaraya University. Prof. Pullaiah has published 49 books, 265 research papers and 35 popular articles. His books include *Flora of Andhra Pradesh* (5 volumes), *Flora of Eastern Ghats* (Vols. 1-4), *Biodiversity in India* (Vols. 1-5), *Herbal Antioxidants* (3 volumes) *Taxonomy of Angiosperms* (presently in 3rd edition), *Plant Development, Plant Reproduction, Plant Tissue Culture, Flora of Guntur district, Flora of Kurnool, Flora of Anantapur district, Flora of Nizamabad, Flora of Ranga Reddi district* etc. He is Principal Investigator of 20 major Research Projects totaling more than a Crore of Rupees funded by DBT, DST, CSIR, UGC, BSI, WWF, GCC etc. Under his guidance 51 students obtained their Ph.D. degrees and 35 students their M.Phil. degrees. He is recipient of Best Teacher Award from Government of Andhra Pradesh, Prof. P. Maheswari Gold Medal and Prof. G. Panigrahi Memorial Lecture of Indian Botanical Society and Prof. Y.D. Tyagi Gold medal of Indian Association for Angiosperm Taxonomy. He was Vice President of Indian Botanical Society. He was Member of Species Survival Commission of International Union for Conservation of Nature and Natural Resources (IUCN). Presently he is President of Indian Association for Angiosperm Taxonomy.

Abiotic Stress and Biotechnology

—Editor—
T. Pullaiah
Department of Botany
Sri Krishnadevaraya University
Anantapur – 515 003
Andhra Pradesh

2013
Regency Publications
A Division of
Astral International Pvt. Ltd.
New Delhi – 110 002

Published by   :   **REGENCY PUBLICATIONS**
                      A Division of
                      Astral International Pvt. Ltd.
                      ISO 9001:2008 Certified Company
                      4760-61/23, Ansari Road, Darya Ganj
                      New Delhi-110 002
                      Ph. 011-43549197, 23278134
                      E-mail: info@astralint.com
                      Website: www.astralint.com

Laser Typesetting   :   **Classic Computer Services**
                           Delhi - 110 035

Printed at   :   **Salasar Imaging Systems**
                    Delhi - 110 035

PRINTED IN INDIA

Preface

Biotechnology has made rapid strides in recent years. It has solved many problems in Agriculture. One of these aspects is Abiotic stress. Two chapters deal with Molecular mechanisms of Abiotic stress and Genetic engineering of plants for abiotic stress tolerance. Methods like Real Time PCR, Subtractive Hybridization and High Throughput DNA sequencing are also given. Other aspects include micropropagation and Production of Secondary Plant products. Biofuel from Algae and Marine actinobacteria in Human health are also included.

We would like to continue the series on Biotechnology. Authors are welcome to send articles on different aspects of Biotechnology for inclusion in future volumes.

T. Pullaiah

Contents

2013, Abiotic Stress and Biotechnology
Editor: T. Pullaiah
Published by: REGENCY PUBLICATIONS, NEW DELHI

Pages 1–28

Chapter 1

Molecular Mechanism Involved in Abiotic Stress Tolerance in Plant

Tapan Kumar Mondal* and Dipnarayan Saha

National Research Centre on DNA Fingerprinting,
National Bureau of Plant Genetic Resources (ICAR),
Pusa Campus, New Delhi – 110 012

ABSTRACT

Agricultural productivity is largely influenced by unprecedented climate changes, leading to heavy crop yield loss and is a serious concern worldwide. The environmental extremes such as high and low temperature, drought, salinity, heavy metals radiation and high and low nutrient etc. are the consequence of rapidly changing climate, which renders havoc in crop yield and productivity. Globally researchers involved in agricultural biotechnology have focused on addressing the challenges of abiotic stress in plants by the expression, over-expression or switching off abiotic stress-related genes. Despite rapid advancements in biotechnological tools and research strategies, several crops are still susceptible to abiotic stresses and are estimated to reduce yield by 15 to 32 per cent in the next fifty years. Thus there is an urgent need to accelerate our research efforts towards engineering genes that would protect and maintain the function and structure of cellular components which can impart tolerance to stress. In this chapter we present the state of art status, principal methods adapted in the control of plants abiotic stresses.

* Corresponding Author: E-mail: mondaltk@yahoo.com

Introduction

"In a world that has the means for feeding its population, the persistence of hunger is a scandal" (FAO, 2006). While overwhelming population growth is the driving force for this comprehensive statement, yet yield of food grain less than the requirement is one of the causes. Today, yield penalty due to climate change is a major concern for 'hunger free world': a dream of noted agriculturist Dr Borlaug. Plant being sessile, a 'good care and management' is prerequisite to harness its full genetic potential for the product they produce. Importantly, abiotic stress due to adverse climate is the major bottlenecks for 'good care and management' to agriculture, which is direct consequences of climate changes at large so that it accounts for 50 per cent of crop yield losses worldwide (Mittlar and Blumwald, 2010). Some of the climatic factors such as water, temperature, light and nutrient are necessary for plant growth and development; become stressful when they exceed certain limits. Other factors, such as, UV-radiation, unfavourable soil condition such as soil acidity, salinity, sodicity, alkalinity, toxic metals, poor irrigation, air pollutant (*e.g.*, ozone) etc. which are not directly involved in plant growth may also become stressful when they exceed the limits. These are the crop productivity-related stresses emerged due to rapid changes in climate, which could override the adaptive potential of plants. Thus understanding of various existing as well as emerging stress, the physiological basis of such abiotic stress, their genetic basis of tolerance, gene manipulation through integrated breeding as well as modern genetic engineering will be immensely useful to overcome the situation in a proactive manner.

Abiotic stresses can be defined as any deviations in normal physiology, development, function of the plants which is detrimental, injurious and can cause a permanent damage to cell, tissue or organ due to non-living cause. Based upon the economic loss they incur, various abiotic stresses may be grouped into i) major stresses such as drought, salinity, cold and soil acidity etc. and ii) emerging stresses such as elevated CO_2, soil nutrient deficiency such as zinc deficiency etc. While the former group of stresses has a bigger impact on crop productivity, thus are well known and catch attention easily, the potential threat from the later group is often neglected but cannot be ignored for long and may be considered equally important. The fact is more pertinent since many of these abiotic stresses occur together and in that process inflict severe yield losses.

Drought or moisture deficiency results when the soil moisture status drops down below critical level and the water absorption by root and water transport within the plant is reduced. It is considered to be the main abiotic factor as it affects 26 per cent of the arable area and causes as high as 50 per cent yield loss in some crop. The tolerance mechanisms of plant to water deficit are mainly through escape, avoidance or tolerance strategies (Turner, 1986).

The detrimental effects of salinity in agricultural soils accounts for 6 per cent of total arable land, mainly found amongst coastal salt marshes or inland desert sands. Secondary salinization occurs from excessive irrigation schedules and tree felling in agricultural land cause water tables to rise and concentrate salts in the crop plant root zone. Approximately 20 per cent of the world's irrigated land, from which one-third of the world's food supply is produced, is presently affected by salinity (Rengasamy, 2006).

Similarly, insufficient or excess of nutrients present in the soil has adverse impacts on plant growth and development (Sunkar, 2010). Numerous genes in roots and shoots play an important role in nutrient acquisition, assimilation and distribution. Micronutrients deficiencies, such as zinc (Zn), iron (Fe) etc. are one of the most widespread crop micronutrient deficiencies which are capable of causing serious yield reductions. These micronutrients are often deficient in human nutrition too, particularly in developing countries and among the vegetarians, thus emphasizing the need of producing food grains with elevated Zn or Fe content.

Acid soils significantly limit crop production worldwide because approximately 50 per cent of the world's potentially arable soils are acidic. The primary limitations on acid soils are due to toxic levels of aluminum (Al) and manganese (Mn), as well as suboptimal levels of phosphorus (P). Aluminum toxicity directly interferes by impairing the root growth. The phosphate ion (Pi) availability is particularly limiting on the highly weathered, acid soils of the tropics and subtropics, due to its fixation with Al and Fe oxides on the surface of clay minerals. Hence, Pi availability is a major factor limiting crop production on acid soils (Kochian *et al.*, 2004).

Other stresses such as the low temperature stress accounts about 15 per cent crop area worldwide and such stress mainly prevails in temperate zone. Heat stress often accompanies drought stress, and the two interact in the way they affect plants. Excessive heat perturbs many cellular and developmental processes and directly affects grain production by reducing fertility and grain quality (Barnaba *et al.*, 2008). Water submergence of plants causes low oxygen availability, which triggers adaptive strategies that alter metabolism, morphology, and physiology. It also causes hypoxia (low-oxygen) stress, which interferes with mitochondrial respiration, normally resulting from water logging condition affecting 13 per cent of the agrarian land of the world (Cramer *et al.*, 2011). Radiation such as ultraviolet-B (UV-B, 280–320 nm) on the earth's surface has been increasing due to increasing depletion of the stratospheric ozone layer. Elevated UV-B is known to negatively affect plant growth and development, which directly/indirectly accelerates the generation and accumulation of excessive ROS (Sunkar, 2010). Ozone (O_3) is another important phytotoxic air pollutant that forms due to the interaction of hydrocarbons or nitrogen oxides in the presence of sunlight. Its concentration in the troposphere has increased significantly at a rate of 1–2 per cent per year during the last century. After entering through stomata, it generates an oxidative burst that leads to an excess of reactive oxygen species (ROS) subsequent leading to foliar injuries besides biochemical and physiological changes (Dizengremel *et al.*, 2009).

Plant Responses to Abiotic Stresses

Plant responses to abiotic stress are complex and are reported to be governed by multiple genes from all the three different genome (nucleus, chloroplast and mitochondria) at different level. Additionally the stresses complex; *i.e.* they do not occur singly and often they are related. For an example drought and high temperature, flooding and salinization are coupled with each other. The responses of plants are both elastic (reversible) and plastic (irreversible). However, there are some commonalities of the plant response in different stresses which is schematically represented in Figure 1.1.

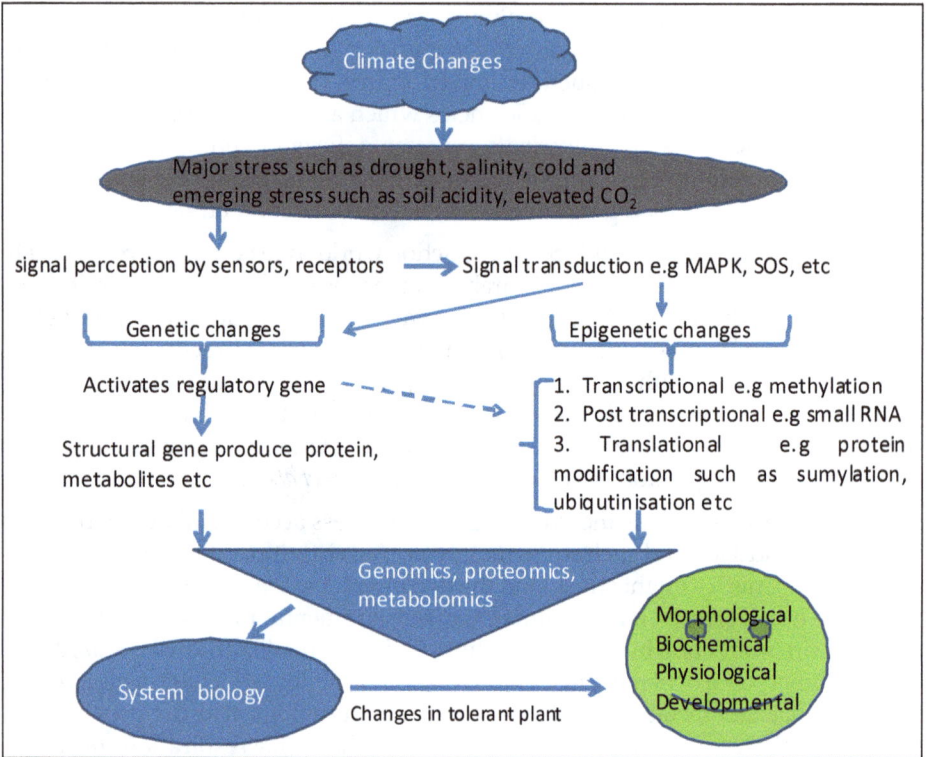

Figure 1.1: Complexity of plant response to abiotic stress. Climate change often induces the primary stresses, such as drought, salinity, cold, heat and chemical pollution. The initial stress signals (*e.g.* osmotic and ionic effects, or temperature, membrane fluidity changes) trigger the downstream signal transduction process and transcription controls which activate stress-responsive mechanisms to re-establish homeostasis, protect and repair damaged proteins and membranes. Some signals through regulatory gene also induce the epigenetic changes. Three different 'omics'; *i.e.* genomics (gene), proteomics and metabolomics together bring down the several visible changes for tolerance to a trait.

Plant adaptation to environmental stresses is controlled by cascades of molecular networks. The ultimate result of stress responsive mechanisms is to re-establish homeostasis and to protect and repair damaged proteins, macromolecules and membranes. The first step of acclimation of a cell is perception of the stress which normally done either by various sensor molecules available in the plant cell or by conformational changes of the protein located in the cell wall. Importantly, the sensor molecule is highly specific for a particular environmental signal. The signal is then transduced from the cell wall and proceeds into deeper cytoplasm through a chain of protein molecules called transcription factor. This group of protein then regulates the expression of desirable proteins or metabolites involved in the response to abiotic stresses. Some time the transcription factor activates a group of genes which in turn produce non-coding RNA molecules which are also actively involved in gene expression related to environmental stresses. Thus gene expressions under abiotic stresses are complex phenomenon, highly co-ordinated among the group of genes and the expressions are inducible in nature (Zhang *et al.*, 2004).

Genetic Engineering Approaches for Abiotic Stress Tolerance

Compared to classical breeding, transgenic development is an alternative technology to improve crop yield. The development of tolerant crops by genetic engineering requires the identification of key genetic determinants underlying stress tolerance in plants and introducing these genes into crops. Introduction of molecular change by genetic engineering takes less time compared to plant breeding methods, only desired gene(s) can be transferred, whereas, conventional breeding approach is associated with simultaneous transfer of undesired gene(s). For re-cultivation of degraded soils and reclamation of industrial sites, stress tolerant plants are required. The advent of plant transformation may have placed within the grasp of the possibility of engineering greater abiotic stress tolerance in plants.

Transformation of gene from any source to plant is perhaps the most powerful techniques for producing improved cultivars. Although enlisting the entire abiotic stress related genes of plant are beyond the scope of this chapter but reader may refer several reviews for the same purpose (Amudha and Balasubramani, 2011). It can be seen from the published literatures that a wide range of gene has been transferred which are discussed below:

Engineering Genes for Stress Signaling

Tolerance response of a plant cell starts from perception of stress signals. However, till now only few stress sensor molecules have been identified (Reguera *et al.*, 2011). The two component trans-member histidine kinase in *Arabidopsis* (AtHK1) has been identified as osmosensor (Urao *et al.*, 1999) and upon transferring into the plant cell; it shows better tolerance in salinity. Transgenic tobacco with receptor-like kinase has shown better tolerance in wound, salt and osmotic stress indicating a role of cross talk between the different stress signaling pathways (Tamura *et al.*, 2003).

MAP-kinase cascades carried information from sensors to cellular responses in all eukaryotes including plants. Activated MAPK is imported to the nucleus where it activates with the specific transcription factors. Transgenic tobacco plants constitutively expressed tobacco MAPKKK (NP1), registered significant tolerance to various abiotic stresses such as heat, freezing, drought and salt (Kovtun *et al.*, 2000). Similarly, it has been documented that ectopically expressed MAPK2 (*MKK2*) in *Arabidopsis* confer tolerance to freezing and salt tolerance (Singh *et al.*, 2010).

Calcium-dependent protein kinase (CDPK) is another group of sensor protein, which plays important role to perceive the signals of abiotic stress. Transgenic rice with over-expressed OsCDPK7 was found to be more tolerant in salt, drought and cold (Saijo *et al.*, 2000). Later, another rice gene CDPK13 were found to be effective in producing cold tolerance (Komatsu *et al.*, 2007). Ca^{++} is well-known secondary messenger in signal transduction. Several other Ca^{++} dependent proteins such as

Calmodulin, Calmodulin like protein, Calcineurin B like protein found to play important role in abiotic stress (White and Broadly, 2003). Transgenic *Arabidopsis* with Calmodulin confer salt tolerance by inducing pyrroline-5-carboxylate synthetase 1, a rate limiting enzyme in proline biosynthesis pathway (Yoo *et al.*, 2005).

Another important pathway that has been discovered to be specific to salt stress is 'salt overlay sensitive' (SOS) pathway. Salt stress induces calcium signal which is perceived by SOS3: a myristolylated calcium binding protein. SOS3 then interact and activates SOS2 which later regulates the expression of SOS1 which is a Na^+/H^+ antiporter. This land mark discovery of Prof J K Zhu and his team at University of California (Zhu, 2002) drew the attention of several workers for producing transgenic plant with SOS pathway genes. Subsequently, it was demonstrated that orthologus sequences of SOS1 and SOS2 in *Arabidopsis*, rice and poplar confer tolerance to salt stress (Wu *et al.*, 2007).

Engineering Genes for Transcriptional Regulation

Expression of the structural gene is governed by the transcriptional rate *i.e.*, binding the transcription factor into an upstream sequence, known as *cis* elements in a sequence-specific manner. Therefore, several workers focus their attention to manipulate transcription factor over-expression to increase the rate of gene expression and enhance protein production.

Ethylene-responsive factors (ERFs) are plant-specific transcription factor that play important role under abiotic stress. Such genes of tomato plant when over-expressed in tobacco, it demonstrated better salt tolerance (Zhang *et al.*, 2004). Transgenic plant from a number of crop with similar gene such as TERF1 (Huang *et al.*, 2004) and JERF3 (Wang *et al.*, 2004) from tomato, ERF like protein of hot pepper (CaERFLP1), wheat ERF1 (TaERF1), barley ERF were found to respond better under salt stress. Recently, it has been shown that transcription factor HARDY with AP2/ERF domain of *Arabidopsis* were transformed into rice, which showed better performance under drought stress (Karaba *et al.*, 2007).

The dehydration responsive elements/C-repeat was identified as *cis* elements which regulates gene expression under salt, drought and cold in *Arabidopsis* (Yamaguchi-Shinozaki and Shinozaki, 1994). Since then, several DREB genes have been cloned and transgenic plants have been produced for better performance in various abiotic stresses (Agrawal *et al.*, 2006). It has been reported that both *rd22BP1* (MYC) and AtMYB2 (MYB) protein of *Arabidopsis* function as transcriptional activators under dehydration and ABA inducible expression of rd22 (Abe *et al.*, 1997). Subsequently, over-expression of *Arabidopsis* MYB2 gene in rice conferred salt stress tolerance in transgenic rice (Malik and Wu, 2005).

Although the NAC domain transcription factors are key regulators of the differentiation of xylem tissues, they also play important role in abiotic stress tolerance in plant. Transgenic rice with NAC1 gene found to be better drought tolerance under field condition (Hu *et al.*, 2006). Recently, it has been found that over-expression of rice OsNAC6 gene also confer tolerance to dehydration and high salt stress (Nakashima *et al.*, 2007). Zinc finger protein is another important transcription factor which has C_2H_2 domain. A soybean C_2H_2 type ZFP, SCOF-1 was isolated under cold stress, which when transferred to *Arabidopsis* and tomato, conferred cold tolerance (Kim *et al.*, 2001). Transgenic *Arabidopsis* with ZAP12, a zinc finger protein found to be tolerant against H_2O_2 stress, suggested its involvement in ROS and abiotic stress signaling in *Arabidopsis* (Davletova *et al.*, 2005).

Engineering for Redox Regulation

Production of reactive oxygen species (ROS) such as superoxide, hydrogen peroxide and free radicals and reactive nitrogen species (RNS) such as NO⁻ etc are the secondary stress in living cells. These reactive species can oxidize the biological macro-molecules such as DNA, protein, carbohydrate, lipids etc. leading to death of the cell. Several scavenging enzymes namely superoxide dismutase, peroxidase, glutathione reductase etc. along with non-enzymatic antioxidants are operative in plants system (Amudha and Balasubramani, 2011). Several attempts have been made to over-express the anti-oxidative enzymes or antioxidants to overcome high temperature (McKersie *et al.*, 2000), salt (Prashanth *et al.*, 2007), drought (Wang *et al.*, 2005), oxidative (Bowler *et al.*, 1991) and others stresses. It has been suggested that Super Oxide Dismutase (SOD) may be a potential candidate gene to engineer plants for multiple stress tolerance. Similarly over-expression of catalase in wheat (Matsumura *et al.*, 2002) for tolerance to chilling and salt, ascorbate peroxidase (Wang *et al.*, 2005) for salt stress, glutathion-S-transferase to overcome UV radiation (Liu and Li, 2002), oxidative stress (Katsuhara *et al.*, 2005), and salinity (Qi *et al.*, 2004) have been reported successful. Glutathion peroxidase (GPX) is the principal cellular enzyme responsible for repairing membrane lipid peroxidation and is generally considered to be the main line of enzymatic defense against oxidative damage. Transgenic tobacco plant with the ectopic expression of tobacco GPX (Roxas *et al.*, 2000) and *Chlamydomonas* GPX (Yoshimura *et al.*, 2004) showed better response when exposed to chilling and salt stress.

Engineering for Osmotic Regulation

Higher accumulation of osmolytes under drought, salt and cold stress is well known mechanism in plant. Glycine betain (N, N, N-trimethyl glycine) is believed to protect membranes and macromolecules from detrimental effects under stresses. It has been demonstrated the various genes in glycine betain biosynthetic pathway such as choline mono-oxygenase (Shirawa *et al.*, 2006), betain aldehyde dehydrogenase (Yang *et al.*, 2005), choline dehydrogenase (Quan *et al.*, 2004) and choline oxidase (Sulpice *et al.*, 2003) enhanced the accumulation of glycine and hence transgenic plant respond to better growth under abiotic stresses. Trehalose, a non-coding disaccharide of glucose play important role in abiotic stress tolerance (Higashiyama, 2002). Several genes in the trehalose synthesis pathways have improved the stress tolerance in plants (Jang *et al.*, 2003, Karim *et al.*, 2007). Polyamines are low-molecular weight polycations which function in one of the most important nitrogenous osmolyte production pathway under osmotic stress. Genes in polyamine biosynthetic pathways such as arginine decarboxylase (Capell *et al.*, 2004), ornithine decarboxylase (Minocha and Sun, 1997) and S-adenosylmethionine decarboxylase (Waie and Rajam, 2003) found to confer better tolerance in various transgenic plants.

The over-expression of osmotin gene in transgenic tobacco resulted in higher accumulation of proline and conferred osmotic stress in transgenic tobacco (Borthakur *et al.*, 2001). Sugar alcohol polyols also play important role for tolerance of abiotic stress. Polyols are both straight chain metabolites such as mannitols and sorbitols or cyclic polyols such as myo-inositol and its methylated derivatives such as ononitol and pinitol accumulated in higher quantity under salinity and drought (Bohnert and Jensen, 1996). Several genes in this biosynthetic pathway have been cloned and over-expressed in plants to found suitable for imparting tolerance under dehydration stress (Singh *et al.*, 2010). L-myo-inositol and its derivatives are commonly associated with cell signaling and membrane biogenesis but found to be associated with salt tolerance (Nelson *et al.*, 1998). A *PINO1* gene encoding the MIP protein from wild halophytic rice *Porteresia coarctata* rendered salt tolerance in transgenic tobacco (Majee *et al.*, 2004). Tobacco plants transformed with *IMT1* encoding myo-inositol-Oomethyl transferase increased the accumulation of D ononitol and found to be efficent under salt and drought condition (Sheveleva *et al.*, 1997).

Ectoine (1, 4, 5, 6-tetrahydro-2-methyl 4-pyrimidine carboxylic acid) act as an enzyme protectant against heat, freezing dehydration (Lippert and Galinski, 1992). Recently transgenic tobacco with ectoine biosynthetic gene from *Halomonas elongata* was found to accumulate higher ectoine and hence rendered efficient salt tolerance (Moghaieb *et al.*, 2006).

Engineering for Cellular Protection

Late embryogenesis proteins (LEA) are associated with tolerance to moisture stress that occurs primarily due to desiccation, heat or as secondary effect of cold stress. They are synthesized and accumulated during late stage of seed development in vegetative tissue under drought, heat, cold, salt stress or abscisic acid (Sivamani *et al.*, 2000). Their functions are known as a molecular chaperon or space filling to prevent cellular collapse at low water potential. LEA protein was first cloned from cotton (Dure and Galau, 1981) and subsequently many LEA genes were characterized from different plant species (Dure, 1992). Transgenic rice with wheat *LEA* genes such as PMA80 and PMA1959 were found to be better under dehydration (Cheng *et al.*, 2002). Similarly, transgenic radish (*Rapahanus sativus* L.) with a type 3 *LEA* gene from *Brassica napus* has been reported to confer tolerance against water deficiency as well as salt stress (Park *et al.*, 2005).

Dehydrins, a group II LEA protein, highly hydrophilic and thermo stable which acts not only as structural stabilizer but also have detergent and chaperone like properties. They are found in different organelle within plant cell *i.e.* chloroplast, mitochondria and nucleolus. It has been found that dehydrin gene from spinach (Kaye *et al.*, 1998) and citrus (Hara *et al.*, 2003) demonstrate better tolerance against freezing stress tolerance in transgenic tobacco. Interestingly, wheat dehydrin (DHN5) when over-expressed in *Arabidopsis* it was found that the transgenic plants performed better under high NaCl concentration or water dehydration. The above work also demonstrated that dehydrin facilitates plant cold acclimation by acting as radical scavenging protein to protect membrane stabilization (Brini *et al.*, 2007).

Engineering for Ionic Balance

Ion homeostasis is a strategy which is adopted by plant to survive both under toxic level as well as deficiency level of ion under various nutrient as well as salt stress. Under salinity condition, it is required to prevent Na^+ accumulation to a high level in cytoplasm or in organelles other than the vacuole, because excess Na^+ inhibits many cellular activities (Zhu, 2002). Several proton pumps, antiporters and ion transporters from various plants found to improve the trait of interest in transgenic background mainly for overcoming the growth either under deficiency or toxic soil (Amudha and Balasubramani, 2011). Most importantly, compartmentalization of Na^+ into vacuoles occurred via vacuolar Na^+/H^+ antiporter (NHX) gene which is well characterized in plant system.

Several transgenic plant with ecotopic expression of *Arabidopsis* NHX (*AtNHX1*) genes such as *Brassica* (Zhang *et al.*, 2001), tomato (Zhang and Blumwald, 2001), wheat (Xue *et al.*, 2004), maize (Xiao-Yan *et al.*, 2004), rice (Fukuda *et al.*, 2004), *Suaeda salsa* (Zhao *et al.*, 2006), cotton (Wu *et al.*, 2004), soybean (Li *et al.*, 2006) have been demonstrated with higher tolerance of salt. Even Na^+/H^+ antiporter from yeast (*Schizosaccharomyces pombe*) increased salt tolerance in rice (Zhao *et al.*, 2006). Another important halotolerence gene (*HAL1*) confered salt tolerance in melon (Bordas *et al.*, 1997), tomato (Zhang *et al.*, 2001), *Arabidopsis* (Yang *et al.*, 2001) and watermelon (Ellul *et al.*, 2003).

A vacuolar H^+-pyrophosphatase encoded by a single gene (*AVP1*) when over expressed in *Arabidopsis* conferred salt tolerance, which has been shown to be associated with auxin transport (Gaxiola *et al.*, 2001). Recently H^+ pyrophosphatase gene from *Thellungiella halophila* was over-expressed in tobacco to find that the transgenic tobacco plants accumulated 25 per cent more Na^+ (Gao *et al.*, 2006).

Aquaporins are the protein channel of membrane which transports the water molecule. They play pivotal role for plant-water relation. It has been shown that transgenic tobacco plant with *NtAQP1* gene showed reduced root hydraulic conductivity and lower water stress tolerance, which suggested the importance of symplastic aquaporin-mediated water transport (Siefritz *et al.*, 2002). Interestingly, it has been found that two *Arabidopsis* plasma membrane aquaporin genes *PIP1:4* and *PIP2:5* enhance water flow and improve seed germination under water stress condition whereas these genes fail to perform under control condition (Jang *et al.*, 2003).

Calcium exchanger (CAX) gene encodes a Ca^+/H^+ antiporter. Interestingly, upon over expressing a soybean CAX1 gene in *Arabidopsis* confered the less accumulation of Na^+, K^+, Li^+ ion in transgenic plant grew under elevated concentrations of Na^+, K^+, Li^+ indicating the fact of enhancing the ion exclusion capability of the plant. Thus they postulated the possible beneficial role of this gene for salt tolerance (Hirschi *et al.*, 1996). Subsequently, several CAX genes of numerous plants have been cloned and used to produce transgenic plants for better Ca^+ sequestration (Singh, 2010).

Understanding of Abiotic Stress in System Biology Prospective

Although several genes have been cloned and functionally validated for their possible role in stress tolerance, yet very few transgenic plants have been released for commercial cultivation, indicating the fact that single gene approaches may not be suitable for tackling abiotic traits. Additionally, it is well understood that plant responses under abiotic stress are complex as abiotic traits are normally polygenic in nature. In addition, the level and duration of stress (acute vs. chronic) may have a significant effect on the complexity of the response (Tattersall *et al.*, 2007; Pinheiro and Chaves, 2011). The complexity of stress tolerance becomes more complex after the recent discovery of the epigenetic changes associated with these stresses. Plant cells are highly complex. For example there are 100,000 cell/cm^2 in the 7th leaf of *Arabidopsis* and plant cell owes three genome *i.e.* nucleolus, chloroplast and mitochondria. A typical plant cell has 30,000 genes which produce manifold higher number of protein which also undergo around 200 different types of post translational modification, yet many more to be discovered. Many classical experiment have shown that often these three genome are involved together to combat the environmental extremes and there are cross-talk among the different genes *i.e.*, often a single gene are involved in more than one trait and more than one genes are expressed under a single trait. Another important aspect of abiotic traits is that multiple factors limits the plant growth and there are several over lapping and interdependency among the various abiotic stress.

In the post genomics era, the three systemic approaches namely genomics, proteomics and metabolomics have increased our understanding. In addition computational biology generated several algorithms, statistical tools which enhanced our understanding and identification of group of genes expressed in given conditions by integrating these three 'omics'. One such important tool, co-expression analysis which identifies the regulatory hubs, is very promising for understanding gene-gene correlations. Another experimental approach known as 'time course' facilitates understanding of the multiple phases in stress responses those distinguish between primary and secondary responses. Thus integrating of the different 'omics' data of 'multiple time course' experiment and their analysis with statistical model enhance the clarity about abiotic stress signaling that allows us with more robust identifications of molecular targets for developing much 'smart' plants to combat the abiotic stress.

Epigenetic Approaches for Abiotic Stress

Epigenetic is the change of characters without any change in the genetic material of a cell. Although significant understanding has been made on molecular genetics of abiotic stress (Yahmaguchi-Shinozaki and Shinozaki, 2006; Chinnusamy and Zhu 2009), yet study of epigenetic mechanism in abiotic stress resistance of plants is in its infancy. Genome imprinting in plants (Kohler and Weinhofer-Molisch 2010), transcriptional gene silencing induced by transgene in plants (Vaucheret *et al.*, 2001) and paramutations (Allemen *et al.*, 2006) are some of the examples of epigenetic changes in plants. Recent advances in molecular genetics and biochemical studies has led to the identification of two epigenetic mechanisms that mediates changes in the properties of a genetic locus without modifying nucleotide sequences. These mechanisms are chromatin modeling/remodeling and small interfering RNA (siRNA) mediated transcriptional gene silencing which have been discussed here.

Chromosome Modeling

Chromatin is composed of DNA and basic histone protein. The N terminal of the histone protein undergoes various post-translational changes. Similarly cytosine residue of DNA undergo methylation as well as demethylation which lead to changes in the chromatin structure resulting the switch on/off of the transcription process depending upon the need of the cell to combat with various external adverseness as heterochromatin is a tightly packed structure that does not permit transcription (Chinnusamy *et al.*, 2008).

In model plant *Arabidopsis* around 50 histone genes have been identified. The N terminal of histone core (H2A, H2B, H3 and H4) offers around 240 position for post-translational modification such as acetylation, methylation, glycosylation, phosphorylation, ADP-ribosylation, carbonylation, biotinylation, sumoylation, fernalization and ubiquitnation (Fuchs *et al.*, 2006). These histone variations and post-translational modification offer enormous combinatorial possibilities for nucleosome assemblies (Li *et al.*, 1999).

Abiotic stress induced histone variants; histone N tail modification and DNA cytosine methylation have been reviewed in detail (Chinnuswamy and Zhu, 2009). For example, submergence stress in rice seedling induced histone H3K4 trimethylation and H3 acetylation in *alcohol dehydrogenase 1* (*ADH1*) and *pyruvate decarboxylase 1* (*PDC1*) genes. These modifications enhanced the expression of *ADH1* and *PDC1* gene under stress (Tsuji *et al.*, 2006). Ectopic expression of a SNF2/BRAHMA-type chromatin remodeling gene *AtCHR12* in *Arabidopsis* arrested the growth of primary buds and stem under drought as well as heat stress than under controlled conditions. Conversely, the growth arrest response under stress was less in the *AtCHR12* knockout mutant than in the wild type plants (Mlynarova *et al.*, 2007).

Small Interfering RNA (siRNA)

Small RNA of 21–24 nt are classified into microRNAs (miRNAs), trans-acting siRNAs (ta-siRNAs), natural antisense siRNAs (nat-siRNAs) and repeat associated siRNAs (rasiRNAs) (Vaucheret, 2006). Recently another category of small RNAs ranging from 30 to 40 nt in size, referred to as long-siRNAs has also been discovered (Katiyar-Agarwal *et al.*, 2007). Among these, the miRNAs are the most abundantly expressed and well-studied class of small RNAs in plants which display near-perfect complementarities with their target mRNAs, and interfere with target gene expression by causing degradation or inhibition of the translation of the mRNA targets in plants (Voinnet, 2009). Some group of small RNAs also results into methylation usually at promoter regions which lead to inactivation of genes depending upon the type of signal transduction (Chen *et al.*, 2010). Several small RNAs that are involved in abiotic stresses are provided in a Table 1.1.

Table 1.1: List of miRNAs that are associated with abiotic stresses.

Stress	Name of the Plant	miRNA	References
Drought	Arabidopsis	miR157, miR167, miR168, miR171, miR408, miR319 and miR397 miR393, miR396	Liu et al., 2008
	Populus	miR1446a-e, miR1444a, miR1447, miR1450	Lu et al., 2008
	Rice	miR393	Zhao et al., 2007
	Medicago truncatula	miR398 and miR408	Trindade et al., 2009
	Phaseolus vulgaris	miR2118	Arenas-Huertero et al., 2009
ABA treatment	Phaseolus vulgaris	miR159.2, miR393,miR1514	Arenas-Huertero et al., 2009
Salt	Arabidopsis	miR396, miR168, miR167, miR165, miR319, miR159, miR394, miR156, miR393, miR171, miR158, miR169	Liu et al., 2008
	Populus trichocarpa	miR530a, miR1445, miR1446a-e, miR1447 and miR171l	Lu et al., 2008
Cold	Arabidopsis	miR165/166, miR169, miR172, miR393, miR396, miR397, miR408	Zhou et al., 2008
	Populus trichocarpa	miR168a,b, miR477a,b	Zhang et al., 2009
Oxidative	Arabidopsis	miR398	Sunkar et al., 2006
Hypoxia	Maize	miR167, miR166, miR171 and miR396	Zhang et al., 2008
	Arabidopsis	19 different miRNA	Moldovan et al., 2009
UV-Radiation	Populus tremula	miR169, miR395, miR472, miR168, miR398 and miR408	Jia et al., 2009
	Arabidopsis	miR156, miR160, miR165/166, miR167 and miR398	Zhou et al., 2007
Sulphate uptake	Arabidopsis	miR395	Takahashi et al., 1997 and Kawashima et al., 2009
Phosphate assimilation	Arabidopsis	miR399	Fujii et al., 2005 and Chiou et al., 2006
Copper assimilation	Arabidopsis	miR397, miR408, miR857	Burkhead et al., 2009
Iron deficiency	Arabidopsis	miR169b, miR169c, miR172c and miR394a	Kong and Yang 2010
Mechanical stress	Populus trichocarpa	miR156, miR162, miR164, miR475, miR480, and miR481	Lu et al., 2005

The above reports suggest that miRNAs are an integral part of stress response regulatory networks in plants. Given the fact that these studies are very recent, the discovery of additional novel and known miRNAs as stress-responsive small RNAs in the near future may emerge out.

Marker Assisted Selection for Abiotic Stress

The recent genome sequence projects on various crop plants have aided the marker assisted selected: starting from QTL identification, NIL development, fine mapping to transfer of QTLs to popular cultivars using precise marker assisted backcrossing (MABC) strategies (Collins *et al.*, 2008). It involves the manipulation of the genomic region that is associated with some particular trait. This can be done efficiently through a robust marker such as SNPs, which combines well with conventional back crossing program and with the ability to differentiate parental chromosomal segments. However the basic bottlenecks for marker assisted selection for QTL includes: i) small effect of several QTLs, ii) interaction of QTL with environment or genetic back ground, iii) poor resolution of QTL on the genetic map, iv) poor selection of parents for mapping populations etc. Despite these difficulties, several QTLs for various abiotic stresses have been identified (Collins *et al.*, 2008), although from the literature survey it is clearly evident that maximum attempts have been made for identification of QTLs related to drought. Subsequently, a number of studies have been reported the QTLs for root architecture and evaluated their effects on yield at different moisture regimes in maize (Tuberosa *et al.*, 2002, 2003; Landi *et al.*, 2007) and rice (MacMillan *et al.*, 2006; Steele *et al.*, 2006, 2007; Yue *et al.*, 2006). Another strategy to avoid low moisture stress is to maintain transpiration which ultimately reduce leaf senescence known as "stay-green" trait. This helps to increase photosynthesis over the crop life cycle (Borrell *et al.*, 2001). For example, four major QTLs namely Stg1 to Stg4, that control stay-green and grain yield have been identified (Harris *et al.*, 2007). Further, near isogenic lines (NILs) for these Stg QTLs have been developed, providing an opportunity of positional cloning of the underlying genes. Water use efficiency is another parameter to indicate drought tolerance which is considered to be a conservative strategy for crops by reducing transpiration. However, under moderately favorable environments, it reduced yield benefit but cause a yield penalty under the most severe conditions (Condon *et al.*, 2004). Several QTL of WUE identified in barley (Teulat *et al.*, 2002), rice (Laza *et al.*, 2006), *Brassica oleracea* (Hall *et al.*, 2005), and wheat (Spielmeyer *et al.*, 2007. Except in few instances, salt tolerance has often been found to be associated with lower accumulation of sodium (Na) in the shoot. Thus, using Na accumulation as a measure of salt tolerance, major and minor QTLs have been mapped in various crop species (Flowers and Flowers, 2005; Jenks *et al.*, 2007).

In a conservative estimation it has been found that 15 million hectares of rain-fed lowland rice growing areas in Asia is affected by submergence stress periodically, causing annual losses of up to U.S. $1 billion (Xu *et al.*, 2006). QTLs for submergence and anoxia tolerance have been identified for many crops. In Rice, *Sub1*, a major QTL was found on chromosome 9, which was used for marker assisted breeding to improve submergence tolerance of Swarna, a cultivar widely grown in flood-prone regions (Neeraja *et al.*, 2007). A submergence tolerant QTL allele together with three genes for tolerance to different biotic stresses was introgressed by maker assisted back cross selection into KDML105, a popular rice cultivar of Thailand (Toojinda *et al.*, 2005).

Low temperature or cold and high temperature or heat stress causes significant yield penalty in crop plants. Often heat stress is associated with drought stress and both of which reduce yield by reducing fertility and grain quality. Two QTLs controlling grain-filling duration, a trait thought to be correlated with heat tolerance was identified and validated in wheat by Yang *et al.* (2002). QTLs responsible for restoration of fertility by rendering pollen heat tolerance (germinability and pollen tube growth); a factor influencing heat-induced sterility was identified in maize (Frova and Sari-Gorla, 1994).

QTLs that are controlling chilling sensitivity have been identified for sorghum seedlings (Knoll and Ejeta, 2008), maize seedling (Hund *et al.*, 2005; Jompuk *et al.*, 2005; Presterl *et al.*, 2007), and in rice at the seedling as well as booting stages (Lou *et al.*, 2007). Interestingly, a QTL allele introgressed from a wild relative (*Solanum hirsutum*) to cultivated potato (*Solanum tuberosum*) that ultimately increased chilling tolerance (Goodstal *et al.*, 2005).

Nutrient is the most important input of growth after water for living organism. Particularly, plant growth and yield are severely impaired in soils containing either toxic or insufficient concentrations of particular minerals (Ismail *et al.*, 2007). Additionally their utilization depends upon the soil acidity (pH, 5.0), a characteristic of over half of the world's arable land, poses a serious limitation to crop production worldwide.

Toxicity of aluminium is associated with soil acidity. Major QTLs controlling Al tolerance have been identified in several crops such as soybean (Bianchi-Hall *et al.*, 2000), rice (Xue *et al.*, 2007), maize (Ninamango-Ca´rdenas *et al.*, 2003), wheat (Raman *et al.*, 2005), oat (Wight *et al.*, 2006), alfalfa (Narasimhamoorthy *et al.*, 2007), rye (Matos *et al.*, 2005) and sorghum (Magalhaes *et al.*, 2007).

Micronutrients play important role for two reasons. First, it is essential and required for the plant's own growth and secondly, higher accumulation of micronutrient in edible parts of the plants emerging as functional food to combat the malnutrition. Zinc (Zn) deficiency is one of the most widespread which is capable of causing severe yield reductions. Two major QTLs of rice, associated for Zn defficiency was identified (Wissuwa *et al.*, 2006). Besides QTL controlling Zn content in bean seeds have also been identified (Gelin *et al.*, 2007).

Similarly boron (B) deficiency or present in toxic concentrations are predominant in many agrarian land of the world. Recently, fine mapping and cloning of the major B tolerance QTL has been reported which pave the way for opportunities to overcome any such limitations in various plants (Miwa *et al.*, 2007). Comparisons with B tolerance QTLs and genomic sequences in the corresponding regions of the closely related *Arabidopsis* genome should assist in the fine-mapping and ultimately the positional cloning of these rapeseed QTLs (Zeng *et al.*, 2008).

Nitrogen (N) as nutrient is highly associated with growth and productivity of the plants. Therefore low nitrogen input cultivars is not only useful to reduce the cost of fertilizer but also due to excessive application of this fertilizer resulting into environmental pollution. Several workers reported the discovery of nitrogen use efficiency QTL in temperate maize (Gallais and Hirel, 2004). They identified a set of

QTLs for N use efficiency, grain yield, and its components at high and low N levels. QTLs for tolerance to low N have also been described in other crops such as *Arabidopsis* (Loudet *et al.,* 2003), rice (Lian *et al.,* 2005), and barley (Mickelson *et al.,* 2003).

Phosphorus is another important nutrient which has low mobility in soil and deficiency of which leads to deform root architect and ultimately reduce the growth of the plants. Thus, QTL associated with phosphorus uptake have been reported in number of crops such as *Arabidopsis* (Reymond *et al.,* 2006), common bean (Ochoa *et al.,* 2006), soybean (Li *et al.,* 2005), rice (Wissuwa *et al.,* 2005), maize (Zhu *et al.,* 2006), and wheat (Su *et al.,* 2006).

Future Prospects

This chapter summarizes the recent efforts of understanding the abiotic stress response, genes involved and QTL discovered to improve abiotic stress tolerance in crop plants. Development of crops befitting the emerging climate change perhaps is the most challenging issue for us. Climate change is not only a serious threat to agriculture but also continued to result into emergence of newer abiotic stresses. Abiotic stresses are complex phenomenon; on the other hand, there are several challenges that have restricted the realization of the full potential of using biotechnology approaches in crop breeding. Nevertheless, with the discovery of new modern techniques, availability of new genome sequence, involvement of multidisciplinary approaches such as GIS technology, computer science, analytical chemistry understanding about gene regulation under abiotic stresses all becomes handy tools. As a result of better understanding, there is a paradigm shift of approaches from a single gene-single trait to integrated breeding involving system biology, marker assisted breeding to genetic engineering which leads the better hope for the development of designer crops with improved features that can use natural resources such as water, soil nutrients, atmospheric carbon and nitrogen more efficiently than ever before. While some of the techniques such as integrated breeding are well-adopted, yet few techniques such as transgenic approaches are though academically robust but due to some social concerns, they are far away from acceptance among the stake-holders. However, a holistic approach including government policy will be required to overcome this situation.

Thus, it can be concluded that improvement of abiotic stress tolerance of agricultural crops can only be achieved by combining traditional breeding and modern biotechnology followed by improvement and adaptation of modern agricultural practices *i.e.,* science, cultivation practices and governmental policies have to come together.

References

Agarwal, P.K., Agarwal, P., Reddy, M.K. and Sopory, S.K. 2006. Role of DREB transcription factors in abiotic and biotic stress tolerance in plants. Plant Cell Rep. **25**: 1263–1274.

Alleman. M., Sidorenko, L., McGinnis, K., Seshadri, V., Dorweiler, J.E., White, J., Sikkink, K. and Chandler, V.L. 2006. An RNA–dependent RNA polymerase is required for paramutation in maize. Nature **442**: 295–8.

Amudha, J. and Balasubramani, G. 2011. Recent molecular advances to combat abiotic stress tolerance in crop plants. Biotechnology and Molecular Biology Review. **6**: 31–58.

Arenas–Huertero, C., Perez, B., Rabanal, F., Blanco–Melo, D., De la Rosa, C., Estrada–Navarrete, G., Sanchez, F., Covarrubias, A.A. and Reyes, J.L. 2009. Conserved and novel miRNAs in the legume *Phaseolus vulgaris* in response to stress. Plant Mol Biol **70**: 385–401.

Barnaba´s, B., Ja¨ger, K. and Fehe´r, A. 2008. The effect of drought and heat stress on reproductive processes in cereals. Plant Cell Environ. **31**: 11–38.

Barthakur, S., Babu, V. and Bansal, K.C. 2001. Overexpression of osmotin induces proline accumulation and confers tolerance to osmotic stress in transgenic tobacco. J Plant Biochem Biotech. **10**: 31–37.

Bianchi–Hall, C.M., Carter, T.E., Bailey, M.A., Mian, M.A.R., Rufty, T.W., Ashley, D.A., Boerma, H.R., Arellano, C., Hussey, R.S. and Parrott, W.A. 2000. Aluminum tolerance associated with quantitative trait loci derived from soybean PI 416937 in hydroponics. Crop Sci **40**: 538–545.

Bohnert, H.J., Nelson, D.E. and Jensen, R.G. 1995 Adaptations to environmental stresses. Plant Cell. **7**: 1099–1111.

Bordas, M., Montesinos, C., Dabauza, M., Salvador, A., Roig, L.A., Serrano, R. and Moreno, V. 1997. Transfer of the yeast salt tolerance gene HAL1 to *Cucumis melo* L. cultivars and *in vitro* evaluation of salt tolerance. Transgenic Res **6**: 41–50.

Borrell, A., Hammer, G. and Van Oosterom, E. 2001. Stay–green: a consequence of the balance between supply and demand for nitrogen during grain filling? Ann Appl Biol. **138**: 91–95.

Bowler, C., Slooten, L., Vandenbranden, S., Rycke, R.D., Botterman, J., Sybesma, C., Montagu, M.V. and Inze, D. 1991. Manganease superoxide dismutase can reduce cellular damage mediated by oxygen radicals in transgenic plants. EMBO J. **10**: 1723–1732.

Brini, F., Hanin, M., Lumbreras, V., Amara, I., Khoudi, H., Hasiairi, A., Pagès, M. and Masmoudi, K. 2007. Overexpression of wheat dehydrine *DHN–5* enhances tolerance to salt and osmotic stress in *Arabidopsis thaliana*. Plant Cell Rep. **26**: 2017–2026.

Burkhead, J.L., Reynolds, K.A.G., Abdel–Ghany, S.E., Cohu, C.M. and Pilon, M. 2009. Copper homeostasis. New Phytol. **182**: 799–816.

Capell, T., Bassie, L. and Christou, P. 2004. Modulation of the polyamine biosynthetic pathway in transgenic rice confers tolerance to drought stress. Proc. Natl. Acad. Sci USA. **101**: 9909–9914.

Chen, M., Meng, Y., Mao, C., Chen, D., and Wu, P. 2010. Methodological framework for functional characterization of plant microRNAs. J Exp. Bot. **61**: 2271–2280.

Cheng, Z., Jayaprakash, T., Huang, X. and Wu, R. 2002. Wheat LEA genes, PMA80 and PMA1959, enhance dehydration tolerance of transgenic rice (*Oryza sativa* L.). Mol. Breed. **16**: 71–82.

Chinnusamy, V. and Zhu, J.K. 2009. Epigenetic regulation of stress responses in plants. Curr. Opin. Plant Biol. **12**: 133–139.

Chinnusamy, V., Gong, Z. and Zhu, J.K. 2008. Abscisic acid-mediated epigenetic processes in plant development and stress responses. J. Integr. Plant Biol. **50**: 1187–1195.

Chiou, T.J., Aung, K., Lin, S.I., Wu, C.C., Chiang, S,F. and Su, C.L. 2006. Regulation of phosphate homeostasis by microRNA in *Arabidopsis*. Plant Cell. **18**: 412–21.

Collins, N.C., Francxois, T. and Roberto, T. 2008. Quantitative trait loci and crop performance under abiotic stress: where do we stand? Plant Physiol. **147**: 469–486.

Condon, A.G., Richards, R.A., Rebetzke, G.J and Farquhar, G.D. 2004. Breeding for high water–use efficiency. J. Exp. Bot. **55**: 2447–2460.

Davletova, S., Schlauch, K., Coutu, J and Mittler, R. 2005. The zinc–finger protein *ZAT12* plays a central role in reactive oxygen and abiotic stress signaling in *Arabidopsis*. Plant Physiol. **139**: 847–856.

Dizengremel, P., Thiec, D.L., Hasenfratz–Sauder, M.P., Vaultier, M.N., Bagard, M. and Jolivet, Y. 2009. Metabolic–dependent changes in plant cell redox power after ozone exposure. Plant Biol. **11**: 35–42.

Dure, L. 1992. The LEA protein of higher plants. In: Verma, D.P.S. (ed.) Control of plant gene expression. CRC Press, Boca Raton, FL, pp 325–335.

Dure, L. and Galau, G.A. 1981. Developmental biochemistry of cotton seed embryogenesis and germination: XII. Regulation of biosynsthesis of principal storage proteins. Plant Physiol. **68**: 187–194.

Ellul, P., Rios, G., Atarés, A., Roig, L.A. Serrano, R and Moreno, V. 2003. The expression of the *Saccharomyces cerevisiae* HAL1 gene increases salt tolerance in transgenic watermelon (*Citrullus lanatus* (Thunb.) Matsum. and Nakai. Theor. Appl. Genet. **107**: 462–469.

Flowers, T.J and Flowers, S.A. 2005. Why does salinity pose such a difficult problem for plant breeders? Agric. Water Manage. **78**: 15–24.

Frova, C. and Sari–Gorla, M. 1994. Quantitative trait loci (QTLs) for pollen thermo-tolerance detected in maize. Mol. Gen. Genet. **245**: 424–430.

Fuchs, J., Demidov, D., Houben, A and Schubert, I. 2006. Chromosomal histone modification patterns–from conservation to diversity. Trends Plant Sci. **11**: 199–208.

Fujii, H., Chiou, T.J., Lin, S.I., Aung, K. and Zhu, J.K. 2005. A miRNA involved in phosphate starvation response in *Arabidopsis*. Curr. Biol. **15**: 2038–2043.

Fukuda, A., Nakamura, A., Tagiri, A., Tanaka, H., Miyao, A., Hirochika, H. and Tanaka, Y. 2004. Function, intracellular localization and the importance in salt tolerance of a vacuolar Na^+/H^+ antiporter from rice. Plant Cell Physiol. **45**: 146–159.

Gallais, A and Hirel, B. 2004. An approach to the genetics of nitrogen use efficiency in maize. J. Exp. Bot. **55**: 295–306.

Gao, F., Gao, Q., Duan, X., Yue, G., Yang, A and Zhang, J. 2006. Cloning of an H+– PPase gene from *Thellungiella halophila* and its heterologous expression to improve tobacco salt tolerance. J. Exp. Bot. **57**: 3259–3270.

Gaxiola, R.A., Li, J., Undurraga, S., Dang, L.M., Allen. G.J., Alper, S.L and Fink, G.R. 2001. Drought and salt–tolerant plants result from overexpression of the AVP1 H+–pump. Proc. Natl. Acad. Sci. USA. **98**: 11444–11449.

Gelin, J.P., Forster, S., Grafton, K.F., McClean, P.E. and Kooas–Cifuentes, G.A. 2007. Analysis of seed zinc and other minerals in a recombinant inbred population of navy bean (*Phaseolus vulgaris* L.). Crop Sci. **47**: 1361–1366.

Goodstal, F.J., Kohler, G.R., Randall, L.B., Bloom, A.J and St, Clair, D.A. 2005. A major QTL introgressed from wild *Lycopersicon hirsutum* confers chilling tolerance to cultivated tomato (*Lycopersicon esculentum*). Theor. Appl. Genet. **111**: 898–905.

Hall, N.M., Griffiths, H., Corlett, J.A., Jones, H.G., Lynn, J and King, G.J. 2005. Relationships between water–use traits and photosynthesis in *Brassica oleracea* resolved by quantitative genetic analysis. Plant Breed. **124**: 557–564.

Hara, M., Terashima, S., Fukaya, T. and Kuboi, T. 2003. Enhancement of cold tolerance and inhibition of lipid peroxidation by citrus dehydrin in transgenic tobacco. Planta. **217**: 290–298.

Harris, K., Subudhi, P.K., Borrell, A., Jordan, D., Rosenow, D., Nguyen, H., Klein, P., Klein, R. and Mullet, J. 2007. Sorghum stay–green QTL individually reduce post–flowering drought–induced leaf senescence. J. Exp. Bot. **58**: 327–338.

Higashiyama, T. 2002. Novel functions and applications of trehalose. Pure Appl. Chem. **74**: 1263–1269.

Hirschi, K.D., Zhen, R.G., Cunningham, K.W., Rea, P.A and Fink, G.R. 1996. CAX1, an H+/Ca2+ antiporter from *Arabidopsis*. Proc. Natl. Acad. Sci. USA. **93**: 8782– 8786.

Hu, H., Dai, M., Yao, J., Xiao, B., Li, X., Zhang, Q. and Xiong, L. 2006. Overexpressing a NAM, ATAF, and CUC (NAC) transcription factor enhances drought resistance and salt tolerance in rice. Proc. Natl. Acad. Sci. USA. **103**: 12987–12992.

Huang, Z., Zhang, Z., Zhang, X., Zhang, H., Huang, D and Huang, R. 2004. Tomato TERF1 modulates ethylene response and enhance osmotic stress tolerance by activating expression of down strem genes. FEBS Lett. **573**: 110–116.

Hund, A., Frascaroli, E., Leipner, J., Jompuk, C., Stamp, P. and Fracheboud, Y. 2005 Cold tolerance of the photosynthetic apparatus: pleiotropic relationship between photosynthetic performance and specific leaf area of maize seedlings. Mol Breed. **16**: 321–331.

Ismail, A.M., Heuer, S., Thomson, M.J. and Wissuwa, M.2007. Genetic and economic approaches to develop rice germplasm for problem soils. Plant Mol. Biol. **65**: 547–570.

Jang, I.C., Oh, S.J., Seo, J.S., Choi, W.B., Song, S.I., Kim, C.H., Kim, Y.S., Seo, H.S., Choi, Y.D., Nahm, N.M and Kim, J.K..2003. Expression of biofuctional fusion of the *Escherichia coli* genes for trehalose–6–phosphate synthase and trehalose–6–phosphate phophatase in transgenic rice plants increases trehalose accumulation and abiotic stress tolerance without stunning growth. Plant Physiol. **131**: 516–524.

Jenks, M.A., Hasegawa, P.M. and Mohan Jain. S. 2007. Advances in molecular breeding toward drought and salt tolerant crops. Springer, Dordrecht, The Netherlands

Jia, X., Ren, L., Chen, Q.J., Li, R and Tang, G. 2009. UV–B–responsive microRNAs in *Populus tremula*. J. Plant Physiol. **166**: 2046–57.

Jompuk, C., Fracheboud, Y., Stamp, P. and Leipner, J. 2005. Mapping of quantitative trait loci associated with chilling tolerance in maize (*Zea mays L.*) seedlings grown under field conditions. J. Exp. Bot. **56**: 1153–1163.

Karaba, A., Dixit, S., Greco, R. Aharoni, A., Trijatmiko, K.R., Marsch–Martinez, N., Krishnan, A., Nataraja, K.N., Udayakumar, M and Pereira, A. 2007. Improvement of water use efficiency in rice by expression of HARDY, an *Arabidopsis* drought and salt tolerance gene. Proc. Natl. Acad. Sci. USA. **104**: 15270–15275.

Karim, S., Aronsson, H., Ericson, H., Pirhonen, M., Leyman, B., Welin, B., Mantyla, Palva, E.T., Van Dijck, P and Holmstrom, K.O. 2007. Improved drought tolerance without undesired side effects intransgenic plants producing trehalose. Plan Mol. Biol. **64**: 371–386.

Katiyar–Agarwal, S., Gao, S., Vivian–Smith, A. and Jin, H. 2007. A novel class of bacteria induced small RNAs in *Arabidopsis*. Genes Dev. **21**: 3123–3134.

Katsuhara, M., Otsuka, T and Ezaki, B. 2005. Salt stress–induced lipid peroxidation is reduced by glutathione S–transferase but this reduction of lipid peroxidase is not enough for a recovery of root growth in *Arabidopsis*. Plant Sci. **169**: 369–373.

Kawashima, C.G., Yoshimoto, N., Maruyama–Nakashita, A, Tsuchiya, Y.N., Saito, K., Takahashi, H and Dalmay, T. 2009. Sulphur starvation induces the expression of microRNA– 395 and one of its target genes but in different cell types. Plant J. **57**: 313–21.

Kaye, C., Neven, L., Hofig, A., Li, Q.B., Haskell, D and Guy, C. 1998. Characterization of a gene for spinach CAP160 and expression of two spinach cold–acclimation proteins in tobacco. Plant Physiol. **116**: 1367–1377.

Knoll, J. and Ejeta, G. 2008. Marker–assisted selection for early–season cold tolerance in sorghum: QTL validation across populations and environments. Theor. Appl. Genet. **116**: 541–553.

Ko¨hler, C and Weinhofer–Molisch, I. 2010. Mechanisms and evolution of genomic imprinting in plants Heredity. **105**: 57–63.

Kochian, L.V., Hoekenga, O.A. and Pi˜neros, M.A. 2004. How do crop plants tolerate acid soils? Mechanisms of aluminum tolerance and phosphorus efficiency. Annu. Rev. Plant Biol. **55**: 459–93.

Komatsu, S., Yang, G., Khan, M., Onodera, H., Toki, S and Yamaguchi, M. 2007. Over-expression of calcium–dependent protein kinase 13 and calreticulin interacting protein 1 confers cold tolerance on rice plants. Mol. Genet. Genomics. **277**: 713–723.

Kong, W.W. and Yang, J.M. 2010. Identification of iron–deficiency responsive microRNA genes and cis–elements in *Arabidopsis*. Plant Physiol. Biochem. **48**: 153–159.

Kovtum, Y., Chiu, W.L., Tena, G and Sheen, J. 2000. Functional analysis of oxidative stress–activated mitogen activated protein kinase cascade in plants. Proc. Natl. Acad. Sci. USA. **97**: 2940–2945.

Kuang, R.B., Liao, H., Yan, X.L and Dong, Y.S. 2005. Phosphorus and nitrogen interactions in field–grown soybean as related to genetic attributes of root morphological and nodular traits. J. Integr. Plant Biol. **47**: 549–559.

Landi, P., Sanguineti, M.C., Liu, C., Li, Y., Wang, T.Y., Giuliani, S., Bellotti, M., Salvi, S and Tuberosa, R. 2007. Root–ABA1 QTL affects root lodging, grain yield, and other agronomic traits in maize grown under well–watered and water–stressed conditions. J. Exp. Bot. **58**: 319–326.

Laza, M.R., Kondo, M., Ideta, O., Barlaan, E and Imbe, T..2006. Identification of quantitative trait loci for d C–13 and productivity in irrigated lowland rice. Crop Sci. **46**: 763–773.

Li, G., Bishop, K.J., Chandrasekharan, M.B. and Hall, T.C. 1999. β-phaseolin gene activation is a two–step process: PvALF–facilitated chromatin modification followed by abscisic acid–mediated gene activation. Proc. Natl. Acad. Sci. USA. **96**: 7104–7109.

Li, W.Y.F., Wong, F.L., Tsai, S.N., Phang, T.H., Shao, G. and Lam, H.N. 2006. Tonoplast–located *GmCLC1* and *GmNHX1* from soybean enhance NaCl tolerance in transgenic bright yellow (BY)–2–cells. Plant Cell Environ. **29**: 1122–1137.

Li, Y.D., Wang, Y.J., Tong, Y.P., Gao, J.G., Zhang, J.S and Chen, S.Y. 2005. QTL mapping of phosphorus deficiency tolerance in soybean (*Glycine max* (L.) Merr.) Euphytica. **142**: 137–142.

Lippert, K and Galinski, E.A. 1992. Enzyme stabilization by ecotoine–type compatible solutes: protection against heating, freezing and drying. App. Micro. Biotech. **37**: 61–65.

Liu, H.H., Tian, X., Li, Y.J., Wu, C.A. and Zheng, C.C. 2008. Microarray–based analysis of stress regulated microRNAs in *Arabidopsis thaliana*. RNA. **14**: 836–43.

Liu, X.F. and Li, J.Y. 2002. Characterization of an ultra–violate inducible gene that encodes glutathione S–transferase in *Arabidopsis thaliana,* Yi Chuan Xue Bao. **29**: 458–460.

Lou, Q.J., Chen, L., Sun, Z.X., Xing, Y.Z., Li, J., Xu, X.Y., Mei, H.W and Luo, L.J..2007. A major QTL associated with cold tolerance at seedling stage in rice (*Oryza sativa* L.). Euphytica. **158**: 87–94.

Loudet, O., Chaillou, S., Merigout, P., Talbotec, J. and Daniel–Vedele, F. 2003. Quantitative trait loci analysis of nitrogen use efficiency in *Arabidopsis.* Plant Physiol. **131**: 345–358.

Lu, S., Sun, Y.H and Chiang, V.L. 2008. Stress–responsive microRNAs in *Populus.* Plant J. 55: 131–51.

Lu, S., Sun, Y.H., Shi, R., Clark, C., Li, L and Chiang, V.L. 2005. Novel and Mechanical Stress–Responsive MicroRNAs in *Populus trichocarpa* That Are Absent from *Arabidopsis. Plant Cell.* **17**: 2186–2203.

MacMillan, K., Emrich, K., Piepho, H.P., Mullins, C.E and Price, A.H. 2006. Assessing the importance of genotype 3 environment interaction for rot traits in rice using a mapping population. I: A soil–filled box screen. Theor. Appl. Genet. **113**: 977–986.

Magalhaes, J.V., Liu, J., Guimara˜es, C.T., Lana, U.G.P., Alves, V.M.C., Wang, Y.H., Schaffert, R.E., Hoekenga, O.A., Pin˜eros, M.A., Shaff, J.E., Klein, P.E., Carneiro, N.P., Coelho, C.M., Trick, H.N and Kochian, L.V. 2007. A gene in the multidrug and toxic compound extrusion (MATE) family confers aluminum tolerance in sorghum. Nat. Genet. **39**: 1156–1161.

Majee, M., Maitra, S., Dastidar, K.G., Pattnaik, S., Chatterjee, A., Hait, N.C., Das, K.P and Majumdar, A.L. 2004. A novel salt–tolerant l–myo–inositol–l–phosphate synthetase from *Porteresia coarctata* (Roxb.) Tateoka, a halophytic wild rice: molecular cloning, bacterial over expression, characterization and functional introgression into tobacco–conferring salt tolerance phenotype. J. Biol. Chem. **279**: 28539–28552.

MaKersie, B.D., Murnaghan, J., Jones, K.S and Bowley, S.R. 2000. Iron–superoxide dismutase expression in transgenic alfalfa increase winter survival without a detectable increase in photosynthesis oxidative stress tolerance. Plant Physiol. **122**: 1427–1437.

Malik, V and Wu, R. 2005. Transcription factor *Atmyb2* increased salt–stress tolerance in rice (*Oryza sativa* L.). Rice Genet. Newslett. **22**: 63–65.

Matos, M., Camacho, M.V., Pe´rez–Flores, V., Pernaute, B., Pinto–Carnide, O and Benito, C. 2005. A new aluminum tolerance gene located on rye chromosome arm 7RS. Theor. Appl. Genet. **111**: 360–369.

Matsumura, T., Tabayashi, N., Kamagata, Y., Souma, C and Saruyama, H. 2002. Wheat catalase expressed in transgenic rice can improve tolerance against low temperatures stress. Physiol. Plant. **116**: 317–327.

Mickelson, S., See, D., Meyer, F.D., Garner, J.P., Foster, C.R., Blake, T.K., Fischer, A.M. 2003. Mapping of QTL associated with nitrogen storage and remobilization in barley (*Hordeum vulgare* L.) leaves. J. Exp. Bot. **54**: 801–812.

Minocha, S.C. and Sun, D.Y. 1997. Stress tolerance in plants through transgenic manipulation of polyamine biosynthesis. Plant Physiol Supp. **114**: 297.

Mittler, R. and Blumwald, E. 2010. Genetic engineering for modern agriculture: challenges and perspectives. Annu. Rev. Plant Biol. **61**: 443–62.

Miwa, K., Takano, J., Omori, H., Seki, M., Shinozaki, K and Fujiwara, T. 2007. Plants tolerant of high boron levels. Sci. **318**: 1417.

Mlynarova, L., Nap, J.P and Bisseling, T. 2007. The SWI/SNF chromatin remodeling gene AtCHR12 mediates temporary growth arrest in *Arabidopsis thaliana* upon perceiving environmental stress. Plant J. **51**: 874–885.

Moghaieb, R.E., Tanaka, N., Saneoka,H., Murooka, Y., Ono, H., Morikawa, H., Nakamura, A., Nguyen, N.T., Suwa, R and Fujita, K. 2006. Characterization of salt tolerance in ectoine–transformed tobacco plants (*Nicotiana tabacum*): photosynthesis, osmotic adjustment and nitrogen portioning. Plant Cell Environ. **29**: 173–182.

Moldovan, D., Spriggs, A., Yang, J., Pogson, B.J., Dennis, E.S., Wilson, I.W. 2009. Hypoxia responsive microRNAs and trans–acting small interfering RNAs in *Arabidopsis*. J. Exp. Bot. **61**: 165–77.

Nakashima, K., Tran, L.S., Van Nguyen, D., Fujita, M., Maruyama, K., Todaka, D., Ito, Y., Hayashi, N., Shinozaki, K. and Yamaguchi–Shinozaki K. 2007. Functional analysis of a NAC–type transcription factor *OsNAC6* involve in abiotic and biotic stress–responsive gene expression in rice. Plant J. **51**: 617–630.

Narasimhamoorthy, B., Bouton, J.H., Olsen, K.M. and Sledge, M.K. 2007. Quantitative trait loci and candidate gene mapping of aluminum tolerance in diploid alfalfa. Theor. Appl. Genet. **114**: 901–913.

Neeraja, C.N., Maghirang–Rodriguez, R., Pamplona, A., Heuer, S., Collard, B.C.Y., Septiningsih, E.M., Vergara, G., Sanchez, D., Xu, K., Ismail, A.M and MacKill, D.J. 2007. A marker–assisted backcross approach for developing submergence-tolerant rice cultivars. Theor. Appl. Genet. **115**: 767–776.

Nelson, D.E., Rammesmayer, G and Bornert, H.J. 1998. Regulation of Cell–specific inositol metabolism and transport in plant salinity tolerance. Plant cell. **10**: 753–764.

Reinberg, D. and Bird, A. 1999. MBD2 is a transcriptional repressor belonging to the MeCP1 histone deacetylase complex. Nat Genet. **23**: 58–61.

Ninamango–Ca´rdenas, F.E., Guimara˜es, C.T., Martins, P.R., Parentoni, S.N., Carneiro, N.P., Lopes, M.A., Moro, J.R. and Paiva, E. 2003. Mapping QTLs for aluminum tolerance in maize. Euphytica. **130**: 223–232.

Ochoa, *I.E.*, Blair, M.W., Lynch, J.P. 2006. QTL analysis of adventitious root formation in common bean under contrasting phosphorus availability. Crop Sci. **46**: 1609–1621.

Park, B.J., Liu, Z., Kanno, A., Kameya, T. 2005. Transformation of radish (*Raphanus sativus* L) via sonication and vacuum infiltration of germinated seeds with *Agrobacterium* harboring a group 3 LEA gene from *B napus*. Plant Cell Repo. **24**. 494–500.

Pinheiro, C., Chaves, M.M. 2011. Photosynthesis and drought: Can we make metabolic connections from available data? J. Exp. Bot. **62**: 869–882.

Prashanth, S.R., Sadhasivam, V and Parida, A. 2007. Overexpression of cytosolic copper/zinc superoxide dismutase from a mangrove plant *Avicennia marina* in Indica Rice var Pusa Basmati–1 confers abiotic stress tolerance. Transgenic Res. **17**: 281–291.

Presterl, T., Ouzunova, M., Schmidt, W., Moller E.M., Rober, F.K., Knaak, C., Ernst, K., Westhoff, P and Geiger, H.H. 2007. Quantitative trait loci for early plant vigour of maize grown in chilly environments. Theor. Appl. Genet. **114**: 1059–1070.

Qi, Y.C., Zhang, S.M., Wang, L.P., Wang, M.D. and Zhang, H. 2004. Overexpression of GST gene accelerates the growth of transgenic *Arabidopsis* under salt stress. Zhi Wu Sheng Li Yu Fen Zi Sheng Wu Xue Xue Bao. **30**: 517–22.

Quan, R., Shang, M., Zhang, H., Zhao, Y., Zhang, J. 2004. Engineering of enhanced glycine betaine synthesis improves drought tolerance in Maize. Plant Biotech J. **2**: 477–486.

Raman, H., Zhang, K.R., Cakir, M., Appels, R., Garvin, D.F., Maron, L.G., Kochian, L.V., Moroni, J.S., Raman, R., Imtiaz, M., Drake–Brockman, F., Waters, I., Martin, P., Sasaki, T., Yamamoto, Y., Matsumoto, H., Hebb, D.M., Delhaize, E and Ryan, P.R. 2005. Molecular characterization and mapping of ALMT1, the aluminium–tolerance gene of bread wheat (*Triticum aestivum* L.). Genome. **48**: 781–791.

Reguera, M., Peleg, Z. and Blumwald, E. 2011. Targeting metabolic pathways for genetic engineering abiotic stress–tolerance in crops. Biochem. Biophys. Acta **1819**: 186–194.

Rengasamy, P. 2006. World salinization with emphasis on Australia. J. Exp. Bot. **57**: 1017–1023.

Reymond, M., Svistoonoff, S., Loudet, O., Nussaume, L and Desnos, T. 2006. Identification of QTL controlling root growth response to phosphate starvation in *Arabidopsis thaliana*. Plant Cell Environ. **29**: 115–125.

Roxas, V.P., Lodhi, S.A., Garrett, D.K., Mahan, J.R. and Allen, R.D. 2000. Stress tolerance in transgenic tobacco seedling that overexpress glutathione S–transferase/glutathione peroxidase. Plant Cell Physiol. **41**: 1229–1234.

Saijo, Y., Hata, S., Kyozuka, J., Shimamoto, K. and Izui, K. 2000. Overexpression of a single Ca+–dependent protein kinase confers both cold and salt/drought tolerance on rice plant. Plant J. **23**: 319–327.

Sheveleva, E.V., Chmara, W., Bohnert, H.J., Jensen, R.G. 1997. Increased salt and drought tolerance by D–ononitol production in transgenic *Nicotiana tabacum* L. Plant Physiol. **115**: 1211–1219.

Shinozaki, K. and Yamaguchi–Shinozaki, K. 2004. Gene networks involved in drought stress response and tolerance. J. Exp. Bot. **58**: 221–227.

Shirasawa, K., Takabe, T., Takabe, T. and Kishitani, S. 2006. Accumulation of glycine betaine in rice plants that overexpress choline monooxygenase from spinach and evaluation of their tolerance to abiotic stress. Ann. Bot. **98**: 565–571.

Siefritz, F., Tyree, M.T., Lovisolo, C., Schubert, A. and Kaldenhoff, R. 2002. PIPI plasma membrane aquaporins in tobacco: from cellular effects to function in plants. Plant Cell. **14**: 869–876.

Singh, A.K., Sopory, S.K., Wu, R and Singla–Pareek, S.L. 2011. Transgenic approaches. In: Abiotic stress adaptation in plants. Physiology, molecular biology and genetic foundation. Springer, pp 417–451.

Sivamani, E., Bahieidin, A., Wraith, J.M., Al–Niemi, T., Dyer, W.E., Ho, T.H.D and Qu,R. 2000. Improved biomass productivity and water use efficiency under water deficit condition in transgenic wheat constitutively expressing the barley HVA1 gene. Plant Sci. **155**: 1–9.

Spielmeyer, W., Hyles, J., Joaquim, P., Azanza, F., Bonnett, D., Ellis, M.E., Moore, C., Richards, R.A. 2007. A QTL on chromosome 6A in bread wheat (*Triticum aestivum*) is associated with longer coleoptiles, greater seedling vigour and final plant height. Theor. Appl. Genet. **115**: 59–66.

Steele, K.A., Price, A.H., Shashidhar, H.E., Witcombe, J.R. 2006. Marker–assisted selection to introgress rice QTLs controlling root traits into an Indian upland rice variety. Theor. Appl. Genet. **112**: 208–221.

Steele, K.A., Virk, D.S., Kumar, R., Prasad, S.C., Witcombe, J.R. 2007. Field evaluation of upland rice lines selected for QTLs controlling root traits. Field Crops Res. **101**: 180–186.

Su, J.Y., Xiao, Y.M., Li, M., Liu, Q.Y., Li, B., Tong, Y.P., Jia, J.Z. and Li, Z.S. 2006 Mapping QTLs for phosphorus–deficiency tolerance at wheat seedling stage. Plant Soil. **281**: 25–36.

Sulpice, R., Tsukaya, H., Nonaka, H., Mustardy, L., Chen, T.H.H., Murata, N. 2003. Enhanced formation of flowers in salt–stress *Arabidopsis* after genetic engineering of the synthesis of glycine betaine. Plant J. **36**: 165–176.

Sunkar, R. 2010. MicroRNAs with macro–effects on plant stress responses. Semin. Cell Dev. Biol. **201021**: 805–811.

Sunkar, R., Kapoor, A and Zhu, J.K.2006. Posttranscriptional induction of two Cu/Zn superoxide dismutase genes in *Arabidopsis* is mediated by downregulation of miR398 and important for oxidative stress tolerance. Plant Cell. **18**: 2051–2065.

Takahashi, H., Yamazaki, M., Sasakura, N., Watanabe, A., Leustek, T., Engler, J.A., Montagu, M.V. and Saito, K. 1997. Regulation of sulphur assimilation in higher plants: a sulfate transporter induced in sulphate–starved roots plays a central role in *Arabidopsis thaliana*. Proc. Natl. Acad. Sci. USA. **94**: 11102–11107.

Takeda, S. and Matsuoka, M. 2008. Genetic approaches to crop improvement: responding to environmental and population changes. Nat. Rev. Genet. **9**: 444–457.

Tamura, T., Hara, K., Yamaguchi, Y., Koizumi, N. and Sano, H. 2003. Osmotic stress tolerance of transgenic tobacco expressing a gene encoding a membrane–located receptor–like protein from tobacco plants. Plant Physiol. **131**: 454–462.

Tattersall, E.A., Grimplet, J., Deluc, L., Wheatley, M.D., Vincent, D., Osborne, C., Ergul, A., Lomen, E., Blank,R.R., Schlauch, K.A., Cushman, J.C. and Cramer. G.R. 2007. Transcript abundance profiles reveal larger and more complex responses of grapevine to chilling compared to osmotic and salinity stress. Funct. Integr. Genomics. **7**: 317–333.

Teulat. B., Merah, O., Sirault, X., Borries, C., Waugh, R. and This, D. 2002. QTLs for grain carbon isotope discrimination in field–grown barley. Theor. Appl. Genet. **106**: 118–126.

Toojinda, T., Tragoonrung, S., Vanavichit, A., Siangliw, J.L., Pa–In, N., Jantaboon, J., Siangliw, M. and Fukai, S. 2005. Molecular breeding for rainfed lowland rice in the Mekong region. Plant Prod. Sci. **8**: 330–333.

Trindade, I., Capitao, C., Dalmay, T., Fevereiro, M.P. and Santos, D.M. 2009. miR398 and miR408 are up–regulated in response to water deficit in *Medicago truncatula*. Planta. **231**: 705–716.

Tsuji, H., Saika, H., Tsutsumi, N., Hirai, A and Nakazono, M. 2006. Dynamic and reversible changes in histone H3–Lys4 methylation and H3 acetylation occurring at submergence–inducible genes in rice. Plant Cell Physiol. **47**: 995–1003.

Tuberosa, R., Salvi, S., Sanguineti, M.C., Landi, P., Maccaferri, M and Conti, S. 2002. Mapping QTLs regulating morpho–physiological traits and yield: case studies, shortcomings and perspectives in drought–stressed maize. Ann. Bot. **89**: 941–963.

Tuberosa. R., Salvi, S., Sanguineti, M.C., Maccaferri, M., Giuliani, S. and Landi, P..2003. Searching for quantitative trait loci controlling root traits in maize: a critical appraisal. Plant Soil. **255**: 35–54.

Turner, N.C. 1986. Crop water deficits: a decade of progress. Adv. Agro. **39**: 1–51.

Urao, T., Yakubov, B., Satoh, R., Yamaguchi–Shinozaki, K., Seki, M., Hirayama, T. and Shinozaki, K. 1999. A transmembrane hybrid–type histidine kinase in *Arabidopsis* functions as an osmosensor. Plant Cell. **11**: 1179–1184.

Vaucheret, H. 2006. Post–transcriptional small RNA pathways in plants: mechanisms and regulations. Genes Dev. **20**: 759–771.

Vaucheret, H., Béclin, C. and Fagard, M. 2001. Post–transcriptional gene silencing in plants. J. Cell Sci. **114**: 3083–3091.

Voinnet, O. 2009. Origin, biogenesis, and activity of plant micro RNAs. Cell. **136**: 669–87.

Waie, B. and Rajam, M.V. 2003. Effect of increased polyamine biosynthesis on stress responses in transgenic tobacco by introduction of human S–adenosyl methionine gene. Plant Sci. **164**: 727–734.

Wang, Y., Ying, Y., Chen, J and Wang, X. 2004. Transgenic *Arabidopsis* overexpressing Mn–SOD enhanced salt–tolerance. Plant Sci. **167**: 671–677.

Wang, Y.J., Wisniewski, M., Meilan, R., Cui, M.G., Webb, R. and Fuchigami, L. 2005. Overexpression of cytosolic ascorbate peroxidase in tomato (*Lycopersicon esculentum* L.) confers tolerance to chilling and salt stress. J. Am. Soc. Hort. Sci. **130**: 167–173.

White, P.J. and Broadley, M.R. 2003. Calcium in plants. Ann. Bot. **92**: 487–511.

Wight, C.P., Kibite, S., Tinker, N.A. and Molnar, S.J. 2006. Identification of molecular markers for aluminium tolerance in diploid oat through comparative mapping and QTL analysis. Theor. Appl. Genet. **112**: 222–231.

Wissuwa, M., Gamat, G and Ismail, A.M. 2005. Is root growth under phosphorus deficiency affected by source or sink limitations? J. Exp. Bot. **56**: 1943–1950.

Wissuwa, M., Ismail, A.M. and Yanagihara, S. 2006. Effects of zinc deficiency on rice growth and genetic factors contributing to tolerance. Plant Physiol. **142**: 731–741.

Wu, C.A., Yang, G.D., Meng, Q.W. and Zheng, C.C. 2004. The cotton *GhNHX1* gene encoding a novel putative tonoplast Na^+/H^+ antiporter plays an important role in salt stress. Plant Cell Physiol. **45**: 600–607.

Wu, Y., Ding, N., Zhao, X., Zhao, M., Chang, Z., Liu, J and Zhang, L. 2007. Molecular characterization of PeSOS1: the putative Na^+/H^+ antiporter of *Populus euphratica*. Plant Mol. Biol. **65**: 1–11.

Xiao–Yan, Y., Fang, Y.A., Wei, Z.K. and Ren, Z.J. 2004. Production and analysis of transgenic maize with improved salt tolerance by the introduction of *AtNHX1* gene. Acta Bot Sinica. **46**: 854–861.

Xu, K., Xu, X., Fukao, T., Canlas, P., Maghirang–Rodriguez, R., Heuer, S., Ismail, A.M., Bailey–Serres, J., Ronald, P.C. and Mackill, D.J. 2006. Sub1A is an ethylene–response–factor–like gene that confers submergence tolerance to rice. Nature. **442**: 705–708.

Xue, Z.Y., Zhi, D.Y., Xue, G.P., Zhang, H., Zhao, Y.X and Xia, G.M. 2004. Enhanced salt tolerance of transgenic wheat (*Ttriticum aestivum* L.) expressing a vacuolar Na^+/H^+ antiporter gene with improved yields in saline soils in the field and a reduced level of leaf Na^+. Plant Sci. **167**: 849–859.

Xue, Y., Jiang, L., Su, N., Wang, J.K., Deng, P., Ma, J.F., Zhai, H.Q. and Wan, J.M. 2007. The genetic basis and fine–mapping of a stable quantitative–trait loci for aluminium tolerance in rice. Planta. **227**: 255–262.

Yamaguchi–Shinozaki, K. and Shinozaki, K. 2006. Transcriptional regulatory networks in cellular responses and tolerance to dehydration and cold stresses. Annu. Rev. Plant Biol. **57**: 781–803.

Yang, J., Sears, R.G., Gill, B.S. and Paulsen, G.M. 2002. Quantitative and molecular characterization of heat tolerance in hexaploid wheat. Euphytica. **126**: 275–282.

Yang, S.X., Zhao, Y.X., Zhang, Q., He, Y.K., Zhang, H and Luo, D. 2001 HAL1 mediate salt adaptation in *Arabidopsis thaliana*. Cell Res. **11**: 142–148.

Yang, X., Liang, Z. and Lu, C. 2005. Genetic engineering of the biosynthesis of glycine betaine enhances photosynthesis against high temperature stress in transgenic tobacco plants. Plant Physiol. **138**: 2299–2309.

Yoo, J.H., Park, C.Y., Kim, J.C., Heo, W.D., Cheong, M.S., Park, H.C., Kim, M.C., Moon, B.C., Choi, M.S., Kang, Y.H., Lee, J.H., Kim, H.S., Lee, S.M., Yoon, H.W., Lim, C.O., Yun, D.J., Lee, S.Y., Chung, W.S. and Cho, M.J. 2005. Direct interaction of a divergent CaM isoform and the transcription factor, MYB2, enhances salt tolerance in *Arabidopsis*. J. Biol. Chem. **280**: 3697–3706.

Yoshimura, K., Miyao, K., Gaber, A., Takeda, T., Kanaboshi, H., Miyasaka, H. and Shigeoka, S. 2004. Enhancement of stress tolerance in transgenic tobacco plants overexpressing *Chlamydomonas* glutathione peroxidase in chloroplast or cytosol. Plant J. 37: 21–33.

Yue, B., Xue, W.Y., Xiong, L.Z., Yu, X.Q., Luo, L.J., Cui, K.H., Jin, D.M., Xing, Y.Z. and Zhang, Q.F..2006. Genetic basis of drought resistance at reproductive stage in rice: separation of drought tolerance from drought avoidance. Genetics. **172**: 1213–1228.

Zeng, C., Han, Y., Shi, L., Peng, L.,Wang, Y., Xu,F. and Meng, J. 2008. Genetic analysis of the physiological responses to low boron stress in *Arabidopsis thaliana*. Plant Cell Environ. **31**: 112–122.

Zhang, H., Huang, Z., Xie, B., Chen, Q., Tian, X., Zhang, X., Zhang, H., Lu, X., Huang, D. and Huang, R. 2004. The ethylene–, jasmonate–, abscisic acid– and NaCl–responsive tomato transcription factor JERF1 modulates expression of GCC box–containing genes and salt tolerance in tobacco. Planta. **220**: 262–270.

Zhang, H.X. and Blumwald, E. 2001. Transgenic salt–tolerant tomato plants accumulate salt in foliage but not in fruit. Nat. Biotech. **19**: 765–768.

Zhang, H.X., Hodson, J.N., Williams, J.P. and Blumwald, E. 2001. Engineering salt–tolerant *Brassica* plants: characterization of yield and seed oil quality in transgenic plants with increased vacuolar sodium accumulation. Proc. Natl. Acad. Sci. USA. **98**: 12832–12836.

Zhang, J.Y., Xu, Y.Y., Huan, Q. and Chong, K. 2009. Deep sequencing of *Brachypodium* small RNAs at the global genome level identifies microRNAs involved in cold stress response. BMC Genomics. **10**: 449 –465.

Zhang. Z., Wei, L., Zou, X., Tao, Y., Liu, Z. and Zheng, Y. 2008. Submergence–responsive microRNAs are potentially involved in the regulation of morphological and metabolic adaptations in maize root cells. Ann. Bot. **102**: 509–19.

Zhao, F., Guo, S., Zhang, H. and Zhao, Y. 2006b. Expression of yeast *SOD2* in transgenic rice results in increased salt tolerance. Plant Sci. **170**: 216–224.

Zhao, F., Wang, Z., Zhang, Q., Zhao, Y. and Zhang, H. 2006a. Analysis of the physiological mechanism of salt–tolerant transgenic rice carrying a vacuolar Na⁺/H⁺ antiporter gene from *Suaeda salsa*. J. Plant Res. **119**: 95–104.

Zhou, B., Li, Y., Xu, Z., Yan, H., Homma, S. and Kawabata, S. 2007. Ultraviolet A-specific induction of anthocyanin biosynthesis in the swollen hypocotyls of turnip (*Brassica rapa*). J. Exp. Bot. **58**: 1771–81.

Zhou, X., Wang, G., Sutoh, K., Zhu, J.K. and Zhang, W. 2008. Identification of cold-inducible microRNAs in plants by transcriptome analysis. Biochim. Biophys. Acta **1779**: 780–788.

Zhu, J.K. 2002. Salt and drought stress signal transduction in plants. Annu. Rev. Plant Biol. **53**: 247–273.

Zhu, J.M., Mickelson, S.M., Kaeppler, S.M. and Lynch, J.P. 2006 Detection of quantitative trait loci for seminal root traits in maize (*Zea mays* L.) seedlings grown under differential phosphorus levels. Theor. Appl. Genet. **113**: 1–10.

2013, Abiotic Stress and Biotechnology
Editor: T. Pullaiah
Published by: REGENCY PUBLICATIONS, NEW DELHI

Pages 29–40

Chapter 2

Rapid *In Vitro* Micropropagation of Chick pea (*Cicer arietinum* L.) from Shoot Tip and Cotyledonary Node Explants

Saheena Parveen[1], T. Ugandhar[2] and K. Jagan Mohan Reddy[2]

*[1]Department of Botany,
S.R.R. Govt Degree and P.G. College, Karimnagar
[2]Department of Botany, Kakatiya University, Warangal – 506 009*

ABSTRACT

A rapid, simple and efficient protocol for *in vitro* multiple shoot induction and plantlet regeneration was achieved from two different explants of *Cicer arietinum* cv (ICCCR) (kranthi). The explants viz. shoot tip and cotyledonary node were cultured on MS medium fortified with Benzyl Amino Purine (BAP) (0.5–3.0 mg/l) and Kinetin (Kn) (0.5–3.0 mg/l) for multiple shoot induction. Multiple shoots proliferation was best observed at 2.0mg/l BAP from the two explants within three weeks of culture. Of the two different cytokinins tested, BAP was found to be more effective than Kinetin for shoot multiplication. The highest number of shoots (12.0 ± 0.3) was achieved on MS medium augmented with 2.0 mg/l BAP. The medium supplemented with (2.0 mg/l) BAP better than all other media concentrations in cotyledonary node explants. Individual shoots

* Corresponding Author: E-mail: reddykankanala@yahoo.co.in

were aseptically excised and sub cultured in the same media for shoot elongation. The elongated shoots were transferred to Indole Butyric Acid (IBA) (0.5–1.0mg/l) and Indole Acetic Acid (IAA) (0.5–1.0mg/l) for root induction. Rooting was observed within two weeks of culture. MS medium supplemented with (1.0 mg/l) IBA proved better with seventy percent rooting after 25 days of implantation. Rooted plantlets were successfully hardened under culture conditions and subsequently established in the field conditions. The recorded survival rate of the plants was 76.3 per cent. Plants looked healthy with no visually detectable phenotypic variations.

Keywords: *Shoot tip, Cotyledonary node, Multiple shoots, Rooting, Hardening.*

Pulse crops, also known as grain legumes, belong to the family Fabaceae, the second largest natural order of flowering plants. Generally, legumes are of a great economic importance as a source of food, fodder as well as for the significant role they play in biological fixation of atmospheric nitrogen. India is the largest producer of pulses in the world and more than a dozen pulse crops are grown on an estimated area of 22–23 million hectares. Of the grain legumes produced in the world, chickpeas stand second as for occupied area (10 million ha) and third in production (7 million t).

Chickpea (*Cicer arietinum* L.) is an important grain legume of the Indian subcontinent, West Asia, Mediterranean region, North and East Africa, Southern Europe and Central America and Australia. Various attributes of chickpea made it the most cultivated pulse crop and the most appreciated protein source among vegetarians all over the world. Chickpea straw has forage value comparable to other straws commonly used for livestock feed. It is able to drive more than 70 per cent of nitrogen from symbiotic dinitrogen fixation, which makes it a promising crop for "alternative agriculture" that is now attracting a considerable attention in the industrialized world. The heavy demand created by the pressure of increasing population in the developing world requires a tremendous scientific effort to meet the requirements of food, fiber, fuel and other necessities of life. Since the conventional techniques employed in crop improvement may not keep pace with the demands of the increasing population and decreasing land resources, the importance of *in vitro* technologies in crop improvement has great relevance. Recent advances made in the field of tissue culture have brought about new emerging technologies for crop improvement.

Micropropagation offers the potential to produce thousands or even millions of plants per annum. Application of tissue culture techniques for genetic upgradation of economically important plants has been reported (Scowcraft and Ryan, 1985). Plant tissue culture offers new ways for the improvement of this crop after many years of recalcitrance. Several researchers have reported on the regeneration of *Cicer arietinum* via direct organogenesis (Kartha *et al.,* 1981; Islam *et al.,* 1995; Barna and Wakhlu 1995; Anju and Chawla 2005). Thus the objective of the present study was to induce maximum number of shoots and regenerate whole plants from shoot tips and cotyledonary node explants of *Cicer arietinum* L. (ICCCR) (kranthi).

A. Shoot Tip Culture

The term shoot culture is now preferred for cultures started from explants bearing an intact shoot meristem, whose purpose is shoot multiplication by the repeated formation of axillary branches. In this technique, newly formed shoots or shoot bases serve as explants for repeated proliferation; severed shoots, explant size, shoot cultures are conventionally started from the apices of lateral or main shoots, up to 20 mm in length, dissected from actively-growing shoots or dormant buds. Larger explants are also sometimes used with advantage: they may consist of a larger part of the shoot apex or be stemming segments bearing one or more lateral buds; sometimes shoots from other *in vitro* cultures are employed. When apical or lateral buds were used almost exclusively as explants, the name 'shoot tip culture' came to be widely used for cultures of this kind. As the use of larger explants has become more common, the term shoot culture has become more appropriate. Large explants have advantages over smaller ones for initiating shoot cultures in them.

Shoot regeneration from shoot apex/shoot tip is direct, relatively simple and is not prone to somaclonal variation and chromosomal abnormalities. This is an elegant methodology of multiplying plants *in vitro* starting from a single shoot with obvious potential when applied to crop plants (Quak, 1977).

Culture of shoot meristem, especially through enhanced branching, permits rapid clonal propagation and a high degree of genetic uniformity of the progeny (Sunitha and Handique, 2000). This mericlone technology has been wide spread practical application in producing virus free plants *in vitro* in recent years (Bhaskaran *et al.*, 1992; Geetha and Shetty, 2000; Sudharshan *et al.*, 2000; Kaur *et al.*, 2000; Emmanuel *et al.*, 2000; Ugandhar *et al.*, 2011; Saheena Parveen *et al.*, 2012) Although mainly used for virus elimination meristem tip culture has also enabled plants to be freed from other pathogens including viroids, mycoplasms, bacteria and fungi, also, meristem freeze preservation as a method of conservation of germplasm has made possible to utilize when needed (Haskins and Katha, 1980). Successful generation of entire plants from frozen meristems of *Arachis hypogaea* and *Cicer arietinum* has been reported (Gosal and Bajaj 1979).

Methodology

The seeds of *Cicer arietinum* L.(ICCC-34) (kranthi) cultivar were obtained from ICRISAT Hyderabad A.P. The seeds were washed thoroughly in tap water 3–5 times and placed in 1 per cent (v/v) Teepol solution (Reckitt Benckiser, India) which was kept under running tap water for 15 min. Then the seeds were disinfected with 0.1 per cent (w/v) mercuric chloride (HgCl$_2$) for 5 min. Finally the seeds were rinsed 3–4 times in sterile distilled water and inoculated on moist cotton in sterile test tubes. To assure uniform and rapid germination of seeds, test tubes were placed in dark at 28°C for 24–48 h. Then the germinated seeds were transferred to light intensity (15 µmol/s2/s), 16 h light per day photoperiod for another 4–7 days and maintained at 25 ± 2°C and 55–60 per cent relative humidity.

Shoot tips, segments of 5–8 mm in length with one or two leaf primordia and Cotyledonary node segments of 5–8 mm in length of 8-d old *in vitro* raised seedlings were selected as explants for direct shoot multiplication.

MS media containing 3.0 per cent sucrose and supplemented with various concentrations cytokinins such as BAP (0.5 – 3.0 mg/l) and Kn (0.5 – 3.0 mg/l) were used. The initial pH of the culture media was adjusted to 5.8 before addition of 0.8 per cent (w/v) agar- agar. The medium was dispensed into culture tubes (25 + 150 mm) each containing 15 ml of the culture medium plugged with non-absorbent cotton and was autoclaved at 121°C for 15 minutes. In each culture tube one shoot tip explant was implanted. The cultures were maintained under 16h light provided with white fluorescent tubes (40 μ mol m^{-2}s^{-2}) at 25 ± 2 °C.

Results

Data on multiple shoot induction from shoot tip explants cultured on MS medium fortified with different concentrations of BAP and Kn alone is presented in Table 2.1. The important part of the present study was the preparation of contamination free explants. This was achieved by using *in vitro* germinated seedlings as an explant source. Sterilization of seeds required 0.1 per cent (w/v) HgCl$_2$ 5 min treatment for maximum germination (98 per cent) and minimum contamination (Narashimhulu and Reddy, 1983). A similar observation was also reported in *Vigna aconitifolia*, confirming the view that the pretreatment of seeds with specific surface sterilizing agents would predetermine the regenerating behaviour of explant tissues (Godbole *et al.*, 1984). The use of direct and large sized explants had higher survival and growth rates than the smaller ones (Hu and Wang 1983).

Table 2.1: Effect of different concentration of BAP, Kn on multiple shoot induction from shoot tip explants of *Cicer arietinum* (ICCC-34) (kranthi).

Growth Regulators (mg/l)	Per cent of Explants Showing Response	No. of Shoots SE*	Average Length of Shoots (cm) SE*
BAP			
0.5	85	3.0 ± 0.4	2.4 ± 0.2
1.0	90	4.0 ± 0.3	5.4 ± 0.3
1.5	95	5.0 ± 0.6	5.2 ± 0.4
2.0	100	6.0 ± 0.3	8.7 ± 0.5
2.5	80	3.5 ± 0.3	6.9 ± 0.4
3.0	70	2.0 ± 0.3	5.3 ± 0.4
Kn			
0.5	70	2.0 ± 0.6	2.9 ± 0.4
1.0	80	3.2 ± 0.7	4.4 ± 0.3
1.5	90	4.0 ± 0.2	5.5 ± 0.4
2.0	95	5.8 ± 0.4	8.3 ± 0.6
2.5	75	4.9 ± 0.5	6.0 ± 0.3
3.0	70	4.0 ± 0.3	4.0 ± 0.4

* Mean ± Standard Error.

Effect of BAP

The meristem containing shoot tip explants were excised from the surface sterilized, *in vitro* grown, 8-d old seedlings and cultured on MS medium augmented with BAP (0.5–3.0 mg/l) for multiple shoot induction. Of all the different concentrations of BAP tested, 2.0 mg/l BAP was found to be more effective in inducing 6.0 ± 0.3 shoots/explant) (Figure 2.1 a). But at high concentration of BAP (3.0 mg/l) considerably the number of shoot induction was found to be reduced.

Effect of Kn

Shoot tip explants were capable of directly developing multiple shoots on MS basal medium containing different concentrations of Kn (0.5 – 3.0 mg/l). Multiple shoot initiation from shoot tip explants was observed within 20 -25 days after inoculation. Highest number of shoots (5.8 ± 0.3) was observed in the medium containing 2.0 mg/l Kn (Table 2.1) (Figure 2.1b).

B. Cotyledonary Node Culture

The cotyledonary node induction is one of the most efficient method of micropropagation in plants since the emerging buds, especially from meristematic organs and tissues posses a great potential for vigorous development (Yadav *et al.,* 1990). Axillary buds have been found to be most suitable for clonal propagation in M*orus nigra* (Yadav *et al.,* 1990) and *Compmiphora wightii* (Durga *et al.,* 1993).

The results of the Cotyledonary node explant cultures on the development of multiple shoots are shown in (Table 2.2). The Cotyledonary node explants of *Cicer arietinum* L. (ICCC-34) (kranthi) cultured on different hormonal combination showed varied results. The axillary buds become active within week after inoculation and new shoots became distinct by the second and third week with leaves and internodes.

The size of the Cotyledonary node explants was found to play an important role in initiation and elongation of shoots. The smaller (1cm) explants could initiate more multiples than the longer (2.0cm) cotyledonary node explants. Although 1.5 cm long segments produced less number of shoot buds, they showed better elongation. Whereas the biggest cotyledonary node segment (2.0 cm) developed only a single shoot from each node.

Effect of BAP

The results on Cotyledonary node culture of *Cicer arietinum* L. (ICCC-34) (kranthi) on MS medium + BAP (0.5-3.0 mg/l) alone are presented in (Table 2.2) and shown in Figure 2.1c. At 1 mg/l BAP the response was 95 per cent with 10.0±0.5 number of shoots and 2.4± 0.2 cm length. The medium containing 2.0 mg/l BAP induced maximum number of shoots (12.0 ± 0.3 shoots/explant)) with (2.0 ±0.29 cm) and also showed good percentage (70 per cent) of response. When BAP concentration was increased above 2.0 mg/l the rate of shoot multiplication and elongation was reduced (Table 2.2).

Effect of Kn

The result on cotyledonary node culture of *Cicer arietinum* L. (ICCC-34) (kranthi) on MS medium + Kn (0.5 – 3.0 mg/l) was observed. High percentage (75) of responding cultures were found at (1.5 mg/l) Kn compared to all other concentrations tested. Whereas more number of shoots were regenerated from Nodal explants at (1.5 mg/) Kn (8.0± 04 shoots/explant) followed by 2.0 mg/l Kn (Table 2.2) (Figure 2.1d).

Table 2.2: Effect of different concentration of BAP, Kn alone in MS medium for multiple shoot induction from cotyledonary node explants of *Cicer arietinum* (ICCC-34) (kranthi).

Growth Regulators (mg/l)	Per cent of Explants Showing Response	No. of Shoots SE*	Average Length of Shoots (cm) SE*
BAP			
0.5	60	09.0 ± 0.6	2.1 ± 0.6
1.0	95	10.0 ± 0.5	2.4 ± 0.2
1.5	90	11.9 ± 0.5	2.3 ± 0.2
2.0	70	12.0 ± 0.3	2.0 ± 0.3
2.5	65	08.0 ± 0.4	1.5 ± 0.3
3.0	50	05.0 ± 0.5	1.0 ± 0.4
Kn			
0.5	50	6.2 ± 1.6	1.3 ± 0.3
1.0	65	7.0 ± 0.5	1.5 ± 0.3
1.5	75	8.0 ± 0.4	2.0 ± 0.3
2.0	70	7.9 ± 0.5	1.8 ± 0.2
2.5	60	6.5 ± 0.4	1.5 ± 0.3
3.0	55	5.0 ± 0.5	1.0 ± 0.5

* Mean ± Standard Error.

Table 2.3: Rooting ability of regenerated shoots from leaf explants culture of *Cicer arietinum.* cv (ICCC-34) cultured on MS medium supplemented with IAA and IBA.

Growth Hormones (mg/l)		Percentage of Response	Average No. of Roots (S.E)*
IAA	IBA		
0	0	23	1.0 ± 0.12
0.5	–	60	2.3 ± 0.37
1	–	70	3.2 ± 0.38
2	–	73	5.6 ± 0.38
–	0.5	54	4.3 ± 0.36
–	1	73	8.3 ± 0.87
–	2	70	6.3 ± 0.36

* Mean ± Standard Error.

In vitro Rooting

Fully elongated healthy shoots were transferred on to half strength MS root induction medium (RIM) (Murashige and Skoog 1962) fortified with different concentration of IAA (0.5 – 2.0 mg/l) and IBA (0.5 – 2.0 mg/l) (Figure 2.1e). Profuse rhizogenesis was observed on 1.5 mg/l IAA, compared to 0.5 -2.0 mg/l) IAA/IBA on

Figure 2.1: *In vitro* micropropagation of *Cicer arietinum* L. cv (ICCCR) (kranthi)

(a) Formation of multiple shoots on MS+BAP (2.0) mg/l from shoot tip, (b) Proliferation of multiple shoots on MS+Kn (2.0mg/l) from shoot tip, (c) Multiple shoot induction on MS+ BAP (2.0mg/l) from cotyledonary node, (d) Multiple shoot induction from cotyledonary node on MS+ Kn (2.0mg/l) (e) Rooting of individual shoots, (f) Hardening of plantlets.

Contd...

Figure 2.1–*Contd...*

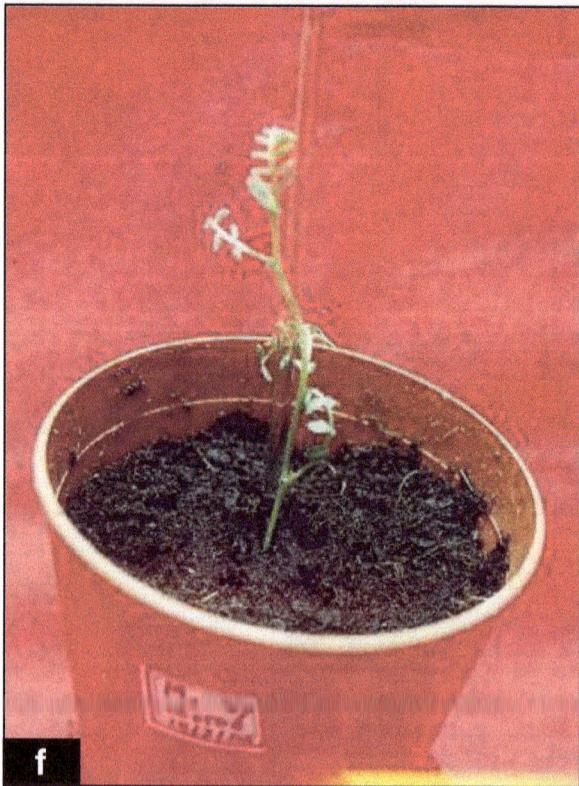

MS medium containing (1.0 mg/l) IBA whereas 73 per cent of plants produced roots with 8.3 ± 0.87 roots/explant (Table 2.3).

Acclimatization

Rooted plantlets were removed from the culture medium and the roots were washed under running tap water to remove agar. Then the plantlets were transferred to polypots containing pre- soaked vermiculite and maintained inside a growth chamber set at 28°C and 70 – 80 per cent relative humidity. After three weeks they were transplanted to poly bags containing mixture of soil + sand + manure in 1: 1: 1 ratio and kept under shade house for a period of three weeks. The potted plantlets were irrigated with Hogland's solution every 3 days for a period of 3 weeks (Figure 2.1 f).

Discussion

The result of present investigation show that the shoot tip explants from mature plants of *Cicer arietinum* L. (ICCC-34) (kranthi) could be induced to produce multiple shoots *in vitro*. Maximum number of shoots was induced on MS medium fortified with various concentrations of BAP and Kn. In recent years, shoot tip explants have been preferred to produce large number of genetically identical clones. Multiple shoot formation from shoot apices was obtained on MS medium supplemented with 2.0 ml/L BA, 0.1/ml NAA in pea (Griga *et al.,* 1986). MS-solid medium fortified with BAP and Kn alone and in combination increased the regeneration potential of shoot apical meristems of soybean, cowpea, peanut, chickpea and bean (Kartha *et al.,* 1981). It was reported that BAP was proved to be an ideal hormone for shoot multiplication of shoot tip culture in grain legumes (Sounder Raj *et al.,* 1989).These results are also in agreement with those on *Tectona grandis* (Gupta *et al.,* 1980). *Abizzia lebbeck* (Gharyl and Maheshwari 1982), *Ziziphus mauritiana* (Sudharshan *et al.,* 2000) and *Vanilla plantifolia* (Geetha and Shetty, 2000).

The capacity of shoot bud differentiation and shoot proliferation from shoot tip explants of *Cicer arietinum* (ICCC) (kranthi) depended on hormonal variation. There was good shoot bud induction and proliferation response only in the presence of cytokinin and no response in the basal medium. Similar results are well documented in several medicinal plants (Pattnaik and Chand, 1996) and *Withania somnifera* (Deka *et al.,* 1999). From our study it was clear that 2.0 mg/L BAP and Kn were significantly more effective for inducing shoot organogenesis.

We were successful in regenerating plants from cotyledonary node cultures on MS medium fortified with different concentrations of cytokinins *i.e.,* BAP and Kn individually. Maximum number of shoots were induced at 1.0 mg/L BAP in comparison to 1.5 mg/l Kn. However the shoot bud proliferation was found to be more on 1.0 mg/l BAP compared to1.5 mg/l. Similarly this was shown in other medicinal herbs including *Ocimum* species (Ahuja *et al.,* 1982: Patnaik and Chand, 1996). Nodal explants were also used to get higher rates of shoot multiplication of several plants (Shekawat and Galston, 1983).

During the present investigations multiple shoots were induced on MS Medium supplemented with various concentrations of cytokinins such as BAP and Kn alone.

Similarly Sudharshan *et al.* (2000) have observed the multiple shoot bud induction from cotyledonary node segments of *Ziziphus mauritiana* on MS medium supplemented with BAP alone. It was also recorded the same results in *Vanilla planifolia* on MS + BAP alone (Geetha and Shetty, 2000). When BAP and Kn concentration was increased (above 2.0 mg/L) the rate of shoot multiplication and elongation was reduced in the present investigation. Similar results were obtained in *Canavalia virosa* (Kathiravan and Ignacimuthu, 1999). Shoot tip and nodal segments were found to be the best explants for multiple shoot formation. In accordance with this in the present study also the shoot tip and nodal explants were found to be suitable for multiple shoot regeneration in *Cicer arietinum* L. (ICCCR) (kranthi).

Conclusion

From the above study, it was concluded that shoot tip and cotyledonary node explants are suitable for clonal propagation of chickpea. Cotyledonary node explants may be used for their higher rate of shoot multiplication. The protocol described in the present study is reproducible and can be used in future for further developments of the crop.

Acknowledgements

The authors are grateful to the Head, Department of Botany, Kakatiya University for his constant encouragement and for providing necessary facilities.

References

Ahuja, A., Verma, M. and Grewal, S. 1982. Clonal propagation of *Ocimum* species by tissue culture. Indian J. Exptl. Biol. **20**: 455–58.

Anju, A., and Chawla, S. 2005. Organogenic plant regeneration via callus induction in chickpea (*Cicer arietinum* L.). Role of genotypes, growth regulators and explants. Indian J. Biotech.4: 251–256.

Barna, K.S. and Wakhlu, A.K. 1995. Modified single node culture method – a new micropropagation method for chickpea. In vitro Plant, **31(3)**: 150–152.

Bhaskaran, S., Rigoldi, M. and Smith, R.H. 1992. Development potential of *Sorghum* shoots apices in culture. J. Plant physical. **140**: 481–484.

Deka, A.C., Kalita, M.C. and Baruah, A. 1999. Micropropagation of a potent herbal medicinal plant, *Withania somnifera*. Environ. Ecology **17(3)**: 594–596.

Durga, M., Brove and Mehta, A.R. 1993. Clonal propagation of mature elite tree *Commiphora wightii*. Plant Cell Tissue Org. Cult. **35**: 237–240.

Emmaneul, S., Ignacimuthu, S. and Kathiravan, K. 2000. Micropropagation of *Wedelia calendulacea* Less. A medical plant. Phytomorphology. **50**: 195–200.

Geetha, S. and Shetty, S.A. 2000. *In vitro* propagation of *Vanilla planifolia* a tropical orchid. Curr. Sci. 79: 886–889.

Geetha, N., Venkatachalam, P. and Lakshmisita, G. 1999. *Agrobacterium* mediated genetic transformation of pigeon pea (*Cajanus cajan* L.) and development to transgenic plants via direct organogenesis. Plant Biotechnology **16**: 213–218

Godbole, D.A., Kunachgi, M., Potdar, N.A., Krishna Murthy, K.U. and Mascarenhas, A. F. 1986. Studies on drought resistant legumes, the moth bean *Vigna aconitifolia* Marechal: Morphogenetic studies. Plant Cell Rep. **3**: 75–78.

Gosal, S.S. and Bajaj, Y.P.S. 1979. Establishment of callus tissue culture and introduction of organogenesis in some grain legumes. Crop Improvement **6**: 154–160.

Griga, M., Kubala Kovaj, M. and Tejk lovie, E. 1986. Somatic embryogenesis in *Vicia faba* L. Plant Cell Tissue Organ Culture **3**: 319–324.

Gupta, P.K., Nadgir, A.L., Mascarenhas, A.F.and Jagannathan, V. 1980. Tissue culture of Forests, trees, clonal multiplication of *Tectona grandis* L (Teak) by tissue culture. Plant Science Lett. **17**: 259–268.

Haskins, R.H. ad Kartha, K.K. 1980. Can J. Bot. **58**: 833–840.

Hu, C.Y. and Wang, P.J. 1983. Meristem, shoot tip and bud cultures. In: P.V. Ammirato, D.A. Evans, W.R. Sharp and Y.Yamada (eds.) Hand book of plant cell culture, Vol.3, crop species MeMillan Publishing Co., New York, pp. 65–90.

Islam, R., Riazuddin, S. and Farooqui, H. 1995. Clonal propagation from seedling nodes and shoot apices of chickpea (*Cicer arietinum* L.). Plant Tissue Cult. **5**: 53–57.

Kartha, K.K., Pahl, K., Leung, N.L. and Mroginski, L.A. 1981. Plant regeneration from meristem of grain legumes soybean, cow pea, peanut, chick pea and bean. Can. J. Bot. **19**: 1671 – 1679.

Kathiravan, K. and Ignacimuthu, S. 1999. Micropropagation of *Canavalia virosa* (Roxb.) Wight & Arn. Phytomorphology **49**: 61–66.

Kaur, R., Mahajan, R., Bhardwaj, S.V. and Sharm, D.R. 2000. Meristem tip culture to eliminate strain berry mottle virus from two cultivars of straw berry. Phytomorphology, **50**: 192–194.

Murashige, T. and Skoog, F. 1962. A revised medium for rapid growth and bioassay with Tobacco tissue culture. Plant Physiol. **159**: 473–497.

Pattnaik, S.K. and Chand, P.L. 1996. *In vitro* propagation of medicinal herb *Ocimum americanum* L. syn *Ocimum canum* Sims (hoary basil) and *Ocimum sanctum* L. (holy basil). Plant cell Rep. **15**: 846–850.

Quak, F. 1977. Meristem Culture and virus free plants. In: Applied and fundamental aspects of plant cell tissue and organ culture (eds) Reinert, J. and Bajaj Y.P.S Spriger – Verlag, Berlin, pp. 598–615.

Scowcraft, W.R. and Ryan, S.A. 1985. Tissue culture and plant breeding. In: Yeoman M. (ed.), Plant Culture Technology. Oxford, Blackwell Scientific.

Shaheena Parveen, Venkateshwarlu, M., Srinivas, D., Jagan Mohan Reddy K. and Ugandhar, T. 2012. Direct *in vitro* shoots proliferation of Chick pea (*Cicer arietinum* L.) from shoot tip explants induced by Thidiazuron. Bioscience Discovery, 3(1): 1–5.

Shekawat, N.S. and Galston, A.W. 1983. Isolation, culture and regeneration of mothbean, *Vigna aconitifolia* leaf protoplasts. Plant Science Letters, **32**: 43–51.

Sounder Raj, V., Tejavathi, D.H., Nijalingappa, B.H.M. 1989. Shoot tip culture in *Dolichos biflorus* L. Curr. Sci. **58**: 1385–1388.

Sudharshan, C., Aboel, M.N. and Hussain, J. 2000. *In vitro* propagation of *Ziziphus mauritiana* Cultivar umrdn by shoot tip ad nodal multiplication. Curr. Sci. **80(2)**: 290–292.

Sunitha, C. and Handique, P.J. 2000. High frequency *In vitro* shoot multiplication of *Plumbago indica* a rare medicinal plant. Curr. Sci. **78**: 1187–1188.

Ugandhar, T., Srilatha, T., Venkateshwarlu, M., Srinivas, D. and Jagan Mohan Reddy, K. 2011. *In vitro* plant regeneration from shoot tip explants of Pigeonpea (*Cajanus cajan* L.). Bioscience Discovery, **2(1)**: 193–195.

Yadav, V., Malan, L. and Jaiswal, V.S. 1990. Micropropagation of *Morus nigra* from shoot tip and nodal explants of mature trees. Scientia Hort B: 61–64.

2013, Abiotic Stress and Biotechnology
Editor: T. Pullaiah
Published by: REGENCY PUBLICATIONS, NEW DELHI

Pages 41–55

Chapter 3

Biofuel from Algae

Purshotam Kaushik* and Abhishek Chauhan**

Department of Botany and Microbiology,
Gurukul Kangri University, Hardwar – 249 404, Uttarakhand

ABSTRACT

The algae like plants require primarily three components to grow, which are: sunlight, carbon dioxide and water. Photosynthesis is an important biochemical process in which plants, algae, and some bacteria convert the energy of sunlight to chemical energy. Microalgae contain lipids and fatty acids as membrane components, storage products, metabolites and sources of energy. Algae contain anywhere between 2 per cent and 40 per cent of lipids/oils by weight. There are several well known methods to extract the oil from oilseeds. The advantages of deriving biofuel from algae include rapid growth rates and a high per acre yield. Moreover, the algal biofuel contains no sulfur, is non-toxic, and is highly biodegradable. Some species of algae namely *Botryococcus braunii, Chlorella, Dunaliella tertiolecta, Gracilaria, Pleurochrysis carterae, Sargassum* are ideally suited to biofuel/biodiesel production due to their high oil content in some species, tapping out near 50 per cent. One of the key reasons why algae are considered as feedstock for oil is their yields *i.e.,* 30 times more energy per acre than land crops such as soybeans, and some estimate even higher yields up to 15000 gallons per acre. Producing biodiesel from algae provides the highest net energy because converting oil into biodiesel is much less energy-intensive than methods for conversion to other fuels (such as ethanol, methane etc.). This characteristic has made biofuel/biodiesel the favourite end product of algae.

Keywords: Algae, Biofuel/biodiesel, Photosynthesis and Oil content.

E-mail: *purshotam.kaushik@gmail.com, **abhimicro19@rediffmail.com

Introduction

In the recent past, there has been a lot of discussion and interest around the viability of first generation biofuels as environment-friendly alternatives to foreign oil, primarily because of their possible competition with food crops and the use of non-sustainable practices for their production. Scientists and research groups have been searching for other sustainable sources for biofuel production, and microalgae seem to be one such alternative with a promising potential.

Microalgae are a diverse group of prokaryotic and eukaryotic photosynthetic microorganisms that can grow rapidly due to their simple structure (Kaushik and Chauhan, 2009). They have been investigated for the production of different biofuels including biodiesel, bio-oil and bio-hydrogen. Microalgal biofuel production is potentially sustainable. It is possible to produce adequate microalgal biofuels to satisfy the fast growing energy demand within the restraints of land and water resources.

Microalgae contain lipids and fatty acids as membrane components, storage products, metabolites and sources of energy. Algae contain anywhere between 2 per cent and 40 per cent of lipids/oils by weight. There is a greater possibiltiy to extract the oil/diesel from these natural resources. The advantages of deriving biodiesel from algae include rapid growth rates and high per-acre yield. Furthermore, algal biofuel contains no sulphur, is nontoxic, and is highly biodegradable. Some species of algae are ideally suited to biodiesel production due to their high oil content in some species, tapping out near 50 per cent.

They are sunlight-driven cell factories that convert carbon dioxide to potential biofuels, foods, feeds and high-value bioactives (Metting and Pyne, 1986; Schwartz, 1990; Kay, 1991; Shimizu, 1996; Borowitzka, 1999; Ghirardi *et al.*, 2000; Akkerman *et al.*, 2002; Banerjee *et al.*, 2002; Melis, 2002; Lorenz and Cysewski, 2003; Metzger and Largeau, 2005; Singh *et al.*, 2005; Spolaore *et al.*, 2006; Walter *et al.*, 2005). In addition, these photosynthetic microorganisms are known to produce intracellular and extracellular metabolites with diverse biological activities such as antibacterial (Falch *et al.*, 1995; Mundt *et al.*, 2001; Rao *et al.*, 2007; Kaushik and Chauhan, 2008; Kaushik *et al.*, 2009), antifungal (MacMillan *et al.*, 2002), cytotoxic (Luesch *et al.*, 2000), algaecide (Papke *et al.*, 1997), immunosuppressive (Koehn *et al.*, 1992) and antiviral activities (Hayashi and Hayashi, 1996; Kaushik and Chauhan, 2009).

Microalgae can provide several different types of renewable biofuels. These include methane produced by anaerobic digestion of the algal biomass (Spolaore *et al.*, 2006); biodiesel derived from microalgal oil (Roessler *et al.*, 1994; Sawayama *et al.*, 1995; Dunahay *et al.*, 1996; Sheehan *et al.*, 1998; Banerjee *et al.*, 2002; Gavrilescu and Chisti, 2005); and photobiologically produced biohydrogen (Ghirardi *et al.*, 2000; Akkerman *et al.*, 2002; Melis, 2002; Fedorov *et al.*, 2005; Kapdan and Kargi, 2006). The idea of using microalgae as a source of fuel is not new (Sawayama *et al.*, 1995), but it is now being taken seriously because of the escalating price of petroleum and, more significantly, the emerging concern about global warming that is associated with burning fossil fuels (Gavrilescu and Chisti, 2005).

Between 1978 and 1996, the Aquatic Species Program (ASP) focused on the production of biodiesel from high lipid-content algae growing in outdoor ponds and using CO_2 from coal-fired power plants to increase the rate of algal growth and reduce carbon emissions (Sheehan *et al.*, 1998). Under optimum growing conditions micro-algae will produce up to 4 lbs./sq. ft./year or 15,000 gallons of oil/acre/year. Microalgae are the fastest growing photosynthesizing organisms. They can complete an entire growing cycle every few days.

Greenfuel Technologies in Cambridge, MA is field testing a closed system that uses the CO_2 in power plant flue gases (13 per cent of flue gases in the test) to feed algae (Novakovic, 2005). Greenhouses can be modified to produce algae all year round. The surface area limitation which applies to ponds could be overcome in a greenhouse by adding a third layer of plastic inside the other two layers over which the pond water could flow in a thin enough film that it would receive enough solar radiation to grow algae. Microalgae is, by a factor of 8 to 25 for palm oil and a factor of 40 to 120 for rapeseed, the highest potential energy yield temperate vegetable oil crop.

Algae Cultivation

Algae can produce 15-300 times more oil per acre than conventional crops, such as rapeseed, palms, soybeans, or jatropha, and they have a harvesting cycle of 1-10 days, which permits several harvests in a very short time frame, increasing the total yield (Chisti, 2007). Algae can also be grown on land that is not suitable for other established crops, for instance, arid land, land with excessively saline soil, and drought-stricken land. This minimizes the issue of taking away pieces of land from the cultivation of food crops (Schenk *et al.*, 2008). Water, carbon dioxide, minerals and light are all important factors in cultivation, and different algae have different requirements. The basic reaction in water is

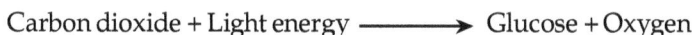

$$\text{Carbon dioxide} + \text{Light energy} \longrightarrow \text{Glucose} + \text{Oxygen}$$

Temperature must remain generally within 20 to 30°C. Microalgae biomass contains approximately 50 per cent carbon by dry weight (Sanchez Miron *et al.*, 2003). All of this carbon is typically derived from carbon dioxide. Producing 100 t of algal biomass fixes roughly 183 t of carbon dioxide. Carbon dioxide must be fed continually during daylight hours. Feeding controlled in response to signals from pH sensors minimizes loss of carbon dioxide and pH variations. Biodiesel production can potentially use some of the carbon dioxide that is released in power plants by burning fossil fuels (Sawayama *et al.*, 1995; Yun *et al.*, 1997). They can grow 20 to 30 times faster than food crops. Algae can be cultivated in open pond system and closed loop system.

Use of Bioreactors

Provide fast growth of algae with the high contents of oil. Photobioreactors have been successfully used for producing large quantities of microalgal biomass (Molina Grima *et al.*, 1999; Tredici, 1999; Pulz, 2001; Carvalho *et al.*, 2006). The term is more commonly used to define a closed system, as opposed to an open tank or pond.

| WATER | ALGAE | CO_2 | NUTRIENTS |

⬇ ⬇ ⬇ ⬇

FEEDING VESSEL

⬇

PHOTOBIOREACTOR

⬇

CENTRIFUGE

⬇

DRYER

⬇

OIL PRESS

⬇ ⬇

| ALGAE OIL | PRESS CAKE |

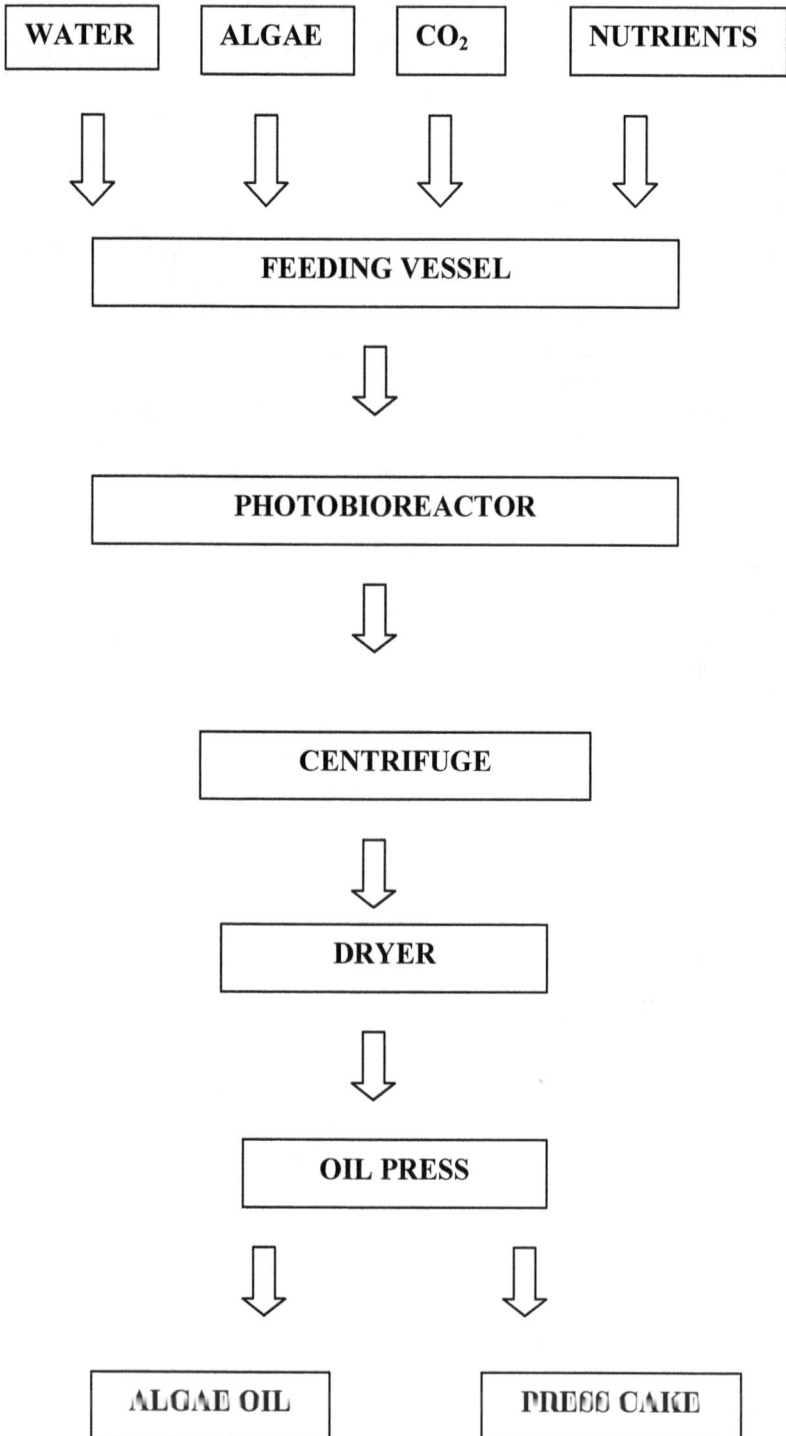

Figure 3.1: The technological circuit of installation of a bioreactor.

Because it is a closed system, the cultivator must provide all nutrients, including CO_2. A photo bioreactor can operate in "batch mode", which involves restocking the reactor after each harvest, but it is also possible to grow and harvest continuously. Continuous operation requires precise control of all elements to prevent immediate collapse. The grower provides sterilized water, nutrients, air, and carbon dioxide at the correct rates. This allows the reactor to operate for long periods. An advantage is that algae that grows in the "log phase" is generally of higher nutrient content than old "senescent" algae. Maximum productivity occurs when the "exchange rate" (time to exchange one volume of liquid) is equal to the "doubling time" (in mass or volume) of the algae.

Harvesting

Algae can be harvested using microscreens, by centrifugation, by flocculation (Bilanovic *et al.,* 1998) and by froth flotation. Interrupting the carbon dioxide supply can cause algae to flocculate on its own, which is called "autoflocculation". "Chitosan", a commercial flocculant, more commonly used for water purification, is far more expensive. The powdered shells of crustaceans are processed to acquire chitin, a polysaccharide found in the shells, from which chitosan is derived via de-acetylation. Water that is more brackish, or saline requires larger amounts of flocculant. Flocculation is often too expensive for large operations. Alum and ferric chloride are other chemical flocculants. In froth flotation, the cultivator aerates the water into a froth, and then skims the algae from the top (Gilbert *et al.,* 1961) Ultrasound and other harvesting methods are currently under development (Bosma *et al.,* 2003; US Patent No. 6524486).

Extration Methods

1. Mechanical Methods

The simplest method is mechanical crushing. Since different strains of algae vary widely in their physical attributes, various press configurations (screw, expeller, piston, etc) work better for specific algae types. Often, mechanical crushing is used in conjunction with chemicals. Mechanical methods are of two types:

(a) Expression/Expeller Press

Algae is dried. It retains its oil content, which then can be "pressed" out with an oil press. Since different strains of algae vary widely in their physical attributes, therefore, various press configurations (screw, expeller, piston, etc.) work better for specific algae types. Many commercial manufacturers of vegetable oil use a combination of mechanical pressing and chemical solvents in extracting oil.

(b) Ultrasonic-Assisted Extraction

Ultrasonic extraction, a branch of sonochemistry, can greatly accelerate extraction processes. Using an ultrasonic reactor, ultrasonic waves are used to create cavitation bubbles in a solvent material, when these bubbles collapse near the cell walls, it creates shock waves and liquid jets that causes those cell walls to break and release their contents into the solvent.

2. Chemical Methods

Algal oil can be extracted using chemicals. Chemical methods include:

(a) Hexane Solvent Method

Hexane solvent extraction can be used in isolation or it can be used along with the oil press/expeller method. After the oil has been extracted using an expeller, the remaining pulp can be mixed with cyclo-hexane to extract the remaining oil content. The oil dissolves in the cyclohexane, and the pulp is filtered out from the solution. The oil and cyclohexane are separated by means of distillation.

(b) Soxhlet Extraction

Soxhlet extraction is an extraction method that uses chemical solvents. Oils from the algae are extracted through repeated washing, or percolation, with an organic solvent such as hexane or petroleum ether, under reflux in a special glassware.

(c) Supercritical Fluid Extraction

In supercritical fluid/CO_2 extraction, CO_2 is liquefied under pressure and heated to the point that it has the properties of both a liquid and a gas, the liquified fluid then acts as the solvent in extracting the oil.

Table 3.1: Oil content of some microalgae (Adopted from Chisti, 2007).

Sl.No.	Microalgae	Oil Content (per cent dry wt)
1.	*Botryococcus braunii*	25–75
2.	*Chlorella* sp.	28–32
3.	*Crypthecodinium cohnii*	20
4.	*Cylindrotheca* sp.	16–37
5.	*Dunaliella primolecta*	23
6.	*Isochrysis* sp.	25–33
7.	*Monallanthus salina*	>20
8.	*Nannochloris* sp.	20–35
9.	*Nannochloropsis* sp.	31–68
10.	*Neochloris oleoabundans*	35–54
11.	*Nitzschia* sp.	45–47
12.	*Phaeodactylum tricornutum*	20–30
13.	*Schizochytrium* sp.	50–77
14.	*Tetraselmis sueica*	15–23

Algal Fuels

The algae product can then be harvested and converted into biodiesel and the algae's carbohydrate content can be fermented into biofuels.

Biodiesel

Microalgae have a large capacity for producing lipids for biodiesel and carbohydrates for bioethanol. The potential for microalgae as biofuel feedstocks is high because of their high rate of productivity, the potentially high percentage of biomass composed of lipids or carbohydrates, and because they lack lignin. The absence of lignin production in most algae is a benefit because processing lignin is currently a major impediment for bioethanol production (Moore, 2009).

Producing biodiesel from algae provides the highest net energy because converting oil into biodiesel is much less energy-intensive than methods for conversion to other fuels (such as ethanol, methane etc.). This characteristic has made biodiesel a favourite end-product from algae. Producing biodiesel from algae requires selecting high-oil content strains, and devising cost effective methods of harvesting, oil extraction and conversion of oil to biodiesel.

Studies show that algae can produce up to 60 per cent of their biomass in the form of oil. Because the cells grow in aqueous suspension where they have more efficient access to water, CO_2 and dissolved nutrients, microalgae are capable of producing large amounts of biomass and usable oil in either high rate algal ponds or photobioreactors. This oil can then be turned into biodiesel which could be sold for use in automobiles. The more efficient this process the larger the profit that is turned by the company. Regional production of microalgae and processing into biofuels will provide economic benefits to rural communities. Other sources of commercial biodiesel include canola oil, animal fat, palm oil, corn oil, waste cooking oil (Felizardo *et al.*, 2006; Kulkarni and Dalai, 2006), and jatropha oil (Barnwal and Sharma, 2005).

Table 3.2: Lipid content of different algae.

Sl.No.	Strains	Per cent Lipid (on a dry basis)
1.	*Chlorella* sp.	14–22
2.	*Scenedesmus* sp.	12–40
3.	*Chlamydomonas* sp.	21
4.	*Euglena* sp.	14–20
5.	*Spirogyra* sp.	11–21
6.	*Dunaliella* sp.	6–8
7.	*Synechoccus* sp.	11
8.	*Prymnesium* sp.	22–38
9.	*Porphyridium* sp.	9–14

Lipids from various sources can be converted to biodiesel through the process of transesterification (Chisti, 2007). Microalgae provide an excellent source of lipids for two major reasons. First, microalgae productivity can be an order of magnitude greater than terrestrial vegetation used for biofuel feedstocks. Second, the lipid content of microalgae can exceed 70 per cent of their dry mass, although algae with lipid content of around 30 per cent is more common (Chisti, 2007). High productivity combined with high lipid content results in a large amount of lipid that can be harvested annually for biodiesel production.

In addition to utilizing the lipids and carbohydrates from microalgal biomass for biofuel production, it is also possible to make use of the other compounds present in microalgal biomass. Microalgal proteins can be used for fish feeds in aquaculture or animal feeds for livestock and poultry (Meng *et al.*, 2009). Algae are currently grown commercially to produce profitable chemicals, such as antioxidants and dietary supplements. These chemicals comprise only a minor component of the microalgae cells, but have a high enough value to make their production profitable (Rosenberg *et al.*, 2008). The cost of producing bioethanol and biodiesel from microalgae can be reduced by exploiting every component of microalgae biomass.

Transesterification of Oil into Biodiesel

According to Chisti (2007) parent oil used in making biodiesel consists of triglycerides (Figure 3.2) in which three fatty acid molecules are esterified with a molecule of glycerol. In making biodiesel, triglycerides are reacted with methanol in a reaction known as transesterification or alcoholysis. Transestrification produces methyl esters of fatty acids, that are biodiesel, and glycerol (Figure 3.2). The reaction occurs stepwise: triglycerides are first converted to diglycerides, then to monoglycerides and finally to glycerol.

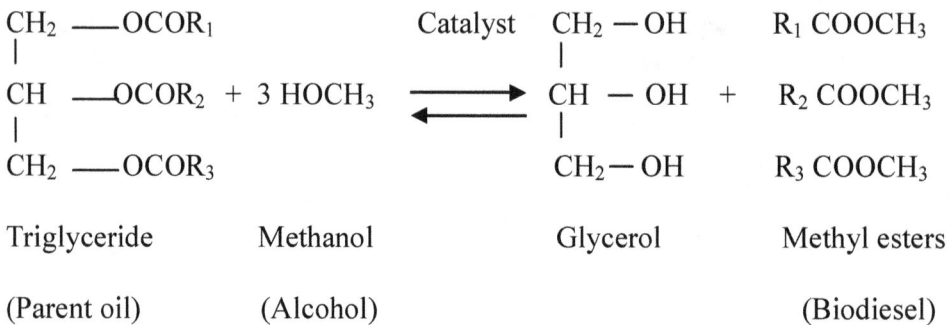

$$
\begin{array}{llll}
CH_2 - OCOR_1 & \text{Catalyst} & CH_2 - OH & R_1\,COOCH_3 \\
| & & | & \\
CH - OCOR_2 + 3\,HOCH_3 \rightleftharpoons & & CH - OH \; + & R_2\,COOCH_3 \\
| & & | & \\
CH_2 - OCOR_3 & & CH_2 - OH & R_3\,COOCH_3 \\
\end{array}
$$

Triglyceride	Methanol	Glycerol	Methyl esters
(Parent oil)	(Alcohol)		(Biodiesel)

Figure 3.2: Transesterification of oil into biodiesel.

Transesterification is catalyzed by acids, alkalis (Fukuda *et al.*, 2001; Meher *et al.*, 2006) and lipase enzymes (Sharma *et al.*, 2001). Alkali-catalyzed transesterification is about 4000 times faster than the acid catalyzed reaction (Fukuda *et al.*, 2001). Biodiesel is recovered by repeated washing with water to remove glycerol and methanol.

Ethanol from Algae

Ethanol from algae is possible by converting the starch (the storage component) and Cellulose (the cell wall component). Put simply, lipids in algae oil can be made into Biodiesel while the carbohydrates can be converted to ethanol. Algae are the optimal source for second generation bioethanol due to the fact that they are high in carbohydrates/polysaccharides and thin cellulose walls. Some prominent strains of algae that have a high carbohydrate content and hence are promising candidates for ethanol production are *Sargassum, Glacilaria, Prymnesium parvum, Euglena gracilis.*

Fermentation Process

Fermentation process to produce ethanol include the following stages:

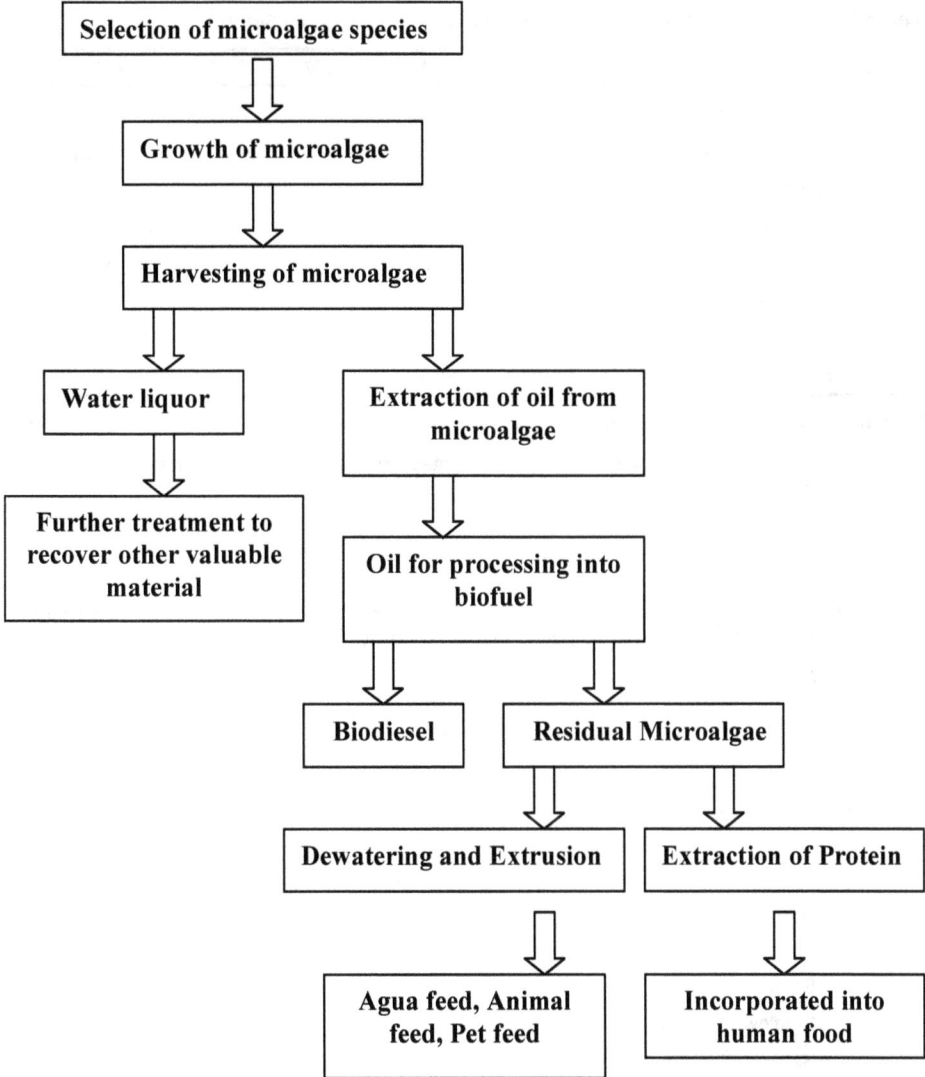

Figure 3.3: A detailed process for production of biodiesel from algae.

1. Growing starch-accumulating, filament-forming, or colony-forming algae in an aqua culture environment;
2. Harvesting the grown algae to form a biomass;
3. Initiating decay of the biomass;
4. Contacting the decaying biomass with a yeast capable of fermenting it to form a fermentation solution; and,
5. Separating the resulting ethanol from the fermentation solution.

Table 3.3: Comparison of biodiesel from microalgal oil and diesel fuel.

Sl.No.	Properties	Biodiesel from Microalgal Oil	Diesel Fuel
1.	Density kg l⁻¹	0.864	0.838
2.	Viscosity Pa s	5.2×10^{-4} (40 °C)	$1.9–4.1 \times 10^{-4}$ (40 °C)
3.	Flash point °C	65-115*	75
4.	Solidifying point °C	−12	−50–10
5.	Cold filter plugging point °C	−11	−3.0 (−6.7 max)
6.	Acid value mg KOH g⁻¹	0.374	0.5 max
7.	Heating value MJ kg⁻¹	41	40–45
8.	HC ratio	1.18	1.18

*Based on data from multiple sources

Source: Department of Biological Sciences and Biotechnology, Tsinghua University, Beijing, China (2004)

Other Algae Fuels

Biogasoline, Methane Straight Vegetable Oil, Hydrocracking to traditional transport fuels and jet fuel are produced from algae.

Oil Yields

The Table 3.4 below presents indicative oil yields from various oilseeds and algae.

Table 3.4: Comparison of average oil yields from algae with that from other oilseeds.

Sl.No.	Oilseeds and Algae	Gallons of Oil per Acre per Year
1	Corn	18
2.	Soybeans	48
3.	Safflower	83
4.	Sunflower	102
5.	Rapeseed	127
6.	Oil Palm	635
7.	Microalgae	5000–15000

Advantages of Biofuel Production

Biofuel production using microalgal farming offers the following advantages

1. The high growth rate of microalgae makes it possible to satisfy the massive demand on biofuels using limited land resources without causing potential biomass deficit.
2. Microalgal cultivation consumes less water than land crops.
3. The tolerance of microalgae to high CO_2 content in gas streams allows high-efficiency CO_2 mitigation.

4. Nitrous oxide release could be minimized when microalgae are used for biofuel production.

5. Microalgal farming could be potentially more cost effective than conventional farming.

Disadvantages of Biofuel Production

On the other hand, one of the major disadvantages of microalgae for biofuel production is the low biomass concentration in the microalgal culture due to the limit of light penetration, which in combination with the small size of algal cells makes the harvest of algal biomasses relatively costly. The large water content of harvested algal biomass also means its drying would be an energy-consuming process. The higher capital costs of and the rather intensive care required by a microalgal farming facility compared to a conventional agricultural farm is another factor that impedes the commercial implementation of the biofuels from microalgae strategy. Nevertheless, these problems are expected to be overcome or minimized by technology development. Given the vast potential of microalgae as the most efficient primary producers of biomass, there is little doubt that they will eventually become one of the most important alternative energy sources. Another problem that arises in microalgae production is that the algae-culture systems can easily and quickly be contaminated with other organisms (Zittelli *et al.*, 2006). Species of algae other than the target species can be introduced to the production system and compete with the target species. This presents a problem because the overall productivity of the system can be reduced. One way to avoid the issue of contamination is to grow microalgae that flourish under extreme conditions. For example, *Arthrospira maxima* grow at very high pH (9.5-11), while *Tertraselmis* grow in extremely saline waters (Dismukes *et al.*, 2008). Another method for avoiding contamination is to grow the microalgae in a closed system, under very controlled conditions. There, however, is no guarantee that contamination will not occur.

Conclusion

The potential benefits of large-scale production of microalgae for biofuels and other products far outweigh the existing and potential issues associated with microalgae production. At the moment, microalgae appear to be the best option for biofuel feedstocks because of their tremendous productivity, ability to use waste products in their production, and the valuable byproducts that can be produced. Research dealing with improving production systems, identifying ideal microalgae for biofuel production, and investigating the sustainability of microalgal biofuel is necessary before large-scale microalgal biofuel operations are established. Microalgal biotechnology appears to possess high potential for biodiesel production because a significant increase in lipid content of microalgae is now possible through heterotrophic cultivation and genetic engineering approaches. Keeping in view the advantages of microalgae, the study can be designed to develop the technologies for the production of biodiesel from microalgae, including the various modes of cultivation for the production of oil-rich microalgal biomass, as well as the subsequent downstream processing for biodiesel production.

Acknowledgements

Sincere thanks are due to Dr. Vedpal S. Malik, a noted microbiologist at Riverdale, MD, USA, for interaction and providing useful information.

References

Akkerman, I., Janssen, M., Rocha, J. and Wijffels, R. H. 2002. Photobiological hydrogen production: photochemical efficiency and bioreactor design. Int. J. Hydrogen Energy. **27**: 1195–208.

Banerjee, A., Sharma, R., Chisti, Y. and Banerjee, U. C. 2002. *Botryococcus braunii*: a renewable source of hydrocarbons and other chemicals. Crit. Rev. Biotechnol. **22**: 245–279.

Barnwal, B.K. and Sharma, M.P. 2005. Prospects of biodiesel production from vegetables oils in India. Renew Sustain Energy Rev. **9**: 363–78.

Bilanovic, D., Sukenik, A. and Shelef, G. 1988. Flocculation of microalgae with cationic polymers. Effects of medium salinity. Elsevier Science Publishers Ltd., England.

Borowitzka, M. A. 1999. Pharmaceuticals and agrochemicals from microalgae. In: Cohen Z. (Ed.). Chemicals from microalgae. Taylor and Francis. p. 313–52.

Bosma, R., Tramper, J. and Wijffels, R.H. 2003. Ultrasound A new technique to harvest microalgae. Journal of Applied Phycology. **15(2–3)**: 143–153.

Carvalho, A.P., Meireles, L.A. and Malcata, F.X. 2006. Microalgal reactors: a review of enclosed system designs and performances. Biotechnol. Prog. **22**: 1490–506.

Chisti, Y. 2007. Biodiesel from microalgae. Biotechnol. Adv., 25: 294–306.

Dismukes, G., Charles, Damian, C., Nicholas, B., Gennady, M. Ananyey and Matthew, C. P. 2008. Aquatic phototrophs: efficient alternatives to land–based crops for biofuels. Current Opinion in Biotechnology **19**: 235–240.

Dunahay, T.G., Jarvis, E.E., Dais, S.S. and Roessler, P.G. 1996. Manipulation of microalgal lipid production using genetic engineering. Appl. Biochem. Biotechnol. **57–58**: 223–31.

Falch, B.S., Konig, G.M., Wright, A.D., Sticher, O., Angerhofer, C.K., Pezzuto, J.M and Bachmann, H.1995. Biological activities of cyanobacteria: Evaluation of extracts and pure compounds. Planta Med. **61**: 321–328.

Fedorov, A.S., Kosourov, S., Ghirardi, M. L. and Seibert, M. 2005. Continuous H_2 photoproduction by *Chlamydomonas reinhardtii* using a novel two–stage, sulfate–limited chemostat system. Appl. Biochem. Biotechnol. **121124**: 403–12.

Felizardo, P., Correia, M.J.N., Raposo, I., Mendes, J.F., Berkemeier, R. and Bordado, J.M. 2006. Production of biodiesel from waste frying oil. Waste Manag. **26(5)**: 487–94.

Fukuda, H., Kondo, A. and Noda H. 2001.Biodiesel fuel production by transesterification of oils. J Biosci Bioeng. 92: 405 16.

Gavrilescu, M. and Chisti, Y. 2005. Biotechnology – a sustainable alternative for chemical industry. Biotechnol. Adv. **23**: 471–99.

Ghirardi, M.L., Zhang, J.P., Lee, J.W., Flynn, T., Seibert, M. and Greenbaum, E. 2000. Microalgae: a green source of renewable H2. Trends Biotechnol. **18**: 506–11.

Gilbert, V., Levin and John, R. 1961. Clendenning, Ahron Gibor, and Frederick D. Bogar. "Harvesting of Algae by Froth Flotation". Research Resources, Inc, Washington, D.C.

Hayashi, T. and Hayashi, K. 1996. Calcium spirulan, an inhibitor of enveloped virus replication, from a blue–green alga, *Spirulina platensis*. J. Nat Prod. **59**: 83–87.

Kapdan, I.K. and Kargi, F.2006. Bio–hydrogen production from waste materials. Enzyme Microb Technol. **38**: 569–82.

Kaushik, P. and Chauhan, A. 2008. *In vitro* antibacterial activity of laboratory grown culture of *Spirulina platensis*. Indian Journal of Microbiology. **48(3)**: 348–352.

Kaushik, P. and Chauhan, A. 2009 Cyanobacteria: Antibacterial Activity, New India Publishing Agency,101, Vikas Surya Plaza, C.U. Block, L.S.C. Mkt., Pitam pura, New Delhi 110088, p. 1–198.

Kaushik, P., Garima, Chauhan, A. and Goyal, P. 2009. Screening of *Lyngbya majuscula* for potential antimicrobial activity and HPTLC analysis of active methanolic extract. Journal of Pure and Applied Microbiology. **3(1)**: 169–174.

Kay, RA.1991. Microalgae as food and supplement. Crit Rev Food Sci Nutr. **30**: 555–73.

Koehn, F.E., Longley, R.E. and Reed, J.K. 1992. Microcolins A and B, new immunosuppressive peptide from the blue–green alga *Lyngbya majuscula*. J. Nat. Prod. **55**: 613–619.

Kulkarni, M.G. and Dalai, A.K. 2006. Waste cooking oil – an economical source for biodiesel: A review. Ind. Eng. Chem. Res. **45**: 2901–13.

Lorenz, R.T. and Cysewski, G.R. 2003. Commercial potential for *Haematococcus microalga* as a natural source of astaxanthin. Trends Biotechnol. **18**: 160–7.

Luesch, H., Yoshida, W.Y., Moore, R.E., Paul, V.J. and Mooberry, S.L.2000. Isolation, structure determination and biological activity of Lyngbyabellin A from the marine Cyanobacterium *Lyngbya majuscula*. J. Nat. Prod. **63**: 611–615.

MacMillan, J.B., Ernst–Russell, M.A., De Roop, J.S. and Molinski, T.F. 2002. Lobocyclamides A–C, lipopeptides from a cryptic cyanobacterial mat containing *Lyngbya confervoides*. J. Org. Chem. **67**: 8210–8215.

Meher, L.C., Vidya, Sagar, D. and Naik, S.N. 2006. Technical aspects of biodiesel production by transesterification: A review. Renew Sustain Energy Rev. **10**: 248–68.

Melis, A. 2002. Green alga hydrogen production: progress, challenges and prospects. Int. J. Hydrogen Energy. **27**: 1217–28.

Meng, Xin., Jianming, Yang, Xin Xu, Lei Zhang, Qingjuan Nie and Mo Xian. 2009. Biodiesel production from oleaginous microorganisms. Renewable Energy. **34**: 1–5.

Metting, B. and Pyne, J.W. 1986. Biologically–active compounds from microalgae. Enzyme Microb Technol. **8**: 386–94.

Metzger, P. and Largeau, C. 2005. *Botryococcus braunii:* a rich source for hydrocarbons and related ether lipids. Appl. Microbiol. Biotechnol., **66**: 486–96.

Microalgae separator apparatus and method, United States Patent 6524486". United States Patent Department.

Molina Grima, E. 1999. Microalgae, mass culture methods. In: Flickinger, M.C., Drew, S.W. (eds.) Encyclopedia of bioprocess technology: fermentation, biocatalysis and bioseparation, vol. 3.Wiley. p. 1753–69.

Molina Grima, E., Acien Fernandez, F.G., Garcia Camacho, F. and Chisti, Y. 1999. Photobioreactors: light regime, mass transfer, and scaleup. J. Biotechnol., **70**: 231–47.

Moore, J. 2009. Microalgae: from Biodiesel to Bioethanol and Beyond, March 23, 2009 Filed under Algae, Biofuels.

Mundt, S., Kreitlow, S., Nowotny, A. and Effmert, U. 2001. Biological and pharmacological investigation of selected cyanobacteria. Int. J. Hyg. Environ. Health. **203**: 327–334.

Novakovic, G.V., Kim, Y., Wu, X., Berzin, I. and Merchuk, J.C. 2005. Air-lift bioreactors for algal growth on flue gas: Mathematical modeling and pilot-plant studies. Ind. Eng. Chem. Res. **44(16)**: 6154–6163.

Papke, U., Gross, E.M. and Francke, W. 1997. Isolation, identification and determination of the absolute configuration of fischerellin B. A new algicide from the freshwater cyanobacterium *Fischerella musscicola*. Tetrahedron Lett. **38**: 379–382.

Pulz, O. 2001. Photobioreactors: production systems for phototrophic microorganisms. Appl. Microbiol. Biotechnol. **57**: 287–93.

Rao, M., Malhotra, S., Fatma, T. and Rattan A. 2007. Antimycobacterial activity from Cyanobacterial extracts and phytochemical screening of methanol extract of *Hapalosiphon*. Pharma. Biol. **45**: 88–93.

Roessler, P.G., Brown, L.M, Dunahay, T.G., Heacox, D.A., Jarvis, E.E. and Schneider, J.C. 1994. Genetic–engineering approaches for enhanced production of biodiesel fuel from microalgae. ACS Symp Ser. **566**: 255–70.

Rosenberg, Julian, N., George, A. Oyler, L. W. and Michael, J. B. 2008. A green light for engineered algae: redirecting metabolism to fuel a biotechnology revolution. Current Opinion in Biotechnology. **19**: 430–436.

Sanchez, Miron A., Ceron Garcia MC, Contreras Gomez A., Garcia Camacho F., Molina Grima E. and Chisti Y. 2003. Shear stress tolerance and biochemical characterization of *Phaeodactylum tricornutum* in quasi steady–state continuous culture in outdoor photobioreactors. Biochem. Eng. J. 16: 287–297.

Sawayama, S., Inoue, S., Dote, Y. and Yokoyama, S.Y. 1995. CO_2 fixation and oil production through microalga. Energy Convers Manag. **36**: 729–31.

Schwartz, R.E. 1990. Pharmaceuticals from cultured algae. J. Ind. Microbiol. **5**: 113–23.

Sharma, R., Chisti, Y. and Banerjee U.C. 2001. Production, purification, characterization, and applications of lipases. Biotechnol. Adv. **19**: 627–62.

Sheehan, J., Dunahay, T., Benemann, J. and Roessler, P. 1998. A look back at the U.S. Department of Energy's Aquatic Species Program–biodiesel from algae. National Renewable Energy Laboratory, Golden, CO; n Report NREL/TP–580–24190.

Shimizu, Y. 1996. Microalgal metabolites: a new perspective. Annu. Rev. Microbiol. **50**: 431–65.

Schenk, P.M., Thomas-Hall, S.R., Stephens, E., Marx, U.C., Mussgnug, J.H., Posten, C., Kruse, O. and Hankamer, B., 2008. Second generation biofuels: high–efficiency microalgae for biodiesel production. Bioenergy Res. **1**: 20–43.

Singh, S., Kate, B.N. and Banerjee, U. C. 2005. Bioactive compounds from cyanobacteria and microalgae: an overview. Crit. Rev. Biotechnol. **25**: 73–95.

Spolaore, P., Joannis–Cassan, C., Duran, E. and Isambert, A. 2006. Commercial applications of microalgae. J. Biosci. Bioeng. **101**: 87–96.

Tredici, M.R. 1999. Bioreactors, photo. In: Flickinger, M.C. and Drew, S.W. (eds.). Encyclopedia of bioprocess technology: fermentation, biocatalysis and bioseparation Wiley. p. 395–419.

Walter, T.L., Purton, S., Becker, D.K. and Collet, C. 2005. Microalgae as bioreactor. Plant Cell Rep. **24**: 629–41.

Yun, Y.S., Lee, S.B., Park, J.M., Lee, C.I. and Yang, J.W. 1997. Carbon dioxide fixation by algal cultivation using wastewater nutrients. J. Chem. Technol. Biotechnol. **69**: 451–5.

Zittelli, G.C., Liliana, R., Natascia, B., Mario, R. and Tredici. 2006. Productivity and photosynthetic efficiency of outdoor cultures of *Tetraselmis suecica* in annular columns. Aquaculture. **261**: 932–943.

2013, Abiotic Stress and Biotechnology
Editor: T. Pullaiah
Published by: REGENCY PUBLICATIONS, NEW DELHI

Pages 57–73

Chapter 4

Application of High Throughput DNA Sequencing for Plant Science Research

*Tapan Kumar Mondal**

National Research Centre on DNA Fingerprinting
National Bureau of Plant Genetic Resources,
Pusa, New Delhi – 110 012

ABSTRACT

High-throughput DNA sequencing technologies are the most recent technologies that are used in biology to sequence the selective or entire genome of an organism. They generate millions of short nucleotide sequence which have various applications in biology particularly in genomics, epigenomics and transcriptomics that are described here.

Introduction

Although the DNA of any organism in this universe is made up of four basic unit or their derivatives, then one might be thinking how such diverse organism starting from tiny bacteria to giant dinosaurs or small grasses to big banyan trees are created by nature. This is because each of them has different genes in their DNA or in more simple sense, it is the difference in sequence of their genome. Thus understanding or knowing the sequence of the entire genome is pre-requisite for modern day genomics

---

* Corresponding Author: E-mail: mondaltk@yahoo.com

study which can be achieved by whole genome sequencing, known as *de novo* sequencing: a terminology which is often used to indicate the sequencing of the entire genome for the first time. Once the genome sequence is known, then the information can be utilized in several ways but primarily for two purposes: 1) to detect the dissimilarity between the two organisms or group of organisms; 2) to identify new or different gene which is why mosquito become small or elephant is so big or a halophyte can survive in high salty water of sea when others can not. Thus sequence level variation can answer many of the questions which scientist can utilize either to identify improved superior cultivar or to create a specific cultivar.

Chemistry of Different Sequencing Technologies

Although different generation sequencing techniques use different chemistry yet there are commonality in the process of sequencing. First, DNAs of interest are broken into small pieces, sequenced and then these small sequences are aligned or joined one by one to form contig or bigger piece of DNA molecule. Finally, the contigs are joined together to get entire long sequence of the genome.

First generation or sanger sequencing method is popularly known as 'chain termination" reaction which is based on DNA polymerase-dependent synthesis of a complementary DNA strand in the presence of regular deoxynucleotides (dNTPs) and modified one (22,32-dideoxynucleotides: ddNTPs) that serve as nonreversible synthesis terminators (Sanger *et al.,* 1977). The DNA synthesis reaction is randomly terminated whenever a modified ddNTP is added to the growing oligonucleotide chain, resulting in truncated products of varying lengths with an appropriate ddNTP at their 32 terminus. The products are separated then by their size in polyacrylamide gel electrophoresis and the terminal ddNTPs are used to reveal the DNA sequence of the template strand. Four different test tubes are required to sequence one piece of DNA for four different ddNTP terminators *i.e.,* ddATP, ddCTP, ddTTP, or ddGTP. However, there is one major limitation of the technique *i.e.,* sequencing of DNA strand require an *in vivo* support which is normally done by cloning or inserting the DNA to be sequenced into bacterial plasmid but this cloning step is prone to host-related biases, is lengthy, and is quite labor intensive. Nevertheless, sanger sequencing is suitable for sequencing of individual gene. Simultaneously, another method known as Maxum-Gilbert was also developed but the principally based on base specific chemical cleavage of DNA stand in four separate reaction, followed by the separation of the fragments in polyacrylamide gel electrophoresis as per the length of the fragments. Later due to technological simplicity, use of non-radioactive material and amenable to higher scale up, sanger sequencing moved forward so much as that it become a highly automated technique.

The cloning to bacterial plasmid, prior to sequencing was eliminated by various second-generation sequencing technology, although, they differ in chemistry of sequencing. For an example, instead of cloning, a highly efficient *in vitro* DNA amplification method is known as 'emulsion PCR' was used in Roche 454, the first next generation sequencing technology came to the market in 2005. The 454 approaches is known as pyrosequencing which is based on sequencing-by-synthesis technique that measures the release of inorganic pyrophosphate (PPi) by chemiluminescence during sequencing reaction. The sequence of DNA template is determined from a "pyrogram," which corresponds to the order of correct nucleotides that had been incorporated. Since chemiluminescent signal intensity is proportional to the amount of pyrophosphate released and hence the number of bases incorporated, the pyrosequencing approach is prone to errors that result from incorrectly estimating the length of homopolymeric sequence stretches (*i.e.*, indels). The advantage of this technique is that it can do the sequencing for longer read length but demerits of having higher cost per unit of sequence. Nevertheless, due to larger sequencing length, this is used for *de novo* sequencing (Morozova and Marra, 2008; Deschamps and Campbell, 2010).

In Illumina/Solexa, the DNA templates are sequenced in a massively parallel fashion using sequencing-by-synthesis approach which employs reversible terminators with removable fluorescent moieties and special DNA polymerases which can incorporate these terminators into growing oligonucleotide chains. The major merit is that it can generate the millions of reads but the demerit is that the very high initial cost of the instruments. Nevertheless, this technique is the most popular, suitable for resequencing which has greater application for studying epigenetic (change of phenotypes without the change in DNA sequence) by chip-seq (identification of protein binding sequence of DNA), small non-coding RNA sequencing and degradome sequencing.

In SOLiDs (**S**upported **O**ligonucleotide **L**igation and **D**etection **S**ystem), the chemistry of sequencing is based on massively parallel sequencing by ligation which is quite similar that of 454 but differ in the sense that after amplification, the sequencing is done in sequential rounds of hybridization and ligation with 16 dinucleotide combinations labeled by four different fluorescent dyes (each dye use to label four dinucleotides). The advantage of this technology is that it has higher sequencing accuracy with the generation of larger amount of sequencing data at minimum time. Due to generation of large number sequencing data, it gives better coverage, a term which is often used to describe the number of times a region of the genome has been sequenced and which is an valuable information for genome sequencing.

'Ion torrent' is another second generation sequencing technology which is based on the fact that adding one nucleotide in the DNA chain always release one H^+ in the medium that causes the change of pH which is then detected by ion sensor somewhat like a pH meter. This technique is also suitable for *de novo* sequencing of the genome with higher accuracy.

Recently in 2011, PacBio, a revolutionary third generation sequencer has come to the market based on SMRT (**S**ingle **M**olecule **R**eal **T**ime) technology which does single molecule DNA sequencing in real time. It has the potentiality of sequencing up to 20,000 bp long DNA within one hour. Therefore this technique is highly recommended for *de novo* sequencing. Another third generation technology is known as 'Helicos', gaining popularity with an advantage of direct RNA sequencing. Therefore the step of making cDNA library (collection of all converted RNA molecule to it complementary DNA) is eliminated which ultimately reduce the cost of sample preparation.

However, based upon the principle of nanotechnology, there are several, "future generation sequencing" or "fourth generation sequencing" technologies which will be available for commercial use in near future. These are based on single molecule real time sequencing but the sequencing length will be potentially unlimited which differs from the third generation sequencing.

Table 4.1: Different technologies used in sequencing.

Technology	Chemistry	Length of Reads	Generation
Automated Sanger sequencer *e.g.*, ABI3730xl	Synthesis in the presence of dye terminators	> 1000 bp	**First generation**
454/RocheFLX system *e.g.*, 454XLR	pyrosequencing	> 1000 bp	**Second generation**
Illumina/Solexa *e.g.*, Hi Seq 2000	Sequencing by synthesis with reversible terminators	>150 bp	**Second generation**
ABI/SOLiD *e.g.*, 5500xl	ABI/SOLiD Based base massively parallel sequencing by ligation	>50 bp	**Second generation**
Ion torent *e.g.*, PGM 318	Change of pH	>400 bp	**Second generation**
PacBio *e.g.*, PacBio *RS*	SMRT (Single molecule real time)	> 10,000 bp	**Third generation**
Helicos	Sequencing by synthesis	> 55 bp	**Third generation**
IBM Nanopore	Transistor mediated DNA sequencing	Potentially unlimited	**Future generation**
Oxford Nanopore *e.g.*, GridON system	Direct electrical detection of single molecule	Potentially unlimited	**Future generation**

Application

Although there are excellent reviews of various aspects of high throughput sequencing (Schadt *et al.*, 2010; Ekblom and Galindo, 2011) yet an introductory account is given below.

Identification of Differentially Expressed Gene

Quantification of differentially expressed mRNA species under different conditions, cell or tissue types have long been of interest to plant scientist.

When the primary objective of a biological study is comparison of gene expression profiling between two samples, mRNA sequencing (RNAseq), is the most appropriate approach for number of reasons. This sort of analysis is particularly relevant for detecting the difference of mRNA population between the wild-type and mutant strains, between treated versus untreated tissue, cancer versus normal cell, and so on.

It is noteworthy to mention that, this identification of differentially expressed genes can be studied in a number of ways such as SSH (selective subtractive hybridization), cDNA-AFLP (amplified fragment length polymorphism), MPSS (massive parallel sequence signature), SAGE (serial analysis of gene expression), DDRT (differential display of reverse transcriptase) and microarray, yet each of them embodied with merits and demerit. Nevertheless, newer techniques become always superior to older one. Due to larger dynamic range and sensitivity of RNA-seq, several additional factors have contributed to the rapid uptake of sequencing for differential expression analysis. For an example, microarrays are simply not available for many non-model organisms (for example, Affymetrix offers microarrays for approximately 30 organisms. Additionally, sequencing gives unprecedented detail about transcriptional features that arrays cannot, such as novel transcribed regions, allele-specific expression, RNA editing and a comprehensive capability to capture alternative splicing. Thus RNAseq became popular choice for number of purpose.

In order to detect the developmental stage specific genes in watermelon, Gau *et al.* (2011) isolated RNA from four fruit developmental stages (immature white, white-pink flesh, red flesh and over-ripe). After RNAseq analysis using 454, they obtained 577,023 high quality ESTs with an average length of 302.8 bp. *De novo* assembly of these ESTs together with 11,786 watermelon ESTs collected from GenBank produced 75,068 unigenes with a total length of approximately 31.8 Mb. Overall 54.9 per cent of the unigenes showed significant similarities to known sequences in GenBank non-redundant (nr) protein database and around two-thirds of them matched proteins of cucumber, the most closely-related species with a sequenced genome. The unigenes were further assigned with gene ontology (GO) terms and mapped to biochemical pathways. Finally they identified 3,023 genes that were differentially expressed during watermelon fruit development and ripening, which provided novel insights into watermelon fruit biology and a comprehensive resource of candidate genes for future functional analysis. Then they generated profiles of several interesting metabolites that are important to fruit quality including pigmentation and sweetness.

Wang *et al.* (2010) used illumina platform to detect the gene responsible for developing cotton fibre by comparing an wild cotton variety and a mutant developed in the same wild genotypes back ground. Overall, they found high degree of transcriptional complexity in early developing fibers and represent a major improvement over the microarrays for analyzing transcriptional changes on a large scale.

Transcriptional dynamics of berry development in *Vitis vinifera* 'Corvina' was analyzed using RNAseq technology (Zenon *et al.*, 2010). More than 59 million sequence reads, 36 to 44 bp in length, were generated from three developmental stages: post setting, ve´raison, and ripening. They reported 17,324 genes expressed during berry development, 6,695 of which were expressed in a stage-specific manner, suggesting differences in expression for genes in numerous functional categories and a significant transcriptional complexity. This exhaustive overview of gene expression dynamics demonstrates the utility of RNA-Seq for identifying single nucleotide polymorphisms and splice variants and for describing how plant transcriptomes change during development.

In order to identify the differentially expressed gene under water logging in cucumber plant, 5.8 million total clean sequence tags per library were obtained with

143013 distinct clean tag sequences. They found most of the genes were linked to carbon metabolism, photosynthesis, reactive oxygen species generation/scavenging, and hormone synthesis/signaling (Qi *et al.,* 2012).

RNA-Seq was used to identify the osmotic stress related gene in sorghum. Root and shoot tissues of sorghum were subjected to polyethylene glycol (PEG)-induced osmotic stress or exogenous ABA and identified 28335 unique gene. Differential gene expression analyses in response to osmotic stress and ABA revealed a strong interplay among various metabolic pathways including abscisic acid and 13-lipoxygenase, salicylic acid, jasmonic acid, and plant defense pathways. Transcription factor analysis indicated that groups of genes may be co-regulated by similar regulatory sequences to which the expressed transcription factors bind (Dugas *et al.,* 2011).

De novo Sequencing of Non-Model Crop Species

This is particularly important for sequencing of a plant for the first time or for a plant which has no sequenced genome. For years, people were doing this in little different way which is known as 'shot gun' technique. However, shot gun technique is less throughput than this *denovo* sequencing. It is noteworthy to mention here that recently pigeon pea (Singh *et al.,* 2012) and Neem (Krishnan *et al.,* 2011) have been sequenced by Indian scientists for the first time. However several plant species have been sequenced which are shwown in Table 4.2.

SNP Discovery

One of the central themes in genomics is to study allelic differences or variations which can be detected by discovering biallelic markers such as single nucleotide polymorphism (SNP) or haplotype *i.e.,* group of SNPs that are associated in a particular trait. Single-nucleotide polymorphisms (SNPs) are the most abundant type of DNA sequence polymorphisms. Their higher availability and stability when compared to simple sequence repeats (SSRs) provide enhanced possibilities for genetic and breeding applications such as cultivar identification, construction of genetic maps, the assessment of genetic diversity, the detection of genotype/phenotype associations, or marker-assisted breeding. In addition, the efficiency of these activities can be improved thanks to the ease with which SNP genotyping can be automated. Short sequences can be compared to a reference genome sequence to detect the variants. These SNPs are of very good molecular marker for developing the genetic map which is useful information for plant breeder. Large scale SNP has been discovered in several crops using various source of sequence. For an example EST as well as transcriptomic sequence were used for grapevine to detect large scale SNPs (Lijavetzky *et al.,* 2007).

Using a multi-tier reduced representation library, Hyten *et al.* (2010) discovered a total of 3,487 SNPs of which 2,795 contained sufficient flanking genomic sequence for SNP assay development. Using Sanger sequencing to determine the validation rate of these SNPs, they found that 86 per cent are likely to be true SNPs. Furthermore, a Golden Gate Assay was designed which contained 1,050 of the 3,487 predicted SNPs. A total of 827 of the 1,050 SNPs produced a working GoldenGate assay (79 per cent).

Table 4.2: Summary of *de novo* sequencing of non-model plant species.

Plant	Purpose	Reference
Eucalyptus grandis	Transcriptome characterization, SNP identification	Novaes *et al.*, 2008
Pachycladon enysii	Reference-guided assembly using diverged reference	Collins *et al.*, 2008
California poppy (*Eschscholzia californica*)	Transcriptome characterization	Wall *et al.*, 2009
Sugarcane (*Saccharum officinarum*)	Transcriptome characterization	Bundock *et al.*, 2009
Red mangrove (*Rhizophora mangle*)	Transcriptome characterization, comparative gene expression	Dassanayake *et al.*, 2009
Roughfruit amaranth (*Amaranthus tuberculatus*)	Transcriptome characterization	Lee *et al.*, 2009
Artemesia annua	Transcriptome characterization	Wang *et al.*, 2009
Chestnut (*Castanea dentata*)	Transcriptome characterization, comparative gene expression	Barakat *et al.*, 2009
Looking-glass mangrove (*Heritiera littoralis*)	Transcriptome characterization, comparative gene expression	Dassanayake *et al.*, 2009
Olive (*Olea europaea* cv. Coratina)	Transcriptome characterization, comparative gene expression	Alagna *et al.*, 2009
Olive (*O. europaea* cv.Tendellone)	Transcriptome characterization, comparative gene expression	Alagna *et al.*, 2009
Amaranthus tuberculatus	Transcriptome characterization, comparative gene expression	Riggins *et al.*, 2010
Wild oat (*Avena barbata*)	Transcriptome characterization, comparative gene expression	Swarbreck *et al.*, 2011
Cucumber (*Cucumis sativus*)	Transcriptome characterization, comparative gene expression	Guo *et al.*, 2010
Eucalyptus grandis × *E. urophylla*	Transcriptome characterization, comparative gene expression	Mizrachi *et al.*, 2010
Garden pea (*Pisum sativum*)	Transcriptome characterization, comparative gene expression	Franssen *et al.*, 2011
Bracken fern (*Pteridium aquilinum*)	Transcriptome characterization	Der *et al.*, 2011
Scabiosa columbaria	Transcriptome characterization, SNP identification	Angeloni *et al.*, 2011
Medicinal Chinese herb (*Salvia miltiorrhiza*)	*De novo* transcriptom sequencing	Wenping *et al.*, 2011
Neem (*Azadirachta indica*)	Whole genome sequence	Krishnan *et al.*, 2011
Pigeon pea (*Cajanus cajan*)	Whole genome sequence	Singh *et al.*, 2011

   To identify a large number of SNPs in elite potato germplasm, Hamilton *et al.* (2011) sequenced normalized cDNA prepared from three commercial potato cultivars: 'Atlantic', 'Premier Russet' and 'Snowden'. For each cultivar, around 2 Gb of sequence was generated which was assembled into a representative transcriptome of ~28-29 Mb for each cultivar. Using the Maq SNP filter that filters read depth, density, and quality, 575,340 SNPs were identified within these three cultivars. In parallel, 2,358 SNPs were identified within existing sanger sequences for three additional cultivars, 'Bintje', 'Kennebec', and 'Shepody'.

Using a stringent set of filters in conjunction with the potato reference genome, Hamilton *et al.* (2011) identified 69,011 high confidence SNPs from these six cultivars for use in genotyping with the Infinium platform. Ninety-six of these SNPs were used with a BeadXpress assay to assess allelic diversity in a germplasm panel of 248 lines; 82 of the SNPs proved sufficiently informative for subsequent analyses. Within diverse North American germplasm, the chip processing market class was most distinct, clearly separated from all other market classes. The round white and russet market classes both include fresh market and processing cultivars. Nevertheless, the russet and round white market classes are more distant from each other than processing are from fresh market types within these two groups.

SSR Discovery

Next-generation sequencing (NGS) technologies have wide application for development of simple sequence repeat (SSR) or microsatellite loci which is powerful molecular marker due to their poly-allelic nature. However, their development has traditionally been a difficult and costly process. NGS technologies allow the efficient identification of large numbers of microsatellites at a fraction of the cost and effort of traditional approaches. This shortfall is taken care by NGS methods due to their ability to produce large amounts of sequence data from which isolation and development of numerous genome-wide and gene-based microsatellite loci become easy (Zalapa *et al.*, 2011). In the future, NGS technologies will massively increase the number of SSRs and other genetic markers available to conduct genetic research in understudied but economically important crops such as cranberry.

Sequence Informed Conservation and Utilization of Plant Growth Regulator (PGR)

NGS technologies are useful for conservation of PGR. Often, high degree of redundancy found among the different *ex situ* collections which wastes a time, resources etc. Most attempts to identify duplicated samples suffered from the difficulty to agree on a common set of markers for a given species, manifold problems to reproduce DNA marker data between different labs. NGS data do not suffer from such shortcomings and therefore represent an ideal information platform to tackle the issue of redundancy. Arguably, sequencing of *ex situ* collections just for the sake of eliminating redundancy would be too expensive an undertaking. Clearly large crop collections cannot be sequenced in one draft. Therefore, developing the core reference set (CRS) i.e 'a set of genetic stocks that are representative of the genetic resources of the crop and are used by the scientific community as a reference for an integrated characterization of its biological diversity' (Glaszmann *et al.*, 2010). Every CRS will serve as a public, standardized and well characterized resource for the scientific community. Well characterized, multiplied, isolated CRS have to be maintained for reference purposes, comparative studies, future reanalysis and integrative genomic analysis (Hawkins *et al.*, 2010).

NGS methods are valuable for low cost sequencing of CRS to develop genome-wide marker which will facilitate the rejection of duplication across the genomes of populations (Bansal *et al.*, 2010, Davey *et al.*, 2011). There are several methods in this regard such as reduced-representation libraries (RRLs) (You *et al.*, 2011; Gompert *et al.*, 2010), complexity reduction of polymorphic sequences (CRoPS) (van Orsouw *et al.*, 2007; Mammadov *et al.*, 2010), restriction-site associated DNA sequencing (RAD-seq) (Baxter *et al.*, 2011) and low-coverage sequencing for genotyping (Huang *et al.*, 2009; Andolfatto *et al.*, 2011; Elshire *et al.*, 2011) are applicable for genetic analysis crop plant particularly to non-model species, to species with high levels of repetitive DNA or to breeding germplasm with low levels of polymorphism-without the need for prior sequence information. These methods can be applied to compare SNP diversity within and between closely related plant species or within wild natural populations (Ossowski *et al.*, 2010; Pool *et al.*, 2010).

Epigenetics

Epigenetic is different from genetics which means changes of phenotype without any change in genetic material That is normally occurred by methylation in cytosine bases of DNA or post-translational modification of amino acid present in histone tails. They play a key role in gene expression and in plant development under stress.

Methylome

One major type of epigenetic modification that has been an important research focus for many years is the methylation of specific cytidine residues in the DNA. Generally, methylation has been studied using bisulphate treatment. Un-methylated cytidine residues in the DNA are converted to uracil after treatment with bisulphite, whereas methylated cytidine are protected from this conversion. Sequencing of these regions can then pinpoint the specific methylated nucleotides. In recent years, NGS has been utilized to characterize DNA methylation patterns genome-wide, using an application known as ultra-deep bisulphite DNA sequencing (BSSeq) (Taylor *et al.*, 2007; Cokus *et al.*, 2008). Methylated fractions of genomes can also be sequenced using a combination of NGS and methyl-DNA immunoprecipitation (MeDIP). By using this approach, detailed methylation maps of the genome will become available (Pomraning *et al.*, 2009), and by comparing these for different samples, one will be able to address the importance of methylation for a large range of ecologically and evolutionary important questions, like the genetic architecture behind differential gene expression because of natural selection.

Becker *et al.* (2011) compared genome-wide DNA methylation among 10 *A. thaliana* lines, derived 30 generations ago from a common ancestor. Epimutations at individual positions were easily detected, and close to 30,000 cytosines in each strain were differentially methylated. In contrast, larger regions of contiguous methylation were much more stable, and the frequency of changes was in the same low range as that of DNA mutations. Like individual positions, the same regions were often affected by differential methylation in independent lines, with evidence for recurrent cycles of forward and reverse mutations. Transposable elements and short interfering RNAs have been causally linked to DNA methylation. In agreement, differentially methylated sites were farther from transposable elements and showed less association with short interfering RNA expression than invariant positions. The biased distribution and frequent reversion of epimutations have important implications for the potential contribution of sequence-independent epialleles to plant evolution.

Mapping of DNA-Binding Proteins and Chromatin

Another important determinant of transcription levels of genes is the structure of DNA packing, together with histone proteins, into nucleosomes (chromatin packing). This chromatin structure can also be studied using a high-throughput sequencing approach (Johnson *et al.*, 2006). The histone proteins themselves, particularly the N-terminal tails, are also subject to a large number of posttranslational modifications, such as methylation and acetylation of specific amino acid residues (Kouzarides, 2007). These can be studied on a large scale using a ChIP-Seq approach with Illumina/Solexa sequencing (Barski *et al.*, 2007). ChIP-Seq technology can also be used to study a large range of other DNA–protein interactions (Hurd and Nelson, 2009) including the identification of binding sites for transcription factors (Bhinge *et al.*, 2007). Using chromatin immunoprecipitation and deep sequencing (ChIP-Seq) Van *et al.* (2010) established the whole-genome distribution patterns of histone H3 lysine 4 mono-, di-, and tri-methylation (H3K4me1, H3K4me2, and H3K4me3, respectively) in *Arabidopsis thaliana* during watered and dehydration stress conditions.

ChIP-on-chip [ChIP (chromatin immuno-precipitation) using microarrays] is a key approach to map *in vivo* binding sites of various DNA-binding proteins across the genome. Here DNA-protein complex is precipitated with the protein specific antibody after which DNA was isolated with protease K digestion for sequence. This approach is termed as 'ChIP-Seq', which should produce a huge windfall, in particular for studies of multi-cellular eukaryotes where whole genome coverage has generally required the use of several arrays.

Small Non-Coding RNA Profiling and the Discovery of Novel Small RNAs

A related application of next-generation sequencing technologies is to the analysis of transcriptomes for the discovery of small noncoding RNA (ncRNAs) and profiling. These ncRNAs are RNA molecules that are not translated into a protein product. This class of RNAs includes transfer RNA (tRNA), ribosomal RNA (rRNA), small nuclear and small nucleolar RNA, and microRNA and small interfering RNA (miRNA and siRNA). Recent research has implicated microRNAs, approximately 21-nucleotide-long RNA molecules, as crucial posttranscriptional regulators of gene expression in both animals and plants (Filipowicz *et al.*, 2008). A related class of noncoding RNAs, siRNAs, 21–24 nt in length, is the predominant class of small RNA molecules in plants (Lu *et al.*, 2005). Definite evidence for the presence of endogenous siRNAs in animals is lacking (Meyers *et al.*, 2006). While miRNAs and siRNAs are similar in size and are both involved in posttranscriptional regulation of gene expression, their biogenesis and exact functions are different (Xie *et al.*, 2004).

High-throughput sequencing of small RNAs provides great potential for the identification of novel small RNAs as well as profiling of known and novel small RNA genes. The MPSS technology has been applied to the sequencing of size-fractionated small RNAs from *A. thaliana* (Lu *et al.*, 2005). This approach involved the generation of 17-nt fragments corresponding to parts of mature small RNA molecules and using bioinformatic analysis to identify the corresponding small RNA genes in the genome (Lu *et al.*, 2005). Several disadvantages of this approach are the high complexity and cost of the MPSS technology, which involves a cloning step, and the short read lengths corresponding to only a portion of small RNA molecules (Meyers *et al.*, 2006).

Nevertheless, using next generation sequencing approach, trait specific miRNA have been identified in several plant species such as cold responsive microRNA in *Populus tomentosa* (Chen *et al.*, 2012), drought-responsive microRNAs in *Medicago truncatula* (Wang *et al.*, 2011), aluminum-responsive microRNAs in *Medicago truncatula* (Chen *et al.*, 2012) and Ethylene-responsive miRNAs in roots of *Medicago truncatula* (Chen *et al.*, 2012).

Degradome

To understand the biological roles of each miRNA, both miRNA cloning and target validation are required. Conventionally 5′ RACE (rapid amplification of cDNA ends) was widely employed for target confirmation and cleavage site mapping (Jones-Rhoades *et al.*, 2006). However, this method is only applicable for fine-scale purposes, and it will be laborious, time-consuming, and costly if it is intended to verify numerous miRNA–target pairs. Fortunately, a high-throughput method combining modified 5′ RACE with the next-generation sequencing technology was developed, and has greatly enhanced our capacity for target validation. Recently, PARE (parallel analysis of RNA ends) has been adopted by several research groups for plant degradome sequencing (Addo-Quaye *et al.*, 2008; German *et al.*, 2008; Li *et al.*, 2010). Excluding the degradation remnants generated during RNA sample preparation, a large portion of degradomes are derived from miRNA-mediated 3′ cleavage products of the target transcripts. Thus, the sequencing technology is a powerful approach for large-scale miRNA–target pair validation *in planta*, thanks to their special features including transcriptome-wide sequencing and quantitative detection. Based on these high-throughput degradome sequencing data, numerous miRNA–target pairs, including previously verified and not yet uncovered pairs, have been confirmed (Addo-Quaye *et al.*, 2008; German *et al.*, 2008; Li *et al.*, 2010).

Conclusion

The technology of sequencing has dramatically changed from a manual mode to fully automated machine generated operation (Shendure and Ji, 2008). Greater speed, lower cost of sequencing, longer sequencing length with lower error rate are the main criteria for advancement of this technology which will yield the genome sequences of many organism in near future. This will help to discover the new genes or developing new varieties with greater yield or suitable to grow in the context of climate changes. Thus this will be a significant step towards the development of hunger free world: a dream of 'green revolution father', Dr Norman E. Borlaug.

References

Addo–Quaye, C., Eshoo, T.W., Bartel, D.P. and Axtell, M.J. 2008. Endogenous siRNA and miRNA targets identified by sequencing of the *Arabidopsis* degradome. Curr Biol. **18**: 758–762.

Alagna, F., Agostino, N.D., Torchia, L., Servili, M., Rao, R., Pietrella, M. and Giuliano, G. 2009. Comparative 454 pyrosequencing of transcripts from two olive genotypes during fruit development. BMC Genomics **10** : 399.

Andolfatto, P., Davison, D., Erezyilmaz, D., Hu, T.T., Mast, J., Sunayama–Morita, T. and Stern, D.L. 2011. Multiplexed shotgun genotyping for rapid and efficient genetic mapping. Genome Res. **21**: 610–617.

Angeloni, F., Wagemaker, C.A.M., Jetten, M.S.M., Opden, H.J.M., Camp, I., Janssen–Megens, E.M., Françoijs, K.J., Stunnenberg H.G. and Ouborg, N.J. 2011. *De novo* transcriptome characterization and development of genomic tools for *Scabiosa columbaria* L. using next generation sequencing techniques. Mol Ecol Res **11**: 662 – 674.

Bansal, V., Harismendy, O., Tewhey, R., Murray, S.S., Schork, N.J., Topol. E.J. and Frazer, K.A. 2010. Accurate detection and genotyping of SNPs utilizing population sequencing data. Genome Res. **20**: 537–545.

Barakat, A., Diloreto, D.S., Zhang, Y., Smith, C., Baier, K., Powell, W.A. and Wheeler, N. 2009. Comparison of the transcriptomes of American chestnut (*Castanea dentata*) and Chinese chestnut (*Castanea mollissima*) in response to the chestnut blight infection. BMC Plant Biol. **9**: 51.

Barski, A., Cuddapah, S., Cui, K., Roh, T.Y., Schones, D.E., Wang, Z., Wei, G., Chepelev, I. and Zhao, K. 2007. High–resolution profiling of histone methylations in the human genome. Cell. **129**: 823–837.

Baxter, S.W., Davey, J.W., Johnston, J.S., Shelton, A.M., Heckel, D.G., Jiggins, C.D. and Blaxter, M.L. 2011. Linkage mapping and comparative genomics using next–generation RAD sequencing of a non–model organism. PLoS ONE **6**: 19315-19326.

Becker, C., Hagmann, J., Müller, J., Koenig, D., Stegle, O., Borgwardt, K. and Weigel, D. 2011. Spontaneous epigenetic variation in the *Arabidopsis thaliana* methylome. Nature. **480**: 245–249.

Bhinge, A.A., Kim, J., Euskirchen, G.M., Snyder, M. and Iyer, V.R. 2007. Mapping the chromosomal targets of STAT1 by sequence tag analysis of genomic enrichment (STAGE). Genome Res. 17: 910–916.

Bundock, P.C., Eliott, F.G., Ablett, G., Benson, A.D., Casu, R.E., Aitken, K.S. and Henry, R.J. 2009. Targeted single nucleotide polymorphism (SNP) discovery in a highly polyploid plant species using 454 sequencing. Plant Biotechnol. J. **7**: 347–354.

Cannon, C.H., Kua, C.S., Zhang, D. and Harting, J.R. 2010. Assembly free comparative genomics of short–read sequence data discovers the needles in the haystack. Mol. Ecol. **19**: 147–161.

Chen, L., Wang, T., Zhao, M., Tian, Q. and Zhang, W.H. 2012. Identification of aluminum–responsive microRNAs in *Medicago truncatula* by genome–wide high–throughput sequencing. Planta. **235**: 375–86.

Chen, L., Wang, T., Zhao, M. and Zhang, W. 2012. Ethylene–responsive miRNAs in roots of *Medicago truncatula* identified by high–throughput sequencing at whole genome level. Plant Sci. **184**: 14–19.

Chen, L., Zhang, Y., Ren, Y., Xu, J., Zhang, Z. and Wang, Y. 2012. Genome–wide identification of cold–responsive. and new microRNAs in *Populus tomentosa* by high–throughput sequencing. Biochem. Biophys. Res. Commun. **417**: 892–986.

Cokus, S.J., Feng, S., Zhang, X., Chen, Z., Merriman, B., Haudenschild, C.D., Pradhan, S., Nelson, S.F., Pellegrini, M. and Jacobsen, S.E. 2008. Shotgun bisulphite sequencing of the *Arabidopsis* genome reveals DNA methylation patterning. Nature. **452**: 215–219.

Collins, L. J., Biggs, P.J., Voelckel, C. and Joly, S. 2008. An approach to transcriptome analysis of non–model organisms using short–read sequences. Genome informat. **21**: 3–14.

Dassanayake, M., Haas, J.S., Bohnert, H.J., Cheeseman, J.M. 2009. Shedding light on an extremophile lifestyle through transcriptomics. New Phytol. **183**: 764 –775.

Davey, J.W., Hohenlohe, P.A., Etter, P.D., Boone, J.Q., Catchen, J.M. and Blaxter, M.L. 2011. Genome–wide genetic marker discovery. and genotyping using next generation sequencing. Nat. Rev. Genet. **12**: 499–510.

Der, J. P., Barker, M.S., Wickett, N.J., Depamphilis, C.W., Wolf, P.G. 2011. *De novo* characterization of the gametophyte transcriptome in bracken fern, *Pteridium aquilinum.* BMC Genomics **12**: 99–111.

Deschamps, S. and Campbell, M.A. 2010. Utilization of next–generation sequencing platforms in plant genomics. and genetic variant discovery. Mol. Breed. **25**: 553–570.

Dugas, D.V., Monaco, M.K., Olsen, A., Klein, R.R., Kumari, S., Ware, D. and Klein, P.E. 2011. Functional annotation of the transcriptome of *Sorghum bicolor* in response to osmotic stress. and abscisic acid. BMC Genomics. **12**: 514–523.

Ekblom, R. and Galindo, J. 2011. Applications of next generation sequencing in molecular ecology of non–model organisms. Heredity **107**: 1–15

Elshire, R.J., Glaubitz, J.C., Sun, Q., Poland, J.A., Kawamoto, K., Buckler, E.S. and Mitchell, S.E. 2011. A robust, simple Genotyping–by–Sequencing (GBS) approach for high diversity species. PLoSONE 2011;6: 19379–19389.

Filipowicz, W., Bhattacharyya, S.N. and Sonenberg, N. 2008. Mechanisms of posttranscriptional regulation by microRNAs: are the answers in sight? Nat. Rev. Genet. **9**: 102–114.

Franssen, S, U., Shrestha, R.P., Bräutigam, A., Bornberg–bauer, E. and Weber, A.P. 2011. Comprehensive transcriptome analysis of the highly complex *Pisum sativum* genome using next generation sequencing. BMC Genomics **12**: 227.

German, M.A., Pillay, M., Jeong, D.H., Hetawal, A., Luo, S., Janardhanan, P., Kannan, V., Rymarquis, L.A., Nobuta, K., German, R., De, Paoli. E., Lu, C., Schroth, G., Meyers, B.C. and Green, P.J. 2008. Global identification of microRNA–target RNA pairs by parallel analysis of RNA ends. Nature Biotech. **26**: 941–946.

Glaszmann, J.C., Kilian, B., Upadhyaya, H.D. and Varshney, R.K. 2010. Accessing genetic diversity for crop improvement. Curr. Opin. Plant Biol. **13**: 167–73.

Gompert, Z., Forister, M.L., Fordyce, J.A., Nice, C.C., Williamson, R.J. and Buerkle, C.A. 2010. Bayesian analysis of molecular variance in pyrosequences quantifies population genetic structure across the genome of Lycaeides butterflies. Mol. Ecol. **19**: 2455–73.

Guo, S., Zheng, Y., Joung, J., Liu, S., Zhang, Z., Crasta, O.R. and Sobral, B.W. 2010. Transcriptome sequencing. and comparative analysis of cucumber flowers with different sex types. BMC Genomics **11**: 384–395.

Guo, S., Liu, J., Zheng, Y., Huang, M., Zhang, M., Gong, G., He, H., Ren, Y., Zhong, S., Fei, Z. and Xu, Y. 2011. Characterization of transcriptome dynamics during watermelon fruit development: sequencing, assembly, annotation. and gene expression profiles. BMC Genomics. **12**: 454–460.

Hamilton, J.P., Hansey, C.N., Whitty, B.R., Stoffel, K., Massa, A.N., Deynze, A.V., Jong, W.S.D., Douches, D.S. and Buell, C.R. 2011. Single nucleotide polymorphism discovery in elite north american potato germplasm. BMC Genomics **12**: 302–313.

Hawkins, R.D., Hon, G.C. and Ren, B. 2010. Next–generation genomics: an integrative approach. Nat Rev Genet **11**: 476–486.

Huang, X., Feng, Q., Qian, Q., Zhao, Q., Wang, L., Wang, A., Guan, J., Fan, D., Weng, Q., Huang, T., Dong, G., Sang, T. and Han, B. 2009. High–throughput genotyping by whole–genome resequencing. Genome Res. **19**: 1068–1076.

Hurd, P.J. and Nelson, C.J. 2009. Advantages of next–generation sequencing versus the microarray in epigenetic research. Brief Funct. Genomic Proteom. **8**: 174–183.

Hyten, D.L., Song, Q., Fickus, E.W., Quigley, E.W., Lim, J., Choi, I.K., Hwang, E.Y., Pastor–Corrales, M. and Cregan, P.B. 2010. High–throughput SNP discovery. and assay development in common bean. Genomics. **11**: 475–482.

John E., Pool, J.E., Hellmann, I., Jensen, J.D. and Nielsen, R. 2010. Population genetic inference from genomic sequence variation. Genome Res. **20**: 291–300.

Johnson, S.M., Tan, F.J., McCullough, H.L., Riordan, D.P. and Fire, A.Z. 2006. Flexibility. and constraint in the nucleosome core landscape of *Caenorhabditis elegans* chromatin. Genome Res. **16**: 1505–1516.

Jones–Rhoades, M.W., Bartel, D.P. and Bartel, B. 2006. MicroRNAs. and their regulatory roles in plants. Annual Rev. Plant Biol. **57**: 19–53.

Kouzarides, T. 2007. Chromatin modifications. and their function. Cell. **128**: 693–705.

Krishnan, N.M., Pattnaik, S., Deepak, S.A., Hariharan, A.K., Gaur, P., Chaudhary, R., Jain, P., Vaidyanathan, S., Krishna, P.G.B. and Panda, B. 2011. *De novo* sequencing. and assembly of *Azadirachta indica* fruit transcriptome. Curr Sci. **10**: 1533–1561.

Lee, R.M., Thimmapuram, J., Kate, A., Gong, T.G., Hernandez, A.G., Wright, C.L., Kim, R.W., Mikel, M.A. and Tranel PJ. 2009. Sampling the waterhemp (*Amaranthus tuberculatus*) genome using pyrosequencing technology. Weed Sci. **57**: 463–469.

Li, Y.F., Zheng, Y., Addo-Quaye, C., Zhang, L., Saini, A., Jagadeeswaran, G., Axtel, M.J., Zhang, W. and Sunkar, R. 2010. Transcriptome–wide identification of microRNA targets in rice. Plant J. **62**: 742–759.

Lijavetzky, D., Cabezas, J.A., Ibáñez, A., Rodríguez, V. and Martínez–Zapater, J.M. 2007. High throughput SNP discovery. and genotyping in grapevine (*Vitis vinifera* L.) by combining a re–sequencing approach. and SNPlex technology BMC Genomics. **8**: 424–432.

Lu, C., Tej, S.S., Luo, S., Haudenschild, C.D., Meyers, B.C. and Green, P.J. 2005. Elucidation of the small RNA component of the transcriptome. Sci. **309**: 1567–1569.

Mammadov, J.A., Chen, W., Ren, R., Pai, R., Marchione, W., Yalçin, F., Witsenboer, H., Greene, T.W., Thompson, S.A. and Kumpatla, S.P. 2010. Development of highly polymorphic SNP markers from the complexity reduced portion of maize (*Zea mays* L.) genome for use in marker–assisted breeding. Theor. Appl. Gen. **121**: 577–588.

Meyers, B.C., Souret, F.F. and Lu C, Green, P.J. 2006. Sweating the small stuff: microRNA discovery in plants. Curr. Opin. Biotechnol. **17**: 139–146.

Mizrachi, E., Hefer C.A., Ranik, M., Joubert, F., Myburg, A.A. 2010. *De novo* assembled expressed gene catalog of a fast–growing *Eucalyptus* tree produced by Illumina mRNA–Seq. BMC Genomics **11**: 681–687.

Morozova, O. and Marra, M.A. 2008. Applications of next–generation sequencing technologies in functional genomics. Genomics **92**: 255–264.

Novaes, E., Drost, D.R., Farmerie, W.G., Pappas, J.P., Grattapaglia, D., Sederoff, R.R. and Kirst, M.2008. High–throughput gene. and SNP discovery in *Eucalyptus grandis,* an uncharacterized genome. BMC Genomics **9**: 312–319.

Ossowski, S., Schneeberger, K., Lucas-Lledó, J.I., Warthmann, N., Clark, R.M., Shaw, R.G., Weigel, D. and Lynch, M. 2010. The rate. and molecular spectrum of spontaneous mutations in *Arabidopsis thaliana*. Sci. **327**: 92–94.

Pomraning, K.R., Smith, K.M. and Freitag, M. 2009. Genome–wide high throughput analysis of DNA methylation in eukaryotes. Methods **47**: 142–150.

Riggins, C.W., Peng, Y., Stewart, J.R.C.N. and Tranel, P.J. 2010. Characterization of *de novo* transcriptome for waterhemp (*Amaranthus tuberculatus*) using GS–FLX 454 pyrosequencing. and its application for studies of herbicide target site genes Pest Manag. Sci. **66**: 1042–1052.

Sanger, F., Nicklen, S. and Coulson, S. A. 1977. DNA sequencing with chain–terminating inhibitors, Proc. Natl. Acad. Sci. USA. **74**: 5463–5467.

Schadt, E.E., Turner, S. and Kasarskis, A. 2010. A window into third–generation sequencing. Human Mol. Genet. **19**: 227–240.

Schmieder, R. and Edwards, R. 2011. Fast identification. and removal of sequence contamination from genomics. and metagenomic datasets. PLoS ONE **6** : e17288.

Shendure, J. and Ji, H. 2008. Next–generation DNA sequencing. Nature Biotech. **26**: 1135 – 1145.

Singh, N.K. *et al.,* 2012. The first draft of the pigeon pea genome sequence. J. Plant Biochem. Biotech. **1**: 98–112.

Taylor, K.H., Kramer, R.S., Davis, J.W., Guo, J., Duff, D.J. and Xu, D. 2007. Ultradeep bisulfite sequencing analysis of DNA methylation patterns in multiple gene promoters by 454 sequencing. Cancer Res. **67**: 8511–8518.

van Dijk, K., Ding, Y., Malkaram, S., Riethoven, J.J., Liu, R., Yang, J., Laczko, P., Chen, H., Xia, Y., Ladunga, I., Avramova, Z. and Fromm, M. 2010. Dynamic changes in genome–wide histone H3 lysine 4 methylation patterns in response to dehydration stress in *Arabidopsis thaliana.* BMC Plant Biol. **10**: 238–245.

van Orsouw, N.J., Hogers, R.C.J., Janssen, A., Yalcin, F., Snoeijers, S., Verstege, E., Schneiders, H., Vander, P., van Oeveren, J., Verstegen, H.,. and van Eijk, M.J.T. 2007. Complexity Reduction of Polymorphic Sequences (CRoPSTM): a novel approach for large-scale polymorphism discovery in complex genomes. PLoSONE 2007;2: 1172–1182.

Wall, P.K., Leebens–Mack, J., Chanderbali, A.S., Barakat, A., Wolcott, E., Liang, H., Landherr, L., Tomsho, L.P., Hu, Y., Carlson, J.E., Ma, H., Schuster, S.C., Soltis, D.E., Soltis, P.S., Altman, N. and dePamphilis, C.W. 2009. Comparison of next generation sequencing technologies for transcriptome characterization. BMC Genomics. **10**: 347–357.

Wang, Q., Liu, F., Chen, X., Ma, X., Zeng, H. and Yang, Z. 2010. Transcriptome profiling of early developing cotton fiber by deep–sequencing reveals significantly differential expression of genes in a fuzzless/lintless mutant. Genomics. **96**: 269–276.

Wang, W. Y., Wang, Q., Zhang, Y. Q. and Guo, D. 2009. Global characterization of *Artemisia annua* glandular trichome transcriptome using 454 pyrosequencing. BMC Genomics **10**: 465–475.

Wang, T., Chen, L., Zhao, M., Tian, Q. and Zhang, W.H. 2011. Identification of drought–responsive microRNAs in *Medicago truncatula* by genome–wide high–throughput sequencing. BMC Genomics **12**: 367–378.

Wenping, H., Yuan, Z., Jie, S., Lijun, Z. and Zhezhi, W. 2011. *De novo* transcriptome sequencing in *Salvia miltiorrhiza* to identify genes involved in the biosynthesis of active ingredients Genomics **98**: 272–279.

Xie, Z., Johansen, L.K., Gustafson, A.M., Kasschau, K.D., Lellis, A.D., Zilberman, D., Jacobsen, S.E. and Carrington, J.C. 2004. Genetic. and functional diversification of small RNA pathways in plants, PLoS Biol. **2**: 642–652

You, F.M., Huo, N., Deal, K.R., Gu, Y.Q., Luo, M.C., McGuire, P.E., Dvorak, J. and. anderson, O.D. 2011. Annotation–based genomewide SNP discovery in the large. and complex *Aegilops tauschii* genome using next–generation sequencing without a reference genome sequence. BMC Genome. **12**: 59–78.

Zalapa, J.E., Cuevas, H., Zhu, H., Steffan, S., Senalik, D.S., Zeldin, E., Mccown, B., Harbut, R. and Simon, P. 2011. Using next–generation sequencing approaches to isolate simple sequence repeat (SSR) loci in the plant sciences. Amer. J. Bot. **99**: 1–16.

Zenoni, S., Ferrarini, A., Giacomelli, E., Xumerle, L., Fasoli, M., Malerba, G., Bellin, D., Pezzotti, M. and Delledonne, M. 2010. Characterization of transcriptional complexity during berry development in *Vitis vinifera* using RNA–Seq. Plant Plysiol. **152**: 1787–1795.

2013, Abiotic Stress and Biotechnology
Editor: T. Pullaiah
Published by: REGENCY PUBLICATIONS, NEW DELHI

Pages 75–126

Chapter 5

Genetic Engineering of Plants for Enhancing Abiotic Stress Tolerance Using Transcription Factors as Tools

Linga R. Gutha[1], Chandra Obul Reddy Puli[2] and Arjula R. Reddy[3]*

[1]Irrigated Agriculture Research and Extension Center,
Washington State University Prosser 24106
N Bunn Road Prosser 99350 WA USA
[2]Department of Plant Sciences, School of Life Sciences,
[3]Agri-Science Park,
Yogi Vemana University, Vemanapuram, Kadapa – 516 003, A.P., India

ABSTRACT

Abiotic stresses limit the growth rate of plants and reduce the economical yield. Several molecular and genomic studies have shown that numerous genes are activated in plant cells to cope with these adverse conditions. Among these genes, transcription factors (TFs) are key regulatory genes activated under such conditions in plants. In recent years, these genes are being intensely studied to unravel the intricate regulation of abiotic stress tolerance in plants. Several abiotic

* Corresponding Author: E-mail: arjulsl@yahoo.com

stress responsive TFs are isolated, and functionally characterized in a number of plants. They are dehydration responsive element binding (DREB), ABA responsive element binding (AREB), MYB/MYC, zinc finger, WRKY, Trihelix, NAC, and NF-Y transcription factors. Further, genetic engineering of plants with genes encoding TFs is considered a powerful strategy to improve plants for abiotic stress tolerance, which is a quantitative trait governed by many dispersed genes but regulated by a few or a small set of TFs. Transgenic plants carrying individual TF genes are reported to show improved tolerance to drought-, cold-, salinity-, and oxidative-stress factors. However, most of the studies were under laboratory conditions and therefore not yet exploited in major field crops to a greater extent. This review will critically analyze the available data on structural features and functional aspects of the abiotic stress responsive TFs from different plant species and the cross talk among them. Future prospects of using potential TFs in developing abiotic stress tolerant-crop plants will be discussed in detail.

Keywords: *Transcription factors, Abiotic stress tolerance, DREB/CBF, AREB/ABF, MYB/MYC, Zinc finger, Homeodomain, NAC, WRKY, Trihelix, Genetic engineering.*

Introduction

Plants as sessile organisms invariably encounter various abiotic stress conditions such as drought, cold and high salinity during their life cycle. Such stress conditions often can challenge the plants simultaneously, and their co-occurrence leads to fatal damage of plant growth under field conditions (Knight and Knight, 2001; Mittler, 2006). However, multiple inherent adaptive processes are activated in plants immediately after perception of the stress signal. These responses range from physical adaptation to the alteration of gene expression. Multitude of genes is activated under stress conditions and their products function in imparting stress tolerance to plants (Ingram and Bartels, 1996; Shinozaki and Yamaguchi-Shinozaki, 1997; Thomashow 1999).

Many major loci for abiotic stress tolerance across multiple genetic backgrounds and stress challenges have been reported (Takeda and Matsuoka, 2008). Induction of such stress responsive loci is primarily modulated by a few TFs and their *cis*-acting regulatory elements at transcriptional level during diverse stress conditions (Hirayama and Shinozaki, 2010). TFs are the sequence-specific DNA-binding proteins that are capable of activating and/or repressing transcription in a temporal and spatial pattern, and play a critical role in almost all biological processes in plants. A single TF can modulate the transcription of several structural and regulatory genes, and modulate the organism's activity for the adaptation to a particular external condition. Hence, plant TFs are studied not only to incisively elucidate the molecular mechanism of stress regulation but also to develop genetically modified crops for improved stress tolerance.

According to the The *Arabidopsis* Genome Initiative (2000), as much as 5.6 per cent of the *Arabidopsis* transcriptome is represented by TFs that fall into about thirty families. About 11 per cent ESTs are represented by TFs that are expressed in a normalized cDNA library of drought stressed seedlings of *indica* rice (Gorantla *et al.*,

2007). These data indicate that TFs represent a significant amount of gene products in a plant genome. Presence of a large number of TFs, and their expression under stress signify their importance in stress response of the plants. In contrast, functional redundancy is also a common phenomenon present among these proteins. Therefore, studying functions of various TFs in the context of families, that share a common DNA binding motif, is the most practical way instead of interpretation for each individual TF function *(Liu et al.,* 1999). There are several TF families reported in plants, and families like EREBP/AP2, bZIP, MYB, bHLH, zinc finger, homeodomain (HD-Zip),WRKY, Trihelix, NAC(NAM, ATAF1-2, CUC2) and Nuclear factor Y (NF-Y) have been studied in relation to abiotic stress response. Currently, functional analysis of the genes encoding transcription factors is being taken in a big way by over expressing them in transgenic plants (Zhang 2003). Furthermore, several members of these TF families have been extensively studied in several plants, and attempts are made to use them in genetic engineering of plants for developing stress tolerance (Table 5.1).

Abscisic acid (ABA), a phytohormone, acts as an internal messenger in signal transduction of plants under abiotic stress conditions. ABA transduces its signal via ABA dependent signaling transduction pathway. There by, ABA regulates the expression of TFs involved in abiotic stress signaling pathways. The expression of ABA responsive genes is correlated with plant responses to environmental stresses, in particular to water deficit stress (Shinozaki and Yamaguchi-Shinozaki, 2000; Finkelstein *et al.,* 2002; Xiong *et al.,* 2002). Transcription factors and their target genes imparting stress tolerance in plants are activated either by ABA-dependent or ABA-independent stress signaling pathways (Chinnusamy *et al.,* 2004; Tuteja, 2007; Agarwal and Jha, 2010). Although many drought inducible genes are inducible by ABA, some stress responsive genes are also up regulated in an ABA independent manner. These findings raise the possibility that there may be an ABA independent drought response pathway in plants (Shinozaki and Yamaguchi-Shinozaki, 1997).

Drought-, salt-, and cold- stress tolerance in plants is a trait independent of a single functional gene but influenced by many genes. Development of stress-resistant genotypes is an efficient approach to improve the plant productivity under adverse conditions, and to expand the agriculture to non-conventional and non-arable lands. These multigenic traits can only be modulated by genetic engineering of crop plants with the useful candidate genes obtained from various sources. Thus, it is a best strategy to manipulate the master-switches, TFs, to regulate the expression of an array of stress responsive genes simultaneously (Liu *et al.,* 2000; Zhang *et al.,* 2000). Although, substantial information on several TFs is available, they are yet to be exploited comprehensively in crop improvement, which is an ever-demanding area of agriculture. In this context, we review here the structural features, functional pattern, and their potential use in genetic engineering of crop plants for improved abiotic stress tolerance.

DREB/CBF Transcription Factors

The DREB/CBF transcription factors, plant specific-AP2 domain-proteins, occupy essential place in the regulation of plant stress response. The dehydration

Table 1.1: Transcription factors over expressed in various plant species to improve abiotic stress tolerance.

Transcription Factor	Promoter	Donor Plant	Host Plant	Stress Tolerance	Target genes	Reference
DREB/CBF transcription Factors						
DREB1B/CBF1	*CaMV 35S*	*Arabidopsis*	*Arabidopsis*	Freezing	–	Jaglo-Ottosen *et al.,* 1998
DREB1A/CBF3	*CaMV 35S*	*Arabidopsis*	*Arabidopsis*	Freezing, dehydration	–	Liu *et al.,* 1998
DREB2A	*CaMV 35S*	*Arabidopsis*	*Arabidopsis*	*	–	Liu *et al.,* 1998
DREB1A/CBF3	*CaMV 35S, rd29A*	*Arabidopsis*	*Arabidopsis*	Drought, high salt, freezing	*Kin1, Cor6.6/Kin2, Cor15a, Cor47/rd17, erd10, rd29A*	Kasuga *et al.,* 1999
DREB1A/CBF3	*CaMV 35S*	*Arabidopsis*	*Arabidopsis*	Freezing	*Cor78, P5CS, SuSy, SPS*	Gilmour *et al.,* 2000
DREB1D/CBF4	*CaMV 35S*	*Arabidopsis*	*Arabidopsis*	Drought, freezing	*Cor15a, Cor78a*	Haake *et al.,* 2002
DREB1B/CBF1	*CaMV 35S*	*Arabidopsis*	Tomato	Water deficit	*Catalase1*	Hseih *et al.,* 2002a
DREB1B/CBF1	*CaMV 35S*	*Arabidopsis*	Tomato	Chilling, oxidative stress	*Catalase1*	Hsieh *et al.,* 2002b
ZmDREB1	*Ubiquitin, In2*	Maize	Maize	Low temperature	–	Chiappetta *et al.,* 2003
OsDREB1A	*CaMV 35S*	Rice	*Arabidopsis*	Freezing, high salt	*Cor15a, rd29A, rd17, AtGolS3, FL05-21-F13, FL05-20-N18, FL06-16-B22*	Dubouzet *et al.,* 2003
OsDREB2A	*CaMV 35S*	Rice	*Arabidopsis*	*	–	Dubouzet *et al.,* 2003
TaDREB1	*CaMV 35S*	Wheat	*Arabidopsis*	*	*Wcs120, rd29A*	Shen *et al.,* 2003 a
AhDREB1	*CaMV 35S*	*Atriplex*	Tobacco	Drought, salt	*rd17, rd29A, LHCP*	Shen *et al.,* 2003b
DREB1A/CBF3	*CaMV 35S, rd29A*	*Arabidopsis*	Tobacco	Drought, low temperature	*NtERD10A, NtERD10B, NtERD10C, NtERD10D*	Kasuga *et al.,* 2004
DREB1A	*rd29A*	*Arabidopsis*	Wheat	Water stress	–	Pellegrineschi *et al.,* 2004
ZmDREB1A	*CaMV 35S*	Maize	*Arabidopsis*	Drought, freezing	*Cor15a, KIN1, KIN2, rd29A, rd17, erd10, Cor15b, AtGolS3, At1g16850*	Qin *et al.,* 2004

Contd...

Table 5.1–*Contd...*

Transcription Factor	Promoter	Donor Plant	Host Plant	Stress Tolerance	Target genes	Reference
DREB1A/CBF3	Ubiquitin1	Arabidopsis	Rice	Drought, high salt	Lip5, Dip1, Jacalin1 and 2, LOX, Hsp70, PP2Ca	Oh et al., 2005
LpCBF3	CaMV 35S	Rye	Arabidopsis	Freezing	Cor15a, rd29A	Xiong and Fei, 2006
GmDREB2	CaMV 35S	Soybean	Arabidopsis	Drought, high salt	rd29A, Cor15a, rd17A	Chen et al., 2007
HARDY	CaMV 35S	Arabidopsis	Arabidopsis/Rice	Drought and salt tolerance	–	Karaba et al., 2007
BjDREB1B	CaMV 35S	Brassica juncea L.	Tobacco	Drought, high salt	NtERD10B	Cong et al., 2008
OsDREB1E	CaMV 35S	Rice	Rice	Water deficit	–	Chen et al., 2008
OsDREB1G	CaMV 35S	Rice	Rice	Water deficit	–	Chen et al., 2008
HsDREB1A	HVA1s, Dhn8	Hordeum spontaneum L.	Bahiagrass	High salt, dehydration	–	James et al., 2008
OsDREB1B	CaMV 35S	Rice	Tobacco	Drought, high salt, oxidative stress, freezing, virus infection	PR1b, PR2,PR3, CHN50, Osmotin	Gutha and Reddy, 2008
OsDREB1F	CaMV 35S	Rice	Rice, Arabidopsis	Salt, drought, low temperature	Cor15a, rd29A, rab18, rd29B	Wang et al., 2008
DREB1A	rd29A	Arabidopsis	Peanut	Drought tolerance	–	Bhatnagar-Mathur et al., 2007
OsDREB1D	CaMV35S	Rice	Arabidopsis	Cold and high-salt tolerance	RD29A, COR15A, KIN1	Zhang et al., 2009
PeDREB2	CaMV35S	Populus euphratica	Tobacco	Salt tolerance	–	Chen et al., 2009
HvDREB1	CaMV35S/rd29A	Hordeum vulgare	Hordeum vulgare	Salt tolerance	Rd29A	Xu et al., 2009

Contd...

Table 5.1–Contd...

Transcription Factor	Promoter	Donor Plant	Host Plant	Stress Tolerance	Target genes	Reference
GmDREB3	rd29A	Soybean	Arabidopsis	Cold, drought, and high salt stresses	–	Chen et al., 2009
ARAG1	CaMV35S	Rice	Rice	Drought tolerance	–	Zhao et al., 2010
DREB2C	CaMV 35S	Arabidopsis	Arabidopsis	Temperature	HsfA3, HSP	Chen et al., 2010
PgDREB2A	CaMV35S	Pennisetum glaucum	Tobacco	Hyperionic and hyperosmotic stresses	NtERD10B, HSP70-3, Hsp18p, PLC3, AP2 domain TF, THT1, LTP1 and heat shock (NtHSF2) and pathogen-regulated (NtERF5) factors	Agarwal et al., 2010
MbDREB1	CaMV 35S	Malus baccata (L.) Borkh.	Arabidopsis	Low temperature, drought, and salt stress	COR15a and rd29B	Yang et al., 2011c
DREB9	CaMV 35S	Arabidopsis	Arabidopsis	Salt and drought tolerant	–	Krishnaswamy et al., 2011
CtCBF3	CaMV 35S	sweet pepper	Tobacco	Cold Stress	NtERD10B and NtERD10C	Yang et al., 2011b
LcDREB3a	CaMV35S	Leymus chinensis	Arabidopsis	Drought and salt tolrance	AtRD29A, AtSAMDC	Xianjun et a., 2011
TaDREB2 and TaDREB3	Duplicated 35S promoter/Maize drought- and salt-inducible Rab17	Wheat	Hordeum vulgare L. cv. Golden Promise	Drought and frost tolerance	CBF D DREB genes and stress responsive LEA D COR DDHN genes	Morran et al., 2011
AtCBF1	CaMV35S	Arabidopsis	Tomato	Low temperature under low Irradiance tolerance	–	Zhang et al., 2011b
SbDREB2	CaMV35S/rd29A	Sorghum	Rice	Drought tolerance	–	Bihani et al., 2011
HARDY	CaMV 35S	Arabidopsis	Trifolium alexandrinum L.	Drought and Salt tolerance	–	Abogadallah et al., 2011

Contd...

Table 5.1–*Contd...*

Transcription Factor	Promoter	Donor Plant	Host Plant	Stress Tolerance	Target genes	Reference
VrCBF1/VrCBF4	CaMV 35S	Vitis vinifera	Arabidopsis	Freezing and drought tolerance	AtCOR15a, AtRD29A, AtCOR6.6 and AtCOR47	Siddiqua and Nassuth, 2011
JcDREB	–	Jatropha curcas	Arabidopsis	Salt and freezing stresses	?	Tang et al., 2011
OsDREB2A	rd29A	Rice	Rice	Dehydration and salt stress tolerance	–	Mallikarjuna et al., 2011
CkDREB	?	Caragana korshinskii	Tobacco	High salinity and osmotic stresses	?	Wang et al., 2011
AREB/ABF transcription factors						
ABF3	CaMV 35S	Arabidopsis	Arabidopsis	Drought	rd29B, rab18, ICK1, ABI1, ABI2, SKOR, ADH1,CHS	Kang et al., 2002
ABF4	CaMV 35S	Arabidopsis	Arabidopsis	Drought	rd29B, rab18, ICK1, ABI1, ABI2, ADH1, CHS	Kang et al., 2002
ABF2	CaMV 35S	Arabidopsis	Arabidopsis	Drought, salt, heat, oxidative stress	ADH1, rd29A, CHS, SUS1	Kim et al., 2004
AREB1	CaMV 35S	Arabidopsis	Arabidopsis	–*	–	Fujita et al., 2005
AREB1ΔQT	CaMV 35S	Arabidopsis	Arabidopsis	Drought	HIS 1-3, AIA, AIL1, rd29B, GBF3, rd20, rab18, KIN2	Fujita et al., 2005
ABF3	Ubiquitin	Arabidopsis	Rice	Drought	Wsi18, rab21, Hsp70, PP2Ca	Oh et al., 2005
ABF3	CaMV 35S	Arabidopsis	Lettuce	Drought, cold	–	Vanjildorj et al., 2005
HvABI5	CaMV35S	Wheat	Tobacco	Freezing, osmotic and salt stresses	–	Kobayashi et al., 2008

Contd...

Table 51–*Contd...*

Transcription Factor	Promoter	Donor Plant	Host Plant	Stress Tolerance	Target genes	Reference
SlARE1	CaMV35S	Tomato	Tomato	Drought and salt stress tolerance	LEA, LTPs	Orellana et al., 2010
SlARE	CaMV35S	Tomato	Tomato/Arabidopsis	Water deficit and high salinity stresses	AtRD29A, AtCOR47, and SlCl7-like dehydrin	Hsieh et al., 2010
ABP9	CaMV 35S	Maize	Arabidopsis	Drought, high salt, freezing temperature and oxidative stresses	COR15A, KIN1, RD29B, AIA1, HIS1-3,ATMYB2, ATMYB78, CBF1, CBF2, AZF2, PP2C (HAI1)/ At5g59220, and HSFA6A	Zhang et al., 2011
GmbZIP1	CaMV 35SRd29A	Soybean	Arabidopsis	ABA, drought and high saltstress	ABI1, ABI2, rd29B, Rab18, KAT1, and KAT2	Gao et al., 2011
MYB/MYC transcription factors						
AtMYC2	CaMV 35S	Arabidopsis	Arabidopsis	Drought	–	Abe et al., 2003
AtMYB2	CaMV 35S	Arabidopsis	Arabidopsis	Drought	–	Abe et al., 2003
OsMyb4	CaMV 35S	Rice	Arabidopsis	Cold, freezing	–	Vanini et al., 2004
CpMyb10	CaMV 35S	Craterostigma	Arabidopsis	Drought, salt	–	Villalobos et al., 2004
OsMyb4	CaMV 35S	Rice	Arabidopsis	Drought	–	Mattana et al., 2005
AtMyb4	CaMV35S	Arabidopsis	Arabidopsis	Drought, salt	ABI1, ABI2, AtPP2CA, HAB1, HAB2	Jung et al., 2008
Myb1E	CaMV35S	Arabidopsis	Arabidopsis	Drought stress	AtADH1, RD22, RD29B, AtEM6	Ding et al., 2009
OsMYB3R-2	Ubiquitin	Rice	Rice	Chilling stress	OsCycBi, OsCPT1	Ma et al., 2009
MYBS3	CaMV35S	Rice	Rice	Cold stress	TPPI, TPP2, WRKY77	Su et al., 2010
OsMyb4	CaMV35S	Rice	Rice	Chilling stress	DREBI CBF	Park et al., 2010
MYB6	CaMV35S	Arabidopsis	Arabidopsis	Drought	PAP1 and F3H	Oh et al., 2011

Contd...

Table 5.1–Contd...

Transcription Factor	Promoter	Donor Plant	Host Plant	Stress Tolerance	Target genes	Reference
TaPIMP1	CaMV35S	Wheat	Tobacco	Drought and salt stresses	–	Liu et al., 2011b
StMYB1R-1	CaMV35S	Potato	Potato	Drought	AtHB-7, RD28, ALDH 22a1, and ERD1-like	Shin et al.,2011
Zinc finger transcription factors						
AtZAT12	CaMV 35S	Arabidopsis	Arabidopsis	Oxidative, high light		Davletova et al., 2005
AtSZF1 and AtSZF2	CaMV 35S	Arabidopsis	Arabidopsis	Salt tolerance	–	Sun et al., 2007
ZFP252	CaMV 35S	Rice	Rice	Drought, salt	–	Xu et al., 2008
ZFP177	CaMV35S	Rice	Tobacco	Temperature stress tolerance	–	Huang et al., 2008
ZFP245	CaMV35S	Rice	Rice	Cold, drought and oxidative stresses	P5CS, PT	Huang et al., 2009
GhZFP1	CaMV35S	Cotton	Tobacco	Salt tolerance	GZIRD21A and GZIPR5	Guo et al., 2009
StZFP1	rd29A	Potato	Tobacco	Salt tolerance	–	Tian et al., 2010
ZFP179	CaMV35S	Rice	Rice	Salt Stress	OsDREB2A, OsP5CS OsProT, and OsLea3	Sun et al., 2010
GhDi19-1 and GhDi19-2	CaMV35S	Cotton	Arabidopsis	Salt/drought stress and ABA	ABF3, ABF4, ABI5 and KIN1	Li et al., 2010a
AlSAP	CaMV35S	Aeluropus littoralis	Tobacco	Salinity, drought, heat and freezing tolerance	–	Saad et al., 2011
AlSAP	CaMV35S	Aeluropus littoralis	Rice	Cold, drought and salt stresses tolerance	?	Saad et al., 2011

Contd...

Table 1.1—Contd...

Transcription Factor	Promoter	Donor Plant	Host Plant	Stress Tolerance	Target genes	Reference
HD-ZIP transcription factors						
ATHB7	CaMV 35S	Arabidopsis	Arabidopsis	Water limiting conditions	–	Hjellstrom et al., 2003
OsBIHD1	CaMV 35S	Rice	Tobacco	Salt, drought	–	Luo et al., 2005
Hahb-4	CaMV35S	Sunflower	Arabidopsis	Drought tolerance	–	Dezar et al., 2005
AtZFHD1	CaMV 35S	Arabidopsis	Arabidopsis	Drought	LEA, Cytochrome P450 etc.	Tran et al., 2007
AtHB11	CaMV35S	Arabidopsis	Tobacco	Drought	NCED3, LOS5/ABA3, CIPK3, CAX3, ABI3, and ERECTA	Yu et al., 2008
AtHB11	CaMV35S	Arabidopsis	Tall fescue	Drought and Salt stress tolerance	–	Cao et al., 2009
WRKY transcription factors						
WRKY25 and WRKY33	CaMV 35S	Arabidopsis	Arabidopsis	Salt tolerance	–	Jiang and Deyholos, 2008
GmWRKY13, GmWRKY 21, GmWRKY54	CaMV 35S	Soybean	Arabidopsis	Salt, cold and drought tolerance	DREB2A and STZ/Zat10	Zhou et al., 2008
TcWRKY53	CaMV35S	Thlaspi caerulescens	Tobacco	Osmotic stress	NtERF5 and NtEREBP-1, NtLEA5	Wei et al., 2008
OsWRKY45	CaMV 35S	Rice	Arabidopsis	Drought resistance	–	Qiu and Yu, 2009
BhWRKY1	CaMV 35S	Boea hygrometrica	Tobacco	Dehydration tolerance	GolS1	Wang et al., 2009
HvWRKY38	CaMV 35S	Barley	Paspalumno tatum Flugge	drought tolerance	–	Xiong et al., 2009
WRKY25	CaMV 35S	Arabidopsis	Arabidopsis	Thermotolerance	–	Li et al., 2009

Contd...

Table 5.1–Contd...

Transcription Factor	Promoter	Donor Plant	Host Plant	Stress Tolerance	Target genes	Reference
OsWRKY11	HSP101	Rice	Rice	Heat and drought tolerance	–	Wu et al., 2009
OsWRKY08	CaMV35S	Rice	Arabidopsis	Osmotic stress tolerance	AtCOR47 and AtRD21	Song et al., 2009
WRKY39	CaMV35S	Arabidopsis	Arabidopsis	Thermotolerance	PR1 and MBF1c	Li et al., 2010
WRKY25, WRKY26, and WRKY33	CaMV35S	Arabidopsis	Arabidopsis	Thermotolerance	HsfA2,HsfB1, Hsp101, and MBF1c	Li et al., 2011
VvWRKY11	CaMV35S	Grapevine	Arabidopsis	Water stress	AtRD29A and AtRD29B	Liu et al., 2011
BcWRKY46	CaMV35S	Brassica campestris ssp. chinensis	Tobacco	Cold, Salt and dehydration	–	Wang et al., 2011
Trihelix transcription Factors						
GmGT-2A, GmGT-2B	CaMV35S	Soybean	Arabidopsis	Salt, freezing and drought stress	NCED3, LTP3,LTP4 and PAD3	Xie et al., 2009
GTL1	CaMV35S	Arabidopsis	Arabidopsis	Drought tolerance	SDD1	Yoo et al., 2010
OsGTγ-1	CaMV35S	Rice	Rice	Salt tolerance	–	Fang et al., 2010
NAC transcription factors						
ANAC019	CaMV 35S	Arabidopsis	Arabidopsis	Drought	–	Tran et al., 2004
ANAC055	CaMV 35S	Arabidopsis	Arabidopsis	Drought	–	Tran et al., 2004
ANAC072	CaMV 35S	Arabidopsis	Arabidopsis	Drought	–	Tran et al., 2004
OsSNAC1	CaMV 35S	Rice	Rice	Drought, salt	UGS1,2,3,4,5,6	Hu et al., 2006
OsNAC6	CaMV35S	Rice	Rice	Drought, high salt	Lipoxygenase, peroxidase, Ser/Thr Kinase motif protein	Nakashima et al., 2007

Contd...

Table 5.1–*Contd...*

Transcription Factor	Promoter	Donor Plant	Host Plant	Stress Tolerance	Target genes	Reference
OsNAC2	CaMV 35S	Rice	Rice	Cold, drought, salt stress	Peroxidase, ornithine aminotransferase, heavy metal-associated protein, sodium/hydrogen exchanger, heat shock protein, GDSL-like lipase, and phenylalanine ammonia lyase	Hu et al., 2008
ATAF1	CaMV35S	Arabidopsis	Arabidopsis	Drought	–	Wu et al., 2009
OsNAC45	CaMV35S	Rice	Rice	Drought and salt tolerance	OsLEA3-1, WPM-1	Zheng et al., 2009
OsNAC5	Maize ubiquitin promoter	Rice	Rice	High salinity and drought	OsLEA3	Takasaki et al., 2010
AhNAC2	CaMV 35S	Arachis hypogaea	Arabidopsis	Drought and salt tolerance	rd294, rd29B, RAB18, AtMYB2, AtMYC2, ERD1, COR47, COR15A, KIN1, AREB1,CBF1 and AMY1	Liu et al., 2011
DgNAC1	CaMV 35S	Chrysanthemum	Tobacco	Salt Tolerance	–	Liu et al., 2011
TaNAC69	HvDhn4s	Wheat	Wheat	Dehydration tolerance	Glyoxalase I family protein,4-hydroxy-phenylpyruvate dioxygenase, ZIM family gene, chitinase	Xue et al., 2011
GmNAC11	CaMV35S	Soybean	Arabidopsis	Salt tolerance	DREB1A, ERD11, cor15A, ERF5, RAB18, and KAT2	Hao et al., 2011

Contd...

Genetic Engineering of Plants for Enhancing Abiotic Stress Tolerance Using Transcription | 87

Table 5.1–Contd...

Transcription Factor	Promoter	Donor Plant	Host Plant	Stress Tolerance	Target genes	Reference
GmNAC20	CaMV35S	Soybean	Arabidopsis	Salt and freezing tolerance	DERB1A/CBF3 and KIN2/cor6.6	Hao et al., 2011
Nuclear Factor (NF-Y) transcription factors						
AtNF-YB1	Enhanced CaMV35S	Arabidopsis	Arabidopsis	Drought tolerance	–	Nelson et al., 2007
ZmNF-YB2	Rice actin promoter and 5' UTR with an embedded intron (OsRACT)	Maize	Maize	Drought tolerance	–	Nelson et al., 2007
NFYA5	CaMV35S	Arabidopsis	Arabidopsis	Drought tolerance	Cytosolic β-amylase, protease inhibitor, glutathione transferase, cold-responsive protein, COR15A	Li et al., 2008

–: Studies on target genes are not made by the authors; *: Transgenic plants have not shown tolerance to any stress.

responsive element binding proteins (DREBs) or C-repeat binding factors (CBFs) are widely reported to be induced in response to drought, cold, and high salinity stresses. DREBs, encoded by a small multigenic family, regulate the target gene expression in response to several stresses through binding a highly conserved, unique *cis*-acting element A/GCCGAC designated as dehydration responsive element (DRE) (Yamaguchi-Shinozaki and Shinozaki, 1994; Jiang *et al.*, 1996; Stockinger *et al.*, 1997; Gao *et al.*, 2002; Knight *et al.*, 2004). Sequence analysis of all the *DREB1* genes showed that they are intronless genes and duplication of them could have been resulted in a small multigene family during species evolution (Haake *et al.*, 2002). Dubouzet *et al.* (2003) reported that *DREB1*s show extensive homology in the AP2 domain, nuclear localization signal (NLS) regions in the N-terminal region, and in a few parts of the C-terminal region like DASW and LWSY motifs. However the homology among *DREB2* proteins is restricted to only AP2 domain and N-terminal region. *DREB2* proteins have ser-/thr- rich domain, and have not shown any significant similarity to *DREB1* proteins except AP2 domain and NLS region. Valine at both 14th and 19th positions is conserved in the AP2 domain of *DREB1* across monocots, whereas the same domain in dicots has valine and glutamic acid conserved at 14th and 19th positions, respectively.

All the CBFs/DREB1s are induced in response to cold, salinity and drought stresses, but *AtCBF4* is induced rapidly in response to drought, but not by low temperature (Dubouzet *et al.*, 2003). They show tissue specific expression in response to a variety of stress factors. Rice *DREB1B* promoter showed organ-, and stress-specific expression pattern of the reporter gene in transgenic *Arabidopsis* seedlings (Gutha and Reddy, 2008). Jaglo-Ottosen *et al.* (1998) reported that *CBF1* is induced in response to freezing in *Arabidopsis*. But, Dubouzet *et al.* (2003) showed that *DREB1A, DREB1B, DREB1C* and *DREB2A* are induced not only by cold stress as in *Arabidopsis* but also by dehydration and salt stresses. Gutha and Reddy (2008) showed that rice *DREB1B* is induced in response to cold, mannitol, NaCl, PEG, methyl viologen, salicylic acid, and ABA stresses.

Genes coding for *DREB1* and *DREB2* are isolated from different species, and over expressed in a variety of plants for functional validation (Table 5.1). Constitutive expression of *DREB1* TFs in plants resulted in expression of numerous stress responsive target genes even under non-stress conditions and made the transgenic plants resistant to a wide range of abiotic stress factors (Jaglo-Ottosen *et al.*, 1998; Liu *et al.*, 1998; Kasuga *et al.*, 1999; Gilmour *et al.*, 2000; Hsieh *et al.*, 2002a; Hsieh *et al.*, 2002b; Oh *et al.*, 2005;Wang *et al.*, 2008). The genes regulated by DREB transcription factors play a major role in cold, drought, high salinity, oxidative stress, pathogen attack and various other cellular processes. *DREB1A/CBF3* induced the overexpression of group2 LEA proteins such as *NtERD10A, NtERD10B, NtERD10C* and *NtERD10D* in transgenic tobacco plants (Kasuga *et al.*, 2004). Constitutive expression of rice *DREB1B* in tobacco resulted in healthy transgenic plants, which have shown improved osmotic, oxidative and cold stress tolerance besides biotic stress tolerance (Gutha and Reddy, 2008). Similarly, constitutive overexpression of *OsDREB1F* in transgenic rice or *Arabidopsis* plants caused no negative growth effect, but enhanced tolerance to drought, salt and cold stresses by activating both ABA-dependent and

ABA-independent downstream genes (Wang *et al.*, 2008). Constitutive overexpression of a *Arabidopsis* Group III member of the DREB/CBF subfamily, HARDY gene in transgenic *Arabidopsis* and rice plants showed pleiotropic effects on vegetative growth, forming small, thick green leaves, abnormally dense root system and improved water use efficiency (WUE) (Karaba *et al.*, 2007). Overexpression of a barley *DREB1* gene in transgenic *Arabidopsis* activated a downstream gene, *rd29A*, under normal growth conditions and led to increased tolerance to salt stress (Xu *et al.*, 2009). Transgenic wheat plants overexpressing a cotton *DREB* gene showed improved tolerance to drought, high salt, and freezing stresses by activating the expression of downstream genes that prevent chlorophyll decomposition and sugar biosynthesis (Gao *et al.*, 2009).

AREB/ABF Transcription Factors

These are, ABA responsive transcription activators, reported to be expressed in response to diverse abiotic stress stimuli in plants. ABA responsive element binding (AREB) proteins shape a small subfamily of bZIP transcription factors, which are characterized by having a basic residues-rich-DNA binding domain coupled to a leucine zipper dimerization domain. According to the categorization given by Jakoby *et al.* (2002), AREB/ABF transcription factors are the members of A-group of plant bZIP transcription factors in which ABI, GBF, DPBF proteins are the other members. AREB proteins bind ABA responsive *cis*-acting element, ABRE/PyACGTGGC, present in promoters of numerous ABA responsive genes and considered as a subset of G box (Choi *et al.*, 2000; Uno *et al.*, 2000). Studies showed that flanking nucleotides of ACGT core motif regulate the binding specificity of these bZIP proteins. These TFs have shown similarity with the embryo specific C/ABRE binding Dc3 promoter binding factors (DBPFs) in the basic region. Based on *in vitro* binding assays, *(Choi et al., 2000; Uno et al.*, 2000) concluded that ABF proteins show a unique binding activity among bZIP transcription factors, and bind both G/ABRE (CACGTG) and C/ABRE (CGCGTC) motifs. It is reported that a single copy of ABRE is not adequate for ABA dependent transcription of stress regulated genes, but they require a coupling element (CE) for functional specificity (Uno *et al.*, 2000). Most of the identified CEs have shown similarity with ABREs, and rarely with DRE. ABRE forms an ABA-responsive *cis*-element (ABRC) complex with the coordination of a coupling element, and regulates expression of ABA dependent genes in plants (Shen and Ho, 1995; Hattori *et al.*, 2002). Many monocots, and *Arabidopsis* have GCGT motif and ABRE motifs as their CEs, respectively however DREs are found to be less effective as CEs (Zhang *et al.*, 2005).

ABFs, cloned from *Arabidopsis* expression library by yeast one-hybrid system, *ABF1*, *ABF2*, *ABF3*, and *ABF4* show much homology with the embryo specific-, ABA responsive-, G/ABRE and C/ABRE-binding DPBF transcription regulators (Choi *et al.*, 2000). Although all the ABFs bind ABRE, they show differential regulation by diverse environmental stress factors with a unique induction pattern. ABFs show several highly conserved regions including the basic region, and also show analogous properties each other in addition to the DNA binding activity. *ABF1* interacts with G/ABRE and also with C/ABRE although slightly, and activates a wide range of target genes (Choi *et al.*, 2000). Apart from ABA and high-salt stress induction, *ABF2* is also induced by sugar. Transcript levels of *ABF2* are increased with the increasing concentration of sucrose (Kim *et al.*, 2004b). *AREB1* and *AREB2* are weakly expressed in unstressed *Arabidopsis*, however they are induced by drought, salt and endogenous ABA treatments. *AREB3* is not induced in both conditions (Uno *et al.*, 2000). *AREB1/ ABF2, AREB2/ ABF4, ABF3/ DPBF5* were mainly expressed in vegetative tissues in response to drought, high salinity, and exogenous ABA applications (Fujita *et al.*, 2005; Nakashima *et al.*, 2006).

Kang and Saltveit (2002) constitutively expressed ABRE binding factors, *ABF3* and *ABF4*, under the influence of *CaMV35S* promoter in *Arabidopsis*. Transgenic *Arabidopsis* plants have shown decreased transpiration and increased drought tolerance. A large number of ABA regulated or other stress responsive genes are over expressed in transgenic plants. Transgenic lettuce of *ABF3* has shown improved stress tolerance without any growth retardation, although transgene is constitutively expressed (Vanjildorj *et al.*, 2005). Kim *et al.* (2004b) developed transgenic plants constitutively expressing *ABF2* under *CaMV35S* promoter, and observed growth retardation, delayed bolting and senescence after maturation in transgenic plants. *Oh et al.* (2005) reported that *ABF3* imparts exclusively drought tolerance in transgenic rice plants, and activates a few target genes without showing any negative effects on plant growth. Further, transgenic *Arabidopsis* plants over expressing *ABF3* are observed to be healthy, and have shown improved chilling, freezing, high temperature, oxidative, and drought stress tolerance (Kim *et al.*, 2004a). Transgenic *Arabidopsis* and tomato plants that overproduce tomato AERB protein SlAREB under CaMV 35 promoter increased tolerance to water deficit and salt stress by regulating the ABA induced stress - related genes *rd29A, Cor47* and *CI7*-like dehydrin genes (Hsieh *et al.*, 2010).

MYB/MYC Transcription Factors

MYB family proteins are a group of functionally diverse transcription activators identified in plants, animals and yeast. These are the proteins showing two or three imperfect helix-turn-helix (HTH) repeats that are denoted as R1, R2, and R3 in their highly conserved DNA binding domain. Large number of MYB transcription factors has been reported in plants than any other eukaryotic group (Martin and Paz-Ares, 1997). Depending upon the conserved C-terminal motifs, MYB proteins are divided into Myb1R, R2R3Myb, and Myb3R groups (Jin and Martin, 1999; Chen *et al.*, 2005a). However, MYB3R-2 with three repeats was reported in rice (Dai *et al.*, 2007). In plants, R2R3Myb group is the largest group containing two imperfect MYB-like repeats in their DNA binding domain and regulate multitude processes like secondary metabolism, cell shape, disease resistance, and environmental stress response (Kranz *et al.*, 1998; Meissner *et al.*, 1999; Feng *et al.*, 2004). R2R3Myb genes represent the largest gene family of TFs reported in plants. About 130 MYB family members are reported in *Arabidopsis thaliana*, and about eighty are reported in maize (Romero *et al.*, 1998; Rabinowicz *et al.*, 1999; Stracke *et al.*, 2001). These proteins show inherent flexibility in their ability to recognize target sites, but their binding preference is influenced by other proteins interacting with them. All MYBs are not the transcription activators but there are also MYBs repressing the target gene expression (Jin and Martin, 1999). MYC transcription factors have helix-loop-helix (bHLH) DNA binding domain. Like bZIP TFs, MYCs have two sub domains, basic region and HLH region required for DNA binding and dimerization, respectively. MYCs are the second largest TF family in plants and in particular to *Arabidopsis*, however these proteins are well studied in reference to diverse biological processes only in mammalian systems (Toledo-Ortiz *et al.*, 2003).

MYBs are reported to be responsive to abiotic stress factors in plants, and are of current research interest although they are studied primarily with reference to flavonoid biosynthetic pathway. MYBs like *Myb7*, *Myb5* and *Myb10* are induced in response to desiccation and ABA stress in a resurrection plant, *Craterostigma plantagineum*, surviving extreme dehydration conditions (Iturriaga *et al.*, 1996). Although MYBs are generally known to show ABA dependent expression, a few MYBs are ABA independent. Chen *et al.* (2005a), reported that *BcMyb1* from *Boea crassifolia* regulate the gene expression in response to dehydration through an ABA independent pathway. *Myc2* is also reported to be acting as transcription factor in jasmonic acid (JA)-ethylene signaling and light regulated gene expression in *Arabidopsis* (Anderson *et al.*, 2004; Boter *et al.*, 2004; Lorenzo *et al.*, 2004; Yadav *et al.*, 2005).

Transgenic *Arabidopsis* plants constitutively expressing both *AtMyc2* and *AtMYyb2* cDNA are dwarf and showed ABA hypersensitiveness than those over expressing either *AtMyc2* or *AtMyb2* cDNA alone. Transgenic plants harboring both *35S:AtMyc2* and *35S:AtMyb2* showed higher expression of *AtADH1* gene than in the individual transgenic plants indicating that the functional cooperation and interaction between *AtMyc2* and *AtMyb2* (Abe *et al.*, 1997; Abe *et al.*, 2003). Microarray studies of transgenic plants showed up regulation of many other genes like *Cor6.6, rd20, rd22,* and *AtADH1* in addition numerous osmotic stress inducible genes having MYB and MYC recognition sequences (Abe *et al.*, 2003). ABA-independent, cold-responsive *OsMyb4* gene was over expressed in *Arabidopsis*, which showed increased cold-, freezing-, and drought stress tolerance in addition to higher accumulation of drought and cold-compatible solutes like glucose, fructose, sucrose, proline, glycine betaine and sinapoyl malate (Vannini *et al.*, 2004; Mattana *et al.*, 2005). Liao *et al.* (2008) overexpressed soybean Myb factors like *Myb76*, *Myb92*, and *Myb177* in *Arabidopsis*. The expression levels of known stress responsive genes *rd29B, DREB2A, P5CS, rd17, erd10,* and *Cor78/rd29A* were enhanced in *GmMYB76*-transgenic plants. But, expression of *rd29B, COR6.6, COR15a,* and *COR78/rd29A* genes was lower, and expression of *DREB2A, rd17,* and *P5CS* were higher in *GmMYB92*-transgenic plants. In *GmMYB177*-transgenic plants, genes including *rd29B, ABI2, DREB2A, rd17, P5CS, erd10, Cor6.6, erd11,* and *Cor78* were upregulated. A R2R3Myb type transcription factor gene *Mdmyb10,* a anthocyanin biosynthesis related gene in apple, was oveexpressed in *Arabidopsis* and analyzed its function to osmotic stress in transgenic plants. Under high osmotic stress, the *Mdmyb10* over-expressing plants showed better toleranace than wild-type plants (Gao *et al.*, 2010). In a recent study it is reported that MYB transcription factor(s) are involved in the activation of of a sugar beet Sodium/proton exchangers *BvNHX1* upon exposure to salt and water stresses (Adler *et al.*, 2010).

Zinc Finger Transcription Factors

Zinc finger transcription factors function in many biological processes. These proteins play an important regulatory role in plant development in addition to the regulation of responses to various other stresses. Structural features of these proteins are extensively reviewed by Takatsuji (1998). The zinc finger is a small protein domain with a helical region that binds to the major groove of the DNA. A zinc atom bound to two cysteines and two histidines holds the structure together. The Cys-2/His-2-type zinc finger TFs, also referred as the classical or TFIIIA-type finger, have the signature $CX_{2-4}CX3FX5LX2HX_{3-5}H$ motif, consisting of about 30 amino acids (Pabo *et al.*, 2001). Depending upon the number and order of Cys and His residues, these are classified into several types such as C_2H_2, C_2C_2, C_2HC, $C_2C_2C_2C_2$, $C_2HCC_2C_2$, and among all these types C_2H_2 type zinc finger proteins are extensively studied in all the eukaryotes (Ciftci-Yilmaz and Mittler, 2008). Plant ZPT-type proteins have one to four fingers each, and can be classified according to the number of fingers. In plants, the first TFIIIA-type zinc finger has been found in a DNA binding protein from petunia, ZPT2-1 (previously named as EPF1) interacting with promoter region of the 5-enolpyruvylshikimate-3-phosphate synthase (EPSPS) gene (Takatsuji *et al.*, 1992). The involvement of petunia *ZPT2-3* in plant responses to various stresses suggested its potential utility to improve drought tolerance (Sugano *et al.*, 2003). Some members of the Cys-2/His-2-type zinc finger protein (ZFP) gene family, *AZF1*, *AZF2*, *AZF3*, and *STZ* are involved in abiotic stress response, and are up-regulated through ABA-dependent or -independent pathway to control downstream genes such as *ENA1*-like in *Arabidopsis*. These four ZPT2-related proteins are shown to act as transcriptional repressors that down-regulate the transactivation activity of other transcription factors. RNA gel-blot analysis showed that expression of *AZF2* and *STZ* is strongly induced by dehydration, high-salt and cold stresses, and ABA treatment.

Transgenic *Arabidopsis* overexpressing *STZ* showed growth retardation and tolerance to drought stress (Sakamoto *et al.*, 2000). Xu *et al.* (2007) reported that *ZF1* (*Zinc Finger Protein 1*) of *Thellungiella halophila* is induced by salinity and drought. Overexpression of a salt- and drought-inducible C_2H_2-type rice Zinc Finger Protein, *ZFP252*, in transgenic rice plants enhanced the expression of stress defense genes, high amount of free proline and soluble sugars, and conferred salt and drought tolerance (Xu *et al.*, 2008). Yang *et al.* (2008) isolated RING zinc finger ankyrin protein (*ZFP1*) from drought-tolerant *Artemisia desertorum* Spreng. Overexpression of the *AdZFP1* gene in transgenic tobacco enhanced tolerance to drought stress. In an another study, a previously unknown DST (drought and salt tolerance) zinc finger transcription factor was shown to function as negative regulator in stomatal closing by modulating the genes related to H_2O_2 homeostasis in guard cells. The rice *DST* knockdown mutant (*dst*) enhanced drought and salt tolerance by reducing the stomatal density and increasing the stomatal closure (Huang *et al.*, 2009). A salt stress-responsive zinc finger protein gene, *ZFP179* was cloned from rice, and transgenic rice plants overexpressing *ZFP179* showed enhanced stress tolerance. Transgenic rice plants have shown expression of several stress defence genes both by ABA dependent and independent manner. It is also speculated that *ZFP179* might enhance ROS-scavenging activity of transgenic plants to remove toxic ROS in plants induced under salt stress (Sun *et al.*, 2010).

Homeodomain-Leucine Zipper TFs

Homeodomain-leucine zipper (HD-Zip) TFs function in different developmental processes in plants in addition to showing the implication in both biotic and abiotic stress conditions (Otting *et al.,* 1990; Masucci and Schiefelbein, 1996; Soderman *et al.,* 1996; Chang *et al.,* 1997; Kerstetter *et al.,* 1997; Tamaoki *et al.,* 1997; Soderman *et al.,* 1999; Ariel *et al.,* 2007). These TFs show both ABA -dependent and -independent activity (Deng *et al.,* 2006). Further, they have HD-DNA binding domain and a closely related leucine zipper motif to the C-terminal region of the homeodomain for homo-, and heterodimer formation. The homeobox is a conserved DNA motif that encodes 61-amino acids sequence known as homeodomain (HD) (Agalou *et al.,* 2008). Hu *et al.* (2008b) suggested the name of these proteins as ZHD (Zincfinger Homeodomain) transcription activators. *Arabidopsis ATHB5, ATHB6, ATHB7,* and *ATHB12* are induced by drought and they act as growth regulators in response to water deficit conditions (Lee and Chun, 1998; Soderman *et al.,* 1999; HjellstrÖM *et al.,* 2003; Johannesson *et al.,* 2003; Olsson *et al.,* 2004). Overexpression of *ATHB6, ATHB7,* and *ATHB12* in *Arabidopsis* resulted in transgenic plants showing a few aberrations in morphological traits such as reduced stem length, altered leaf shape irrespective of water levels present in the plant. These observations indicated that morphological changes may be playing a role as a mediator for growth responses in water limiting conditions in plants (HjellstrÖM *et al.,* 2003; Olsson *et al.,* 2004). In a resurrection plant, *Craterostigma plantagineum,* the transcripts of *HB6* and *HB7* (type I HD-Zips) accumulated during dehydration whereas the transcripts of *HB3, HB4,* and *HB5* are reduced by dehydration in both leaves and roots (Deng *et al.,* 2002). Transgenic tobacco and *Arabidopsis* plants ectopically expressing *CpHB7* also displayed reduced sensitivity towards ABA during seed germination and stomatal closure. Gene expression analysis in transgenic plants have shown that some ABA-responsive genes are either induced or repressed indicating that this HD-Zip proteins modify ABA-responsive gene expression (Deng *et al.,* 2006). *ZFHD1* is activated in response to drought, high salinity, and ABA stresses in *Arabidopsis.* Constitutive overexpression of this TF in *Arabidopsis* imparted improved drought stress tolerance to transgenic plants, and also led to overexpression of several target genes (Tran *et al.,* 2007).

In *Arabidopsis,* homeodomain containing proteins are central regulators of plant growth and development (Abe *et al.,* 2001; McConnell *et al.,* 2001; Abe *et al.,* 2003), however a few of them have been reported to play important role in abiotic stress tolerance (Himmelbach *et al.,* 2002; Johannesson *et al.,* 2003; Zhu *et al.,* 2004) and not contain START domain. The dramatic role of *AtHDG11* in enhancing drought was identified via activation tagging and subsequent analyses in *Arabidopsis* and tobacco (Yu *et al.,* 2008). Activation tagging of *HDG11* gene in *Arabidopsis* mutant and constitutive overexpression in transgenic tobacco improved multiple characteristics related to drought tolerance-phenotype, including extensive root system, reduced stomatal density, high levels of ABA and proline, increased tolerance to oxidative stress, and high levels of superoxide dismutase. The constitutive expression of the gene was not associated with retarded growth or any other observable deleterious phenotypic effects, and the gene was also shown to transactivate a number of other genes involved in the drought stress response including *ERECTA.*

WRKY Transcription Factor

WRKY proteins constitute a large family of plant specific transcription factors comprised of 74 members in *Arabidopsis*, 102 in *indica* rice, and 103 in *japonica* rice genome (Dong *et al.*, 2003; Song *et al.*, 2010). These TFs consist of a well conserved DNA binding region, called as the WRKY domain with a highly conserved WRKYGQK amino acid signature at their N terminus and a $CX_{4-5}CX_{22-23}HXH$ zinc binding motif in the C-terminus region (Eulgem *et al.*, 2000; Yamasaki *et al.*, 2005). WRKY domain mostly binds to the cognate *cis*-acting element W box (C/T) TGAC (T/C) in the promoter (Eulgem *et al.*, 2000), and some alternative binding sites like SURE (sugar-responsive *cis*-element) (Rushton *et al.*, 1995). Based on the nature of DNA binding domain and number of domains, these TFs are categorized into four groups (Eulgem *et al.*, 2000; Ross *et al.*, 2007). Group-I proteins comprise of a C_2-H_2 zinc finger motif and two WRKY domains, group-II proteins contains single WRKY domain including C_2-H_2 zinc finger motif, group-III proteins contains two domains of WRKY with C_2-HC zinc finger motif and group-IV proteins contain only a WRKY domain, but lack zinc finger motif (Xie *et al.*, 2005; Ross *et al.*, 2007). These factors function both as positive or negative regulators of gene expression. Role of *WRKY* genes in various biological processes has been discussed in several reviews (Eulgem *et al.*, 2000; Singh *et al.*, 2002; Ulker and Somssich, 2004; Eulgem and Somssich, 2007; Agarwal *et al.*, 2011). In addition a number of *WRKY* family genes were found to be induced by various abiotic stresses including freezing (Huang and Duman, 2002), oxidative stress (Rizhsky *et al.*, 2002), drought and salinity (Seki *et al.*, 2002), drought and heat shock (Rizhsky *et al.*, 2002), cold and drought stress (Mare *et al.*, 2004; Wei *et al.*, 2008).

Although these TFs are induced by numerous abiotic stresses and have been isolated from a number of plants, the precise mechanism of individual *WRKY* genes during extreme environmental conditions remains unclear. Several studies have suggested that WRKY proteins function, either directly binding to the W box region or indirectly interacting with other stress responsive transcriptions factors. Constitutive overexpression of three of these TFs, *WRKY13*, *WRKY27* and *WRKY54* in transgenic *Arabidopsis* conferred differential tolerance to abiotic stresses. Transgenic plants carrying *WRKY54* have shown higher expression of stress responsive transcription factors like *DREB2A*, *rd29B* and *erd10*, and down regulation of *STZ*. Interestingly, *WRKY13* promoted the lateral root development in transgenic *Arabidopsis* plants through activation of auxin signalling and downstream expression of *ARF6*, and *ABI1* genes (Zhou *et al.*, 2008). In contrast, overexpression of a *WRKY53* from *Thlaspi caerulescens* in transgenic tobacco decreased the tolerance to osmotic stress and down regulated the two ERF family genes, *NtERF5* and *NtEREBP-1*. These results indicated that *WRKY53* might regulate the plant osmotic stress response by interacting with ERF-type transcription factors (Wei *et al.*, 2008). Constitutive overexpression of an abiotic stress inducible rice gene *WRKY45* in transgenic *Arabidopsis* showed improved drought tolerance with smaller stomatal openings, reduced transpiration and induction of genes involved in ABA signaling pathway (Qiu and Yu, 2009). Overexpression of rice gene *WRKY11* under heat inducible promoter *HSP101* imparted significant drought and heat tolerance as indicated by the slower leaf wilting, increased survival rate of green plants and slower water loss from the detached leaves (Wu *et al.*, 2009). All these studies suggest the essentiality of WRKY transcription factors in plant developmental process for a better adaptation during extreme environmental conditions. A WRKY transcription factor in *Arabidopsis*, *WRKY 39* has been reported to confer thermotolerance through calmodulin mediated calcium signaling pathway by linking the salicylic acid and jasmonic acid signaling pathways (Li,., 2010). Similarly, three of the *Arabidopsis* genes *WRKY25*, *WRKY26*, and *WRKY33* synergistically interact among themselves and regulate heat induced ethylene-activated signaling pathways and thereby fine tuning the plant thermotolerance (Li *et al.*, 2009; Li *et al.*, 2011). Recently, transgenic *Arabidopsis* seedlings overexpressing grapevine *WRKY11* showed higher tolerance to water stress by regulating the expression of two stress responsive genes, *rd29A* and *rd29B* (Liu *et al.*, 2011). All these studies suggest that *WRKYs* plays a role in abiotic stress tolerance, and might be useful for improvement of stress tolerance.

Trihelix Transcription Factors

Trihelix TFs are found uniquely in plant species that contain the highly degenerated and the deduced consensus core sequence 5'-G-Pu-(T/A)-A-A-(T/A)-3' called trihelix DNA-binding motif that binds GT elements; thus these factors also called as GT factors. These transcription factors usually contain one or two trihelix DNA binding domains, suggesting that they might be playing a major role in fine tuning of gene expression at the post- translation level (Smalle *et al.*, 1998). The cDNAs encoding GT binding proteins, rice *GT-2* and tobacco *GT1a/B2F* were first isolated using GT2 sequence and Box II sequence by affinity screening method (Dehesh *et al.*, 1990; Gilmartin *et al.*, 1992; Perisic and Lam, 1992). Since then, several genes encoding trihelix transcription factors have been cloned from *Arabidopsis*, tobacco, soybean, and other plant species (Gilmartin and Chua, 1990; Dehesh *et al.*, 1992; Le Gourrierec *et al.*, 1999; Ayadi *et al.*, 2004). The genome of *Arabidopsis* have 28 members (Riechmann *et al.*, 2000), rice have 22 members (Riano-Pachon *et al.*, 2007), and soybean have 13 members of trihelix family proteins (Tian *et al.*, 2004). Members of trihelix transcription factors further divided into three subgroups based on the binding of functionally distinct types of GT elements. A GT-1 and GT-3 type factor consists of a conserved single trihelix domain at N-terminal and C-terminal regions. But, GT-1 binds to Box II core sequence in the promoter, whereas GT-3a binds to the site-I (Kuhn *et al.*, 1993; Ayadi *et al.*, 2004). GT-2 type subfamily consisting of two trihelix domains, the one at the N-terminal region preferentially binds to the GT3-bx and the C-terminal one to the GT2-bx (Hiratsuka *et al.*, 1994).

These TFs play important role in light-regulated responses and other developmental processes like endoreduplication and petal development (Brewer *et al.*, 2004; Breuer *et al.*, 2009). Although the roles of the trihelix factors are gradually disclosed, the regulatory function of this kind of TFs in abiotic stress response remains largely unknown. More recently (Xie *et al.*, 2009), have cloned two abiotic stress induced *GT-2* type factors namely *GT-2A*, *GT-2B* from soybean, and their overexpression in transgenic *Arabidopsis* plants improved tolerance to salt, freezing and drought stress by altering ABA-regulated genes *NCED3, LTP3, LTP4* and *PAD3*. Similarly in *Arabidopsis*, a member of a GT-2 type gene (*GTL1*) enhanced drought tolerance and water use efficiency by regulating the expression of *Stomatal Density and Distribution1* (*SDD1*) gene. GTL1is induced by drought stress, and it binds to GT element of the *SDD1* promoter leading to repression of SDD1 expression, reduced number of stomata and improved water use efficiency (Yoo *et al.,* 2010). Very recently, Fang *et al.* (2010) identified a novel *GTγ-1*gene, member of GTγ subfamily in rice, which is strongly induced by salt stress and slightly induced by drought, cold stresses and ABA treatment. Overexpression of *GTγ-1*gene in rice enhanced the salt stress tolerance. Further, more studies are needed to determine the role of trihelix transcription factors under abiotic stress tolerance.

NAC Transcription Factors

Members of *NAC (NAM/ATAF1/2/CUC2)* are plant specific transcription factors. There are about 171 non redundant NAC genes in *Arabidopsis,* 151 in rice, 205 in soybean and 152 in tobacco (Rushton *et al.,* 2008; Mochida *et al.,* 2009; Nuruzzaman *et al.,* 2010). Based on the sequence similarity, NAC proteins were classified in to two groups and 18 subgroups (Ooka *et al.,* 2003). N-terminal region of NAC proteins consists of a five highly conserved domains (A-E) with 160 residues. Putative NLS have been detected in sub domains C and D. Sub domains D and E consist a DNA Binding Domain (Aida *et al.,* 1997; Kikuchi *et al.,* 2000; Duval *et al.,* 2002). C-terminal region consists of highly variable transcriptional regulatory regions that act as a transcriptional activation region (Xie *et al.,* 2000; Duval *et al.,* 2002; Hegedus *et al.,* 2003; Lu *et al.,* 2007; Nakashima *et al.,* 2007; Pinheiro *et al.,* 2009). Recent studies revealed that transcriptional activity of *NACs* mainly depends on the interaction between transcriptional repressor NARD domain (located in the N-terminal region *d* sub domain) and activation domain (C-terminal region). NACs acts as transcriptional repressors if the NARD domain function is more, and they act as transcriptional activators if the activation domain function is more (Hao *et al.,* 2010). Initially, most of the NAC TFs have been isolated and investigated extensively with reference to the plant growth and development perspectives. But, in the recent years increasing evidence implicated that these proteins also play a vital role in defense processes, senescence and abiotic stress conditions (Aida *et al.,* 1997; Seki *et al.,* 2002; Olsen *et al.,* 2005; Balazadeh *et al.,* 2008; Balazadeh *et al.,* 2010; Balazadeh *et al.,* 2011). A preliminary transcriptome analysis of 171 non redundant *NACs* in rice suggested that 45 genes are upregulated in response any one of the abiotic stress, and among them eleven genes were up regulated by at least three abiotic stresses (Nuruzzaman *et al.,* 2010).

NAC TF's role in abiotic stress tolerance was first reported in *Arabidopsis.* Overexpression of three cDNAs, *ANAC019, ANAC055,* and *ANAC072* individually in transgenic *Arabidopsis* enhanced tolerance to drought stress and altered expression of many known stress responsive genes. Probably, drought tolerance in transgenic plants is mediated by the detoxification of the toxic aldehydes through glyoxalase pathway (Tran *et al.,* 2004). Transgenic *Arabidopsis* plants over expressing *ONAC063* exhibited enhanced tolerance to high-temperature, high-salinity, and high-osmotic stresses. Transgenic plants also showed higher expression of downstream stress responsive genes almost similar to those upregulated by *ANAC019, ANAC055* or

ANAC072. More recently, a similar type of mechanism was operated in transgenic bread wheat overexpressing *TaNAC69-1* driven by either a constitutive or a drought-inducible promoter. Transgenic wheat plants showed improved drought tolerance and water-use efficiency by regulating the expression levels of several known stress responsive genes like *chitinase, ZIM,* and *glyoxalase I* (Xue *et al.,* 2011). Genetic engineering for improved root traits such as increased development of lateral roots and proliferation are very important for higher drought tolerance. It was also evidenced that overexpression of a salt and ABA induced *AtNAC2* in *Arabidopsis* resulted in improved drought and salt tolerance. Transgenic plants showed enhanced lateral root development by expressing genes involved in ethylene and auxin signaling pathways (He *et al.,* 2005). Similarly, overexpression of a soybean gene *NAC20* in transgenic *Arabidopsis* improved lateral root development and tolerance to both salt and freezing stresses. Transgenic plants showed altered expression levels of genes in *DREB/CBF-COR* pathway and auxin signalling pathway (Hao *et al.,* 2011). An *Arabidopsis NAC* gene *ORS1* regulates the expression of senescence induced genes probably through a regulatory network involving cross-talk with salt- and H_2O_2-dependent signalling pathways (Balazadeh *et al.,* 2011). Further, several NAC transcription factors have been transformed into several other crops species. A *Chrysanthemum NAC1* gene in to tobacco (Liu *et al.,* 2011d), and groundnut *NAC2* gene into *Arabidopsis* were introduced for improved drought and salt tolerance. These results also showed that NAC transcription factors are appropriate candidates for application in genetic engineering strategies aimed at improving abiotic stress tolerance in crop plants.

Nuclear Factor (NF-Y) Transcription Factors

Nuclear factor Y (*NF-Y*) is a heterotrimeric complex with high affinity and sequence specificity for the CCAAT box, a ubiquitous eukaryotic promoter element. NF-Ys are composed of three unique subunits: *NF-YA, NF-YB,* and *NF-YC,* with one copy of each subunit. *Arabidopsis* genome has numerous *NF-Y* subunit genes, also known as heme activator proteins (HAPs). Of these, 10 genes encode *NF-YA (HAP2),* 13 genes encode *NF-YB (HAP3),* and 13genes encode *NF-YC (HAP5)* (Edwards *et al.,* 1998; Gusmaroli *et al.,* 2002; Siefers *et al.,* 2009). However, there is only single gene for each subunit in mammals and yeast. It has been shown that *NF-Y* factors play a major role in embryo development (Lee *et al.,* 2003). At least 37 genes for *NF-Y* subunits have been identified in wheat, and nine of them were expressed in response to drought stress (Stephenson *et al.,* 2007). Furthermore, Nelson *et al.* (2007) identified nuclear factor Y (NF-Y) family protein, *AtNF-YB1* using functional genomics approach from drought tolerant *Arabidopsis* lines.

Constitutive overexpression of *AtNF-YB1,* in *Arabidopsis* showed improved performance under drought conditions without negatively affecting plant development. Consequently, an orthologous maize TF gene, *ZmNF-YB2,* was constitutively expressed in maize. Transgenic lines subjected to field-based drought stress at the late vegetative stage exhibited superior health, higher chlorophyll amount and photosynthetic rates, lower leaf temperatures, higher stomatal conductance, and less yield reduction than wild type. Interestingly, a member of the *Arabidopsis* NF-YA family gene *NFYA5* has a potential regulatory target of the small RNA miR169a, which is down- regulated by drought stress. Expression of this drought and ABA inducible gene was regulated at transcriptional level. Transgenic *Arabidopsis* plants overexpressing *NFYA5* gene under constitutive promoter displayed reduced leaf water loss, and were more resistant to drought stress.. In transgenic plants, drought stress induced several stress responsive genes involved in oxidative stress. Hence, *NFYA5* is regulated by drought stress at both transcriptional and post-transcriptional levels *via* a miRNA. This dual regulation is consistent with the critical importance of *NFYA5* for drought resistance (Li *et al.*, 2008). These results suggest the potential applications of this gene as a target for genetic engineering of plant for improved drought tolerance.

Scope of Using TFs in Engineering Crop Plants for Improved Abiotic Stress Tolerance

Genetic engineering of plants through transfer of global regulators, TFs, which set off the simultaneous transcription of the several down stream genes, to confer a wide range of stress tolerance to transgenic plants is a good strategy instead of transferring single genes providing limited stress tolerance to plants (Jaglo-Ottosen *et al.*, 1998; Liu *et al.*, 1998; Bajaj *et al.*, 1999; Ramanjulu and Bartels, 2002; Shinozaki *et al.*, 2003; Zhang *et al.*, 2004). Although a massive quantity of information has been generated on interaction of abiotic stresses and specific TFs, the information is still to be exploited in economically important crop plants. All the examples discussed here indicate that TFs regulate many downstream target genes also in heterologous plants and even under non-stress conditions, and prepare the plants to face any adverse condition at any time during its growth. Transcription factors like DREB/CBF, bZIP, MYBs/MYCs, Zinc fingers, Homeodomain (HD), NAC, WRKY, Trihelix and NF-Ys individually or in combination can be successfully used to engineer abiotic stress tolerance in a number of crop plants. Exploitation of all this information in crop plants is an attractive research subject and must be the ultimate destination of biotechnological applications.

Conclusion and Perspectives

Unlike in the laboratory, where the each stress is considered in isolation to other stress factors for easy interpretation of the stress-specific effect, plants have to defeat the stress combinations concurrently or separated temporally under field conditions and must present an integrated response (Knight and Knight, 2001). Transformation of plants with a single gene proved to be an ineffective strategy to improve the abiotic stress tolerance controlled by multiple genes unlike the disease or insect resistance. Manipulation of regulatory genes like TFs that control the expression of set of genes in concert will be a practical option in these conditions to improve stress tolerance of plants. The transcription factors discussed here in detail are the genes expressed in response abiotic stress conditions in plants and provide broad-spectrum tolerance to plants to face the inopportune environmental conditions by stimulating many signaling pathways and employing various mechanisms synchronously. In many cases genes were over expressed in plants, and studied their expression pattern in relation to abiotic stresses and the target genes, and obtained promising results towards the improvement of abiotic stress tolerance in plants. But, transgenic plants showed mild to severe growth retardation and reduced fertility that could be the stress-related-phenotypes or the result of lopsided metabolism in transgenic plants due to the overexpression of TFs. However, these effects were mitigated to some extent by limiting the expression of TF to only stress conditions by expressing the gene under control of conditional promoters in plants. Overexpression of TFs in plants is the ray of hope for improving the complex abiotic stress tolerance in plants to reduce the instability in crop yields, and to expand the agriculture to non arable lands to feed the ever-increasing world population by sustainable agriculture.

Acknowledgements

This work in A.R.R. laboratory was supported by the DBT, India and Rockefeller Foundation, USA. Work in the C.O.R.P laboratory was supported by CSIR and DST, India. L.R.G is thankful to CSIR, India for Senior Research Fellowship.

References

Abe, H., Yamaguchi–Shinozaki, K., Urao, T., Iwasaki, T., Hosokawa, D. and Shinozaki, K. 1997. Role of *Arabidopsis* MYC and MYB homologs in drought– and abscisic acid–regulated gene expression. Plant Cell **9 (10)**: 1859–1868.

Abe, M., Takahashi, T. and Komeda, Y. 2001. Identification of a cis–regulatory element for L1 layer–specific gene expression, which is targeted by an L1–specific homeodomain protein. Plant J. **26(5)**: 487–94.

Abe, H., Urao, T., Ito, T., Seki, M., Shinozaki, K., and Yamaguchi–Shinozaki, K. 2003. *Arabidopsis* AtMYC2 (bHLH) and AtMYB2 (MYB) function as transcriptional activators in abscisic acid signaling. Plant Cell **15(1)**: 63–78.

Abogadallah, G.M., Nada, R.M., Malinowski, R. and Quick, P. 2011. Overexpression of HARDY, an AP2/ERF gene from *Arabidopsis*, improves drought and salt tolerance by reducing transpiration and sodium uptake in transgenic *Trifolium alexandrinum L.* Planta **233(6)**: 1265–76.

Adler, G., Blumwald, E. and Bar–Zvi, D. 2010. The sugar beet gene encoding the sodium/proton exchanger 1 (BvNHX1) is regulated by a MYB transcription factor. Planta **232(1)**: 187–195.

Agalou, A., Purwantomo, S., Overnas, E., Johannesson, H., Zhu, X., Estiati, A., de Kam, R.J., Engstrom, P., Slamet–Loedin, I.H., Zhu, Z., Wang, M., Xiong, L., Meijer, A.H. and Ouwerkerk, P.B. 2008. A genome–wide survey of HD–Zip genes in rice and analysis of drought–responsive family members. Plant Mol. Biol. **66(1–2)**: 87–103. doi 10.1007/s11103–007–9255–7.

Agarwal, M., Hao, Y., Kapoor, A., Dong, C.H., Fujii, H., Zheng, X. and Zhu, J.K. 2006. A R2R3 type MYB transcription factor is involved in the cold regulation of CBF genes and in acquired freezing tolerance. J. Biol. Chem. **281(49)**: 37636–37645. doi M605895200 [pii]10.1074/jbc.M605895200.

Agarwal, P. and Jha, B. 2010. Transcription factors in plants and ABA dependent and independent abiotic stress signalling. Biologia Plantarum **54(2)**: 201–212. doi 10.1007/s10535–010–0038–7.

Agarwal, P., Agarwal, P.K., Joshi, A.J., Sopory, S.K. and Reddy, M.K. 2010. Overexpression of PgDREB2A transcription factor enhances abiotic stress tolerance and activates downstream stress–responsive genes. Mol. Biol. Rep. **37(2)**: 1125–35.

Agarwal, P., Reddy, M. P. and Chikara, J. 2011. WRKY: its structure, evolutionary relationship, DNA–binding selectivity, role in stress tolerance and development of plants. Mol. Biol. Rep., **38**: 3883–3896.

Aida, M., Ishida, T., Fukaki, H., Fujisawa, H. and Tasaka, M. 1997. Genes involved in organ separation in *Arabidopsis:* an analysis of the cup–shaped cotyledon mutant. Plant Cell **9(6)**: 841–857.

Anderson, J.P., Badruzsaufari, E., Schenk, P.M., Manners, J.M., Desmond, O.J., Ehlert, C., Maclean, D.J., Ebert, P.R. and Kazan, K. 2004. Antagonistic Interaction between Abscisic Acid and Jasmonate–Ethylene Signaling Pathways Modulates Defense Gene Expression and Disease Resistance in *Arabidopsis*. Plant Cell **16 (12)**: 3460–3479.

Ariel, F.D., Manavella, P.A., Dezar, C.A. and Chan, R.L. 2007. The true story of the HD–Zip family. Trends Plant Sci. **12 (9)**: 419–426.

Artus, N.N., Uemura, M., Steponkus, P., Gilmour, S., Lin, C. and Thomashow, M.F. 1996. Constitutive expression of the cold–regulated *Arabidopsis thaliana* COR15a gene affects both chloroplast and protoplast freezing tolerance. Proc. Natl. Acad. Sci. U S A **93(23)**: 13404–13409.

Ayadi, M., Delaporte, V., Li, Y.F. and Zhou, D.X. 2004. Analysis of GT–3a identifies a distinct subgroup of trihelix DNA–binding transcription factors in *Arabidopsis*. FEBS Lett. *562:* 147–154.

Bajaj, S., Targolli, J., Liu, L.F., Ho, T.H.D. and Wu, R. 1999. Transgenic approaches to increase dehydration–stress tolerance in plants. Mol. Breeding **5(6)**: 493–503.

Baker, S.S., Wilhelm, K.S. and Thomashow, M.F. 1994. The 52–region of *Arabidopsis thaliana* cor15a has cis–acting elements that confer cold–, drought– and ABA–regulated gene expression. Plant Molecular Biol. **24(5)**: 701–713.

Balazadeh, S., Kwasniewski, M., Caldana, C., Mehrnia, M., Zanor, M.I., Xue, G.P. and Mueller–Roeber, B. 2011. ORS1, an HO–responsive NAC transcription factor, controls senescence in *Arabidopsis thaliana*. Mol. Plant **4(2)**: 346–360. doi ssq080 [pii]10.1093/mp/ssq080.

Balazadeh, S., Riano–Pachon, D.M and, Mueller–Roeber, B. 2008. Transcription factors regulating leaf senescence in *Arabidopsis thaliana*. Plant Biol. (Stuttg) **10** Suppl 1: 63–75. doi PLB088 [pii]10.1111/j.1438–8677.2008.00088.x.

Balazadeh, S., Siddiqui, H., Allu, A.D., Matallana–Ramirez, L.P., Caldana, C., Mehrnia, M., Zanor, M.I., Kohler, B. and Mueller–Roeber, B. 2010. A gene regulatory network controlled by the NAC transcription factor ANAC092/AtNAC2/ORE1 during salt–promoted senescence. Plant J. 62(2): 250–264. doi TPJ4151 [pii]10.1111/j.1365–313X.2010.04151.x.

Ben Saad, R., Zouari, N., Ben Ramdhan, W., Azaza, J., Meynard, D., Guiderdoni, E. and Hassairi, A. 2010. Improved drought and salt stress tolerance in transgenic tobacco overexpressing a novel A20/AN1 zinc–finger "AlSAP" gene isolated from the halophyte grass *Aeluropus littoralis*. Plant Mol. Biol. **72(1–2)**: 171–190. doi 10.1007/s11103–009–9560–4.

Bhatnagar–Mathur, P., Devi, M.J., Reddy, D.S., Lavanya, M., Vadez, V., Serraj, R., Yamaguchi–Shinozaki, K. and Sharma, K.K. 2007. Stress–inducible expression of At DREB1A in transgenic peanut (*Arachis hypogaea* L.) increases transpiration efficiency under water–limiting conditions. Plant Cell Rep. **27**: 411–24.

Bihani, P., Char, B. and Bhargava, S. 2011. Transgenic expression of sorghum DREB2 in rice improves tolerance and yield under water limitation. J. Agril. Sci. 149: 95–101.

Boter, M., Ruiz–Rivero, O., Abdeen, A. and Prat, S. 2004. Conserved MYC transcription factors play a key role in jasmonate signaling both in tomato and *Arabidopsis*. Genes Dev. **18**: 1577–1591.

Breuer, C., Kawamura, A., Ichikawa, T., Tominaga–Wada, R., Wada, T., Kondou, Y., Muto, S., Matsui, M. and Sugimoto, K. 2009. The trihelix transcription factor GTL1 regulates ploidy–dependent cell growth in the *Arabidopsis* trichome. Plant Cell **21**: 2307–2322.

Brewer, P.B., Howles, P.A., Dorian, K., Griffith, M.E., Ishida, T., Kaplan–Levy, R.N., Kilinc, A. and Smyth, D.R. 2004. PETAL LOSS, a trihelix transcription factor gene, regulates perianth architecture in the *Arabidopsis* flower. Development **131**: 4035–4045.

Cai, H., Tian, S., Liu, C. and Dong, H. 2011. Identification of a MYB3R gene involved in drought, salt and cold stress in wheat (*Triticum aestivum* L.). Gene **485**: 146–52.

Cao, Y.J., Wei, Q., Liao, Y., Song, H.L., Li, X., Xiang, C.B. and Kuai, B.K. 2009 Ectopic overexpression of AtHDG11 in tall fescue resulted in enhanced tolerance to drought and salt stress. Plant Cell Rep. **28(4)**: 579–588. doi 10.1007/s00299–008–0659–x.

Chang, C., Jacobs, Y., Nakamura, T., Jenkins, N., Copeland, N. and Cleary, M. 1997 Meis proteins are major *in vivo* DNA binding partners for wild–type but not chimeric Pbx proteins. Mol. Cell. Biol. **17 (10)**: 5679–5687.

Chao, Y., Kang, J., Sun, Y., Yang, Q., Wang, P., Wu, M., Li, Y., Long, R. and Qin, Z. 2009. Molecular cloning and characterization of a novel gene encoding zinc finger protein from *Medicago sativa* L. Mol. Biol. Rep. **36**: 2315–21.

Chen, B.J., Wang, Y., Hu, Y.L., Wu, Q. and Lin, Z.P. 2005. Cloning and characterization of a drought–inducible MYB gene from *Boea crassifolia*. Plant Sci. **168(2)**: 493–500.

Chen, M., Wang, Q.Y., Cheng, X.G., Xu, Z.S., Li, L.C., Ye, X.G., Xia, L.Q. and Ma, Y.Z. 2007. GmDREB2, a soybean DRE–binding transcription factor, conferred drought and high–salt tolerance in transgenic plants. Biochem. Biophys. Res. Commun. **353**: 299–305. doi: S0006–291X(06)02653–2 [pii]10.1016/j.bbrc.2006.12.027.

Chen, J.Q., Meng, X.P., Zhang, Y., Xia, M. and Wang, X.P. 2008. Over–expression of OsDREB genes lead to enhanced drought tolerance in rice. Biotechnology Let. **30(12)**: 2191–2198.

Chen, R., Ni, Z., Nie, X., Qin, Y., Dong, G. and Sun, Q. 2005b. Isolation and characterization of genes encoding Myb transcription factor in wheat (*Triticum aestivem* L.). Plant Sci. **169(6)**: 1146–1154.

Chen, J., Xia, X. and Yin, W. 2009. Expression profiling and functional characterization of a DREB2–type gene from *Populus euphratica*. Biochem. Biophys. Res. Commun. **378**: 483–7.

Chen, H., Hwang, J.E., Lim, C.J., Kim, D.Y., Lee, S.Y. and Lim, C.O. 2010. *Arabidopsis* DREB2C functions as a transcriptional activator of HsfA3 during the heat stress response. Biochem. Biophys. Res. Commun. **15**: 401: 238–44.

Chen, W., Provart, N.J., Glazebrook, J., Katagiri, F., Chang, H.S., Eulgem, T., Mauch, F., Luan, S., Zou, G., Whitham, S.A., Budworth, P.R., Tao, Y., Xie, Z., Chen, X., Lam, S., Kreps, J.A., Harper, J.F., Si–Ammour, A., Mauch–Mani, B., Heinlein, M., Kobayashi, K., Hohn, T., Dangl, J.L., Wang, X. and Zhu, T. 2002. Expression profile matrix of *Arabidopsis* transcription factor genes suggests their putative functions in response to environmental stresses. Plant Cell **14(3)**: 559–574.

Chiappetta, L., Tomes, D., Xu. D., Sivasankar, S., Sanguineti, M.C. and Tuberosa, R. 2003. DREB1 overexpression improves tolerance to low temperature in maize in the wake of the double helix: from the green revolution to the gene revolution. Italy, pp. 533–544.

Chinnusamy, V., Ohta, M., Kanrar, S., Lee, B.H., Hong, X., Agarwal, M. and Zhu, J.K. 2003 ICE1: a regulator of cold–induced transcriptome and freezing tolerance in *Arabidopsis*. Genes Dev. **17**: 1043–1054. doi: 10.1101/gad.1077503U–10775R [pii].

Chinnusamy, V., Schumaker, K. and Zhu, J.K. 2004. Molecular genetic perspectives on cross–talk and specificity in abiotic stress signalling in plants. J. Exp. Bo.t **55(395)**: 225–236. doi 10.1093/jxb/erh005erh005 [pii].

Choi, D.W., Rodriguez, E.M. and Close, T.J. 2002. Barley Cbf3 gene identification, expression pattern, and map location. Plant Physiol. **129(4)**: 1781–1787. doi 10.1104/pp.003046.

Choi, H., Hong, J., Ha, J., Kang, J. and Kim, S.Y. 2000. ABFs, a family of ABA–responsive element binding factors. J. Biol. Chem. **275(3)**: 1723–1730.

Ciftci–Yilmaz, S. and Mittler, R. 2008. The zinc finger network of plants. Cell Mol. Life Sci. **65(7–8)**: 1150–1160. doi 10.1007/s00018–007–7473–4.

Collinge, M. and Boller, T. 2001. Differential induction of two potato genes, Stprx2 and StNAC, in response to infection by *Phytophthora infestans* and to wounding. Plant Mol. Biol. **46(5)**: 521–529.

Cominelli, E., Galbiati, M., Vavasseur, A., Conti. L., Sala, T., Vuylsteke, M., Leonhardt, N., Dellaporta, S.L. and Tonelli, C. 2005 A guard cell specific MYB transcription factor regulates stomatal movements and plant drought tolerance. Curr. Biol. **15(13)**: 1196–1200.

Cong, L., Chai, T.Y. and Zhang,Y.X. 2008. Characterization of the novel gene BjDREB1B encoding a DRE–binding transcription factor from *Brassica juncea* L. Biochem. Biophys. Res. Commun. **371**: 702–706. doi: S0006–291X(08)00808–5 [pii]10.1016/j.bbrc.2008.04.126.

Cook, D., Fowler, S., Fiehn, O. and Thomashow, M.F. 2004. A prominent role for the CBF cold response pathway in configuring the low–temperature metabolome of *Arabidopsis*. Proc. Natl. Acad. Sci. U S A **101(42)**: 15243–15248. doi 10.1073/pnas.0406069101.

Dai, X., Xu, Y., Ma, Q., Xu, W., Wang, T., Xue, Y. and Chong, K. 2007. Overexpression of an R1R2R3 MYB Gene, OsMYB3R–2, increases tolerance to freezing, drought, and salt stress in transgenic *Arabidopsis*. Plant Physiol. **143(4)**: 1739–1751.

Dehesh, K., Bruce, W.B. and Quail, P.H. 1990. A trans–acting factor that binds to a GTmotif in a phytochrome gene promoter. Science **250**: 1397–1399.

Dehesh, K., Hung, H., Tepperman, J.M. and Quall, P.H. 1992. GT–2: A transcription factor with twin autonomous DNA–binding domains of closely related but different target sequence specificity. EMBO J. **11**: 4131–4144.

Deng, X., Phillips, J., Meijer, A.H., Salamini, F. and Bartels, D. 2002. Characterization of five novel dehydration–responsive homeodomain leucine zipper genes from the resurrection plant *Craterostigma plantagineum*. Plant Mol. Biol. **49(6)**: 601–610.

Deng, X., Phillips, J., Bräutigam, A., Engström, P., Johannesson, H., Ouwerkerk, P.B., Ruberti, I,, Salinas, J., Vera, P., Iannacone, R., Meijer, A.H. and Bartels, D. 2006. A homeodomain leucine zipper gene from *Craterostigma plantagineum* regulates abscisic acid responsive gene expression and physiological responses. Plant Mol. Biol. **61(3)**: 469–89.

Davletova, S., Schlauch, K., Coutu, J. and Mittler, R. 2005. The zinc–finger protein Zat12 plays a central role in reactive oxygen and abiotic stress signaling in *Arabidopsis*. Plant Physiol. **139(2)**: 847–856. doi pp.105.068254 [pii]10.1104/pp.105.068254.

Devaiah, B.N., Karthikeyan, A,S. and Raghothama, K.G. 2007. WRKY75 transcription factor is a modulator of phosphate acquisition and root development in *Arabidopsis*. Plant Physiol. **143(4)**: 1789–1801. doi pp.106.093971 [pii]10.1104/pp.106.093971.

Dezar, C.A., Gago, G.M., Gonzalez, D.H., Chan, R.L. (2005) Hahb–4, a sunflower homeobox–leucine zipper gene, is a developmental regulator and confers drought tolerance to *Arabidopsis thaliana* plants. Transgenic Res. 14(4): 429–440.

Ding, Z., Li, S., An, X., Liu, X., Qin, H. and Wang, D. 2009. Transgenic expression of MYB15 confers enhanced sensitivity to abscisic acid and improved drought tolerance in *Arabidopsis thaliana*. J. Genet. Genomics. **36**: 17–29.

Dong, J.X., Chen, C.H. and Chen, Z.X. 2003. Expression profiles of the *Arabidopsis* WRKY gene superfamily during plant defense response. Plant Mol. Biol. **51**: 21–37.

Dubouzet, J.G., Sakuma, Y., Ito, Y., Kasuga, M., Dubouzet, *E.G.*, Miura, S., Seki, M., Shinozaki, K. and Yamaguchi–Shinozaki, K. 2003. OsDREB genes in rice, *Oryza sativa* L., encode transcription activators that function in drought–, high–salt– and cold–responsive gene expression. Plant J. **33(4)**: 751–763. doi 1661 [pii].

Duval, M., Hsieh, T.F., Kim, S.Y. and Thomas, T.L. 2002. Molecular characterization of AtNAM: a member of the *Arabidopsis* NAC domain superfamily. Plant Mol. Biol. **50(2)**: 237–248.

Edwards, D., Murray, J.A. and Smith, A.G. 1998. Multiple genes encoding the conserved CCAAT–box transcription factor complex are expressed in *Arabidopsis*. Plant Physiol. **117(3)**: 1015–1022.

Eulgem, T., Rushton, P.J., Robatzek, S. and Somssich, *I.E.* 2000. The WRKY superfamily of plant transcription factors. Trends Plant Sci. **5(5)**: 199–206. doi S1360–1385(00)01600–9 [pii].

Eulgem, T. and Somssich, *I.E.* 2007. Networks of WRKY transcription factors in defense signaling. Curr Opin Plant Biol. **10**: 366–371.

Fang, Y., Xie, K., Hou, X., Hu, H. and Xiong, L. 2010. Systematic analysis of GT factor family of rice reveals a novel subfamily involved in stress responses. Mol. Genet. Genomics **283(2)**: 157–169. doi 10.1007/s00438–009–0507–x.

Feng, C., Andreasson, E., Maslak, A., Mock, H.P., Mattsson, O. and Mundy, J. 2004. *Arabidopsis* MYB68 in development and responses to environmental cues. Plant Sci. **167(5)**: 1099–1107.

Finkelstein, R.R., Gampala, S.S. and Rock, C.D. 2002. Abscisic acid signaling in seeds and seedlings. Plant Cell **14** Suppl: S15–45.

Fowler, S.G., Cook, D. and Thomashow, M.F. 2005. Low temperature induction of *Arabidopsis* CBF1, 2, and 3 is gated by the circadian clock. Plant Physiol. **137(3)**: 961–968. doi 10.1104/pp.104.058354.

Frank, W., Phillips, J., Salamini, F. and Bartels, D. 1998. Two dehydration–inducible transcripts from the resurrection plant *Craterostigma plantagineum* encode interacting homeodomain–leucine zipper proteins. Plant J. **15(3)**: 413–421. doi 10.1046/j.1365–313X.1998.00222.x.

Fujita, M., Fujita, Y., Maruyama, K., Seki, M., Hiratsu, K., Ohme–Takagi, M., Tran, L.S., Yamaguchi–Shinozaki, K. and Shinozaki, K. 2004. A dehydration–induced NAC protein, RD26, is involved in a novel ABA–dependent stress–signaling pathway. Plant J. **39(6)**: 863–876. doi 10.1111/j.1365–313X.2004.02171.xTPJ2171 [pii].

Fujita, Y., Fujita, M., Satoh, R., Maruyama, K., Parvez, M.M,, Seki, M., Hiratsu, K., Ohme–Takagi, M., Shinozaki, K. and Yamaguchi–Shinozaki, K. 2005. AREB1 is a transcription activator of novel ABRE–dependent ABA signaling that enhances drought stress tolerance in *Arabidopsis*. Plant Cell **17(12)**: 3470–3488. doi 10.1105/tpc.105.035659.

Furini, A., Koncz, C., Salamini, F. and Bartels, D. 1997. High level transcription of a member of a repeated gene family confers dehydration tolerance to callus tissue of *Craterostigma plantagineum*. EMBO J. **16(12)**: 3599–3608. doi 10.1093/emboj/16.12.3599.

Ganesan, G., Sankararamasubramanian, H.M., Narayanan, J.M., Sivaprakash, K.R. and Parida, A. 2008. Transcript level characterization of a cDNA encoding stress regulated NAC transcription factor in the mangrove plant *Avicennia marina*. Plant Physiol. Biochem. **46(10)**: 928–934. doi S0981–9428(08)00090–9 [pii]10.1016/j.plaphy.2008.05.002.

Gao, S.Q., Chen, M., Xia, L.Q. Xiu, H.J., Xu, Z.S., Li, L.C., Zhao, C.P., Cheng, X.G. and Ma, Y.Z. 2009. A cotton (*Gossypium hirsutum*) DRE–binding transcription factor gene, GhDREB, confers enhanced tolerance to drought, high salt, and freezing stresses in transgenic wheat. Plant Cell Rep. **28(2)**: 301–11.

Gao, F., Xiong, A., Peng, R., Jin, X., Xu, J., Zhu, B., Chen, J. and Yao, Q. 2010. OsNAC52, a rice NAC transcription factor, potentially responds to ABA and confers drought tolerance in transgenic plants. Plant Cell Tissue Organ Cult. **100(3)**: 255–262. doi 10.1007/s11240–009–9640–9.

Gao, M.J., Allard, G., Byass, L., Flanagan, A.M. and Singh, J. 2002. Regulation and characterization of four CBF transcription factors from *Brassica napus*. Plant Mol. Biol. **49(5)**: 459–471.

Gao, S.Q., Chen, M., Xu, Z.S., Zhao, C.P., Li, L., Xu, H.J., Tang, Y.M., Zhao, X. and Ma, Y.Z. 2011. The soybean GmbZIP1 transcription factor enhances multiple abiotic stress tolerances in transgenic plants. Plant Mol. Biol. **75**: 537–553.

Gilmartln, P.M. and Chua, Nr.H. 1990. Localization of a phytochrome responsive element within the upstream region of a pea rbcS–3A. Mol. Cell. Biol. **10**: 5565–5568.

Gilmartin, P.M., Memelink, J., Hiratsuka, K., Kay, S.A. and Chua, N.H. 1992. Characterization of a gene encoding a DNA binding protein with specificity for a light–responsive element. Plant Cell **4**: 839–849.

Gilmour, S.J., Sebolt, A.M., Salazar, M.P., Everard, J.D. and Thomashow, M.F. 2000. Overexpression of the *Arabidopsis* CBF3 transcriptional activator mimics multiple biochemical changes associated with cold acclimation. Plant Physiol. **124(4)**: 1854–1865.

Gilmour, S.J., Zarka, D.G., Stockinger, E.J., Salazar, M.P., Houghton, J.M. and Thomashow, M.F. 1998. Low temperature regulation of the *Arabidopsis* CBF family of AP2 transcriptional activators as an early step in cold–induced COR gene expression. Plant J. **16(4)**: 433–442.

Gorantla, M., Babu, P.R., Reddy, L.V.B., Reddy, A.M.M., Wusirika, R., Bennetzen, J.L. and Reddy, A.R. 2007. Identification of stress–responsive genes in an indica rice (*Oryza sativa* L.) using ESTs generated from drought–stressed seedlings. J. Exp. Bot. **58(2)**: 253–265. doi 10.1093/jxb/erl213.

Guo, S.Q., Huang, J., Jiang, Y. and Zhang, H.S. 2007. Cloning and characterization of RZF71 encoding a C2H2–type zinc finger protein from rice. Yi Chuan **29(5)**: 607–613. doi 0253–9772(2007)05–607–07 [pii].

Guo, Y.H., Yu, Y.P., Wang, D., Wu, C.A., Yang, G.D., Huang, J.G. and Zheng, C.C. 2009. GhZFP1, a novel CCCH–type zinc finger protein from cotton, enhances salt stress tolerance and fungal disease resistance in transgenic tobacco by interacting with GZIRD21A and GZIPR5. New Phytol. **183(1)**: 62–75. doi NPH2838 [pii]10.1111/j.1469–8137.2009.02838.x.

Gupta, K., Agarwal, P.K., Reddy, M.K. and Jha, B. 2010. SbDREB2A, an A–2 type DREB transcription factor from extreme halophyte *Salicornia brachiata* confers abiotic stress tolerance in *Escherichia coli.* Plant Cell Rep. **29**: 1131–1137.

Gusmaroli, G., Tonelli, C. and Mantovani, R. 2002. Regulation of novel members of the *Arabidopsis thaliana* CCAAT–binding nuclear factor Y subunits. Gene **283(1–2)**: 41–48. doi S0378111901008332 [pii].

Gutha, L.R. 2008. Molecular cloning and functional characterization of rice dehydration responsive element binding (DREB1) transcription factors and their promoters. Dissertation, University of Hyderabad.

Gutha, L.R. and Reddy, A.R. 2008. Rice DREB1B promoter shows distinct stress–specific responses, and the overexpression of cDNA in tobacco confers improved abiotic and biotic stress tolerance. Plant Mol. Biol. **68(6)**: 533–555.

Haake, V., Cook, D., Riechmann, J.L., Pineda, O., Thomashow, M.F. and Zhang, J.Z. 2002. Transcription factor CBF4 is a regulator of drought adaptation in *Arabidopsis.* Plant Physiol. **130(2)**: 639–648. doi 10.1104/pp.006478.

Han, Q., Zhang, J., Li, H., Luo, Z., Ziaf, K., Ouyang, B., Wang, T. and Ye, Z. 2011. Identification and expression pattern of one stress–responsive NAC gene from *Solanum lycopersicum*. Mol. Biol. Rep. doi 10.1007/s11033–011–0911–2.

Hao, Y.J., Song, Q.X., Chen, H.W., Zou, H.F., Wei, W., Kang, X.S., Ma, B., Zhang, W.K., Zhang, J.S.and Chen, S.Y. 2010. Plant NAC–type transcription factor proteins contain a NARD domain for repression of transcriptional activation. Planta **232(5)**: 1033–1043. doi 10.1007/s00425–010–1238–2.

Hao, Y.J., Wei, W., Song, Q.X., Chen, H.W., Zhang, Y.Q., Wang, F., Zou, H.F., Lei, G., Tian, A.G., Zhang, W.K., Ma, B., Zhang, J.S. and Chen, S.Y. 2011. Soybean NAC transcription factors promote abiotic stress tolerance and lateral root formation in transgenic plants. Plant J. **68(2)**: 302–313.

Hattori, T., Totsuka, M., Hobo, T., Kagaya, Y. and Yamamoto–Toyoda, A. 2002. Experimentally Determined Sequence Requirement of ACGT–Containing Abscisic Acid Response Element. Plant Cell Physiol. **43(1)**: 136–140.

He, X.J., Mu, R.L., Cao, W.H., Zhang, Z.G., Zhang, J.S. and Chen, S.Y. 2005. AtNAC2, a transcription factor downstream of ethylene and auxin signaling pathways, is involved in salt stress response and lateral root development. Plant J. **44(6)**: 903–916.

Hegedus, D., Yu, M., Baldwin, D., Gruber, M., Sharpe, A., Parkin, I., Whitwill, S. and Lydiate, D. 2003. Molecular characterization of *Brassica napus* NAC domain transcriptional activators induced in response to biotic and abiotic stress. Plant Mol. Biol. **53(3)**: 383–397.

Himmelbach, A., Hoffmann, T., Leube, M., Höhener, B. and Grill, E. 2002. Homeodomain protein ATHB6 is a target of the protein phosphatase ABI1 and regulates hormone responses in *Arabidopsis*. EMBO J. **21(12)**: 3029–38.

Hiratsuka, K., Wu, X., Fukuzawa, H. and Chua, N.H. 1994. Molecular dissection of GT–1 from *Arabidopsis*. Plant Cell **6(12)**: 1805–1813. doi 10.1105/tpc.6.12.1805.

Hirayama, T. and Shinozaki. K. 2010. Research on plant abiotic stress responses in the post–genome era: past, present and future. Plant J. **61(6)**: 1041–1052. doi TPJ4124 [pii]10.1111/j.1365–313X.2010.04124.x.

HjellstrÖM, M., Olsson, A.S.B., EngstrÖM, P. and SÖDerman, E.M. 2003. Constitutive expression of the water deficit–inducible homeobox gene ATHB7 in transgenic *Arabidopsis* causes a suppression of stem elongation growth. Plant Cell Environ. **26(7)**: 1127–1136. doi 10.1046/j.1365–3040.2003.01037.x.

Hong, J.P. and Kim, W.T. 2005. Isolation and functional characterization of the Ca–DREBLP1 gene encoding a dehydration–responsive element binding–factor-like protein 1 in hot pepper (*Capsicum annuum* L. cv. Pukang). Planta **220**: 875–888.

Hsieh, T.H., Lee, J.T., Charng, Y.Y. and Chan, M.T. 2002a. Tomato plants ectopically expressing *Arabidopsis* CBF1 show enhanced resistance to water deficit stress. Plant Physiol. **130(2)**: 618–626. doi 10.1104/pp.006783.

Hsieh, T.H., Lee, J.T., Yang, P.T., Chiu, L.H., Charng, Y.Y., Wang, Y.C. and Chan, M.T. 2002b. Heterology expression of the *Arabidopsis* C–repeat/dehydration response element binding factor 1 gene confers elevated tolerance to chilling and oxidative stresses in transgenic tomato. Plant Physiol. **129**: 1086–1094. doi: 10.1104/pp.003442.

Hsieh, T.H., Li, C.W., Su, R.C., Cheng, C.P., Sanjaya, Tsai, Y.C. and Chan, M.T. 2010. A tomato bZIP transcription factor, SlAREB, is involved in water deficit and salt stress response. Planta. **231**: 1459–73.

Hu, H., Dai, M., Yao, J., Xiao, B., Li, X., Zhang, Q. and Xiong, L. 2006. Overexpressing a NAM, ATAF, and CUC (NAC) transcription factor enhances drought resistance and salt tolerance in rice. Proc. Natl. Acad. Sci. U S A **103(35)**: 12987–12992. doi 10.1073/pnas.0604882103.

Hu, H., You, J., Fang, Y., Zhu, X., Qi, Z. and Xiong, L. 2008a. Characterization of transcription factor gene SNAC2 conferring cold and salt tolerance in rice. Plant Mol. Biol. **67(1–2)**: 169–181. doi 10.1007/s11103–008–9309–5.

Hu, W., dePamphilis, C.W. and Ma, H. 2008b. Phylogenetic analysis of the plant-specific zinc finger–homeobox and mini zinc finger gene families. J. Integr. Plant Biol. **50(8)**: 1031–1045.

Huang, T. and Duman, J.G. 2002. Cloning and characterization of a thermal hysteresis (antifreeze) protein with DNA–binding activity from winter bittersweet nightshade, *Solanum dulcamara*. Plant Mol. Biol. **48**: 339–350.

Huang, J., Wang, J.F., Wang, Q.H. and Zhang, H.S. 2005. Identification of a rice zinc finger protein whose expression is transiently induced by drought, cold but not by salinity and abscisic acid. DNA Seq. **16**: 130–136.

Huang, J., Wang, M.M., Jiang, Y., Bao, Y.M., Huang,X,. Sun, H., Xu, D.Q., Lan, H.X. and Zhang, H.S. 2008. Expression analysis of rice A20/AN1–type zinc finger genes and characterization of ZFP177 that contributes to temperature stress tolerance. Gene **420(2)**: 135–144.

Huang, J., Yang, X., Wang, M.M., Tang, H.J., Ding, L.Y., Shen, Y. and Zhang, H.S. 2007. A novel rice C2H2–type zinc finger protein lacking DLN–box/EAR–motif plays a role in salt tolerance. Biochim Biophys. Acta **1769(4)**: 220–227. doi S0167–4781(07)00049–8 [pii]10.1016/j.bbaexp.2007.02.006.

Huang, X.Y., Chao, D.Y., Gao, J.P., Zhu, M.Z., Shi, M. and Lin, H.X. 2009. A previously unknown zinc finger protein, DST, regulates drought and salt tolerance in rice via stomatal aperture control. Genes Dev. **23(15)**: 1805–1817. doi 23/15/1805 [pii]10.1101/gad.1812409.

Ingram, J. and Bartels, D. 1996. The molecular masis of dehydration tolerance in plants. Annu. Rev. Plant Physiol. Plant Mol. Biol. **47**: 377–403. doi 10.1146/annurev.arplant.47.1.377.

Islam, M.S. and Wang, M.H. (2009) Expression of dehydration responsive element–binding protein–3 (DREB3) under different abiotic stresses in tomato. BMB Rep. 42(9): 611–616.

Ito, Y., Katsura, K., Maruyama, K., Taji, T., Kobayashi, M., Seki, M., Shinozaki, K. and Yamaguchi–Shinozaki, K. 2006. Functional Analysis of Rice DREB1/CBF–type transcription factors involved in cold–responsive gene expression in transgenic rice. Plant and Cell Physiol. **47**: 141"153.

Iturriaga, G., Leyns, L., Villegas, A., Gharaibeh, R., Salamini, F. and Bartels, D. 1996. A family of novel myb–related genes from the resurrection plant *Craterostigma plantagineum* are specifically expressed in callus and roots in response to ABA or desiccation. Plant Mol. Biol. **32(4)**: 707–716.

Jaglo–Ottosen, K.R., Gilmour, S.J., Zarka, D.G., Schabenberger, O., Thomashow, M.F. 1998. *Arabidopsis* CBF1 overexpression induces COR genes and enhances freezing tolerance. Science 280(5360): 104–106.

Jaglo, K.R., Kleff, S., Amundsen, K.L., Zhang, X., Haake,V., Zhang, J.Z., Deits, T. and Thomashow, M.F. 2001. Components of the *Arabidopsis* C–repeat/dehydration–responsive element binding factor cold–response pathway are conserved in *Brassica napus* and other plant species. Plant Physiol. **127(3)**: 910–917.

Jakoby, M., Weisshaar, B., Dröge–Laser, W., Vicente–Carbajosa, J., Tiedemann, J., Kroj, T. and Parcy, F. 2002. bZIP transcription factors in *Arabidopsis*. Trends Plant Sci. **7(3)**: 106–111.

James, V.A., Neibaur, I. and Altpeter, F. 2008. Stress inducible expression of the DREB1A transcription factor from xeric, *Hordeum spontaneum* L. in turf and forage grass (*Paspalum notatum* Flugge) enhances abiotic stress tolerance. Transgenic Res. **17**: 93–104. doi: 10.1007/s11248–007–9086–y.

Jensen, M.K., Hagedorn, P.H., de Torres–Zabala, M., Grant, M.R., Rung, J.H., Collinge, D.B. and Lyngkjaer, M.F. 2008. Transcriptional regulation by an NAC (NAM–ATAF1,2–CUC2) transcription factor attenuates ABA signalling for efficient basal defence towards *Blumeria graminis* f. sp. *hordei* in *Arabidopsis*. Plant J. **56(6)**: 867–880. doi TPJ3646 [pii]10.1111/j.1365–313X.2008.03646.x.

Jeong, J.S., Kim, Y.S., Baek, K.H., Jung, H., Ha, S.H., Do, Choi, Y., Kim, M., Reuzeau,C. and Kim, J.K. 2010. Root–specific expression of OsNAC10 improves drought tolerance and grain yield in rice under field drought conditions. Plant Physiol. **153(1)**: 185–197. doi pp.110.154773 [pii]10.1104/pp.110.154773.

Jiang, C., Iu, B. and Singh, J. 1996. Requirement of a CCGAC cis–acting element for cold induction of the BN115 gene from winter *Brassica napus*. Plant Mol. Biol. **30(3)**: 679–684.

Jiang, Y. and Deyholos, M.K. 2009. Functional characterization of *Arabidopsis* NaCl–inducible WRKY25 and WRKY33 transcription factors in abiotic stresses. Plant Mol. Biol. **69(1–2)**: 91–105. doi 10.1007/s11103–008–9408–3.

Jin, H. and Martin, C. 1999. Multifunctionality and diversity within the plant MYB–gene family. Plant Mol. Biol. **41(5)**: 577–585.

Johannesson, H., Wang, Y, Hanson, J. and Engström, P. 2003. The *Arabidopsis thaliana* homeobox gene ATHB5 is a potential regulator of abscisic acid responsiveness in developing seedlings. Plant Mol. Biol. **51(5)**: 719–729.

Jung, C., Seo, J.S., Han, S.W., Koo, Y.J., Kim, C.H., Song, S.I., Nahm, B.H., Choi, Y.D. and Cheong, J.J. 2008. Overexpression of AtMYB44 enhances stomatal closure to confer abiotic stress tolerance in transgenic *Arabidopsis*. Plant Physiol. **146(2)**: 623–635. doi pp.107.110981 [pii]10.1104/pp.107.110981.

Kam, J., Gresshoff, P.M., Shorter, R. and Xue, G.P. 2008. The Q–type C2H2 zinc finger subfamily of transcription factors in *Triticum aestivum* is predominantly expressed in roots and enriched with members containing an EAR repressor motif and responsive to drought stress. Plant Mol. Biol. **67(3)**: 305–322.

Kang, H.M. and Saltveit, M.E. 2002. Antioxidant enzymes and DPPH–radical scavenging activity in chilled and heat–shocked rice (*Oryza sativa* L.) seedlings radicles. J Agric Food Chem. **50(3)**: 513–518. doi jf011124d [pii].

Karaba, A., Dixit, S., Greco, R., Aharoni, A., Trijatmiko, K.R., Marsch–Martinez, N., Krishnan, A., Nataraja, K.N., Udayakumar, M. and Pereira, A. 2007 Improvement of water use efficiency in rice by expression of HARDY, an *Arabidopsis* drought and salt tolerance gene. Proc. Natl. Acad. Sci. U S A. **104**: 15270–5.

Kasuga, M., Liu, Q., Miura, S., Yamaguchi–Shinozaki, K. and Shinozaki, K. 1999. Improving plant drought, salt, and freezing tolerance by gene transfer of a single stress–inducible transcription factor. Nat. Biotechnol. **17(3)**: 287–291. doi 10.1038/7036.

Kasuga, M., Miura, S., Shinozaki, K. and Yamaguchi–Shinozaki, K. 2004. A combination of the *Arabidopsis* DREB1A gene and stress–inducible rd29A promoter improved drought– and low–temperature stress tolerance in tobacco by gene transfer. Plant Cell Physiol. **45(3)**: 346–350.

Kerstetter, R., Laudencia–Chingcuanco, D., Smith, L. and Hake, S. 1997. Loss–of–function mutations in the maize homeobox gene, knotted1, are defective in shoot meristem maintenance. Development **124 (16)**: 3045–3054.

Kikuchi, K., Ueguchi–Tanaka, M., Yoshida, K.T., Nagato, Y., Matsusoka. M. and Hirano, H.Y. 2000. Molecular analysis of the NAC gene family in rice. Mol. Gen. Genet. **262(6)**: 1047–1051.

Kim, H.J., Kim, Y.K., Park, J.Y. and Kim, J. 2002. Light signalling mediated by phytochrome plays an important role in cold–induced gene expression through the C–repeat/dehydration responsive element (C/DRE) in *Arabidopsis thaliana*. Plant J. **29(6)**: 693–704.

Kim, J.B., Kang, J.Y. and Kim, S.Y. 2004a. Over–expression of a transcription factor regulating ABA–responsive gene expression confers multiple stress tolerance. Plant Biotechnol. J. **2(5)**: 459–466. doi PBI090 [pii]10.1111/j.1467–7652.2004.00090.x.

Kim, S., Kang, J.Y., Cho, D.I., Park, J.H. and Kim, S.Y. 2004b. ABF2, an ABRE–binding bZIP factor, is an essential component of glucose signaling and its overexpression affects multiple stress tolerance. Plant J. **40(1)**: 75–87. doi 10.1111/j.1365–313X.2004.02192.xTPJ2192 [pii].

Kim, Y.H., Yang, K.S., Ryu, S.H., Kim, K.Y., Song, W.K., Kwon, S.Y., Lee, H.S., Bang, J.W. and Kwak, S.S. 2008. Molecular characterization of a cDNA encoding DRE–binding transcription factor from dehydration–treated fibrous roots of sweet potato. Plant Physiol. Biochem **46(2)**: 196–204.

Kobayashi, F., Maeta, E., Terashima, A. and Takumi, S. 2008. Positive role of a wheat HvABI5 ortholog in abiotic stress response of seedlings. Physiol. Plant. **134**: 74–86.

Knight, H. and Knight, M.R. 2001. Abiotic stress signalling pathways: specificity and cross–talk. Trends Plant Sci. **6(6)**: 262–267.

Knight, H., Veale, E.L., Warren, G.J. and Knight, M.R. 1999. The sfr 6 mutation in *Arabidopsis* suppresses low–temperature induction of genes dependent on the CRT/DRE sequence motif. Plant Cell **11(5)**: 875–886. doi 10.1105/tpc.11.5.875.

Knight, H., Zarka, D.G., Okamoto, H., Thomashow, M.F. and Knight, M,R. 2004. Abscisic acid induces CBF gene transcription and subsequent induction of cold–regulated genes via the CRT promoter element. Plant Physiol. **135(3)**: 1710–1717.

Kranz, H.D., Denekamp, M., Greco, R., Jin, H., Leyva. A., Meissner, R.C., Petroni, K., Urzainqui, A., Bevan, M., Martin, C., Smeekens, S., Tonelli, C., Paz–Ares, J. and Weisshaar, B. 1998. Towards functional characterisation of the members of the R2R3–MYB gene family from *Arabidopsis thaliana*. Plant J. **16 (2)**: 263–276.

Krishnaswamy, S., Verma, S., Rahman, M.H. and Kav, N.N. 2011. Functional characterization of four APETALA2–family genes (RAP2.6, RAP2.6L, DREB19 and DREB26) in *Arabidopsis.* Plant Mol. Biol. **5**: 107–127.

Kuhn, R.M., Casper, T., Dehesh, K. and Quail, P.H. 1993. DNA binding factor GT–2 from *Arabidopsis.* Plant Mol. Biol. **23**: 337–348.

Lata, C., Bhutty, S., Bahadur, R.P., Majee, M. and Prasad, M. 2011. Association of an SNP in a novel DREB2–like gene SiDREB2 with stress tolerance in foxtail millet [*Setaria italica* (L.)]. J. Exp. Bot. **62(10)**: 3387–3401.

Lee, H., Fischer, R.L., Goldberg, R.B. and Harada, J.J. 2003. *Arabidopsis* LEAFY COTYLEDON1 represents a functionally specialized subunit of the CCAAT binding transcription factor. Proc. Natl. Acad. Sci. U S A **100(4)**: 2152–2156. doi 10.1073/pnas.04379091000437909100 [pii].

Lee. Y.H. and Chun, J.Y. 1998 A new homeodomain–leucine zipper gene from *Arabidopsis thaliana* induced by water stress and abscisic acid treatment. Plant. Mol. Biol **37**: 377–384.

Lee, H., Xiong, L., Ishitani, M., Stevenson, B.and Zhu, J.K. 1999. Cold–regulated gene expression and freezing tolerance in an *Arabidopsis thaliana* mutant. Plant J. **17(3)**: 301–308.

Le Gourrierec, J., Li, Y.F. and Zhou, D.X. 1999. Transcriptional activation by *Arabidopsis* GT–1 may be through interaction with TFIIA–TBP–TATA complex. Plant J. **18**: 663–668.

Li, G., Tai, F.J., Zheng, Y., Luo, J., Gong, S.Y., Zhang, Z. and Li, X.B. 2010a. Two cotton Cys2/His2–type zinc–finger proteins, GhDi19–1 and GhDi19–2, are involved in plant response to salt/drought stress and abscisic acid signaling. Plant Mol. Biol. **74(4–5)**: 437–452. doi 10.1007/s11103–010–9684–6.

Li, S., Fu, Q., Chen, L., Huang, W. and Yu, D. 2011. *Arabidopsis thaliana* WRKY25, WRKY26, and WRKY33 coordinate induction of plant thermotolerance. Planta **233(6)**: 1237–1252. doi 10.1007/s00425–011–1375–2.

Li, S., Fu, Q., Huang, W. and Yu, D. 2009. Functional analysis of an *Arabidopsis* transcription factor WRKY25 in heat stress. Plant Cell Rep. **28(4)**: 683–693. doi 10.1007/s00299–008–0666–y.

Li, S., Zhou, X., Chen, L., Huang, W. and Yu, D. 2010b. Functional characterization of *Arabidopsis thaliana* WRKY39 in heat stress. Mol. Cells **29(5)**: 475–483. doi 10.1007/s10059–010–0059–2.

Li, W.X., Oono, Y., Zhu, J., He, X.J., Wu, J.M., Iida, K., Lu, X.Y., Cui, X., Jin, H. and Zhu, J.K. 2008. The *Arabidopsis* NFYA5 transcription factor is regulated transcriptionally and posttranscriptionally to promote drought resistance. Plant Cell **20(8)**: 2238–2251.

Liao, Y., Zou, H.F., Wang, H.W., Zhang, W.K., Ma, B., Zhang. J.S. and Chen, S.Y. 2008 Soybean GmMYB76, GmMYB92, and GmMYB177 genes confer stress tolerance in transgenic *Arabidopsis* plants. Cell Res. **18 (10)**: 1047–60.

Lin, R., Zhao, W., Meng, X., Wang, M. and Peng, Y. 2007. Rice gene OsNAC19 encodes a novel NAC–domain transcription factor and responds to infection by *Magnaporthe grisea*. Plant Sci. **172(1)**: 120–130.

Liu, H., Yang, W., Liu, D., Han, Y., Zhang, A. and Li, S. 2011a. Ectopic expression of a grapevine transcription factor VvWRKY11 contributes to osmotic stress tolerance in *Arabidopsis*. Mol. Biol. Rep. **38(1)**: 417–427. doi 10.1007/s11033–010–0124–0.

Liu, H., Zhou, X., Dong, N., Liu, X., Zhang, H. and Zhang, Z. 2011b. Expression of a wheat MYB gene in transgenic tobacco enhances resistance to *Ralstonia solanacearum*, and to drought and salt stresses. Funct. Integr. Genomics **11(3)**: 431–443. doi 10.1007/s10142–011–0228–1.

Liu, L., White, M.J. and MacRae, T.H. 1999. Transcription factors and their genes in higher plants functional domains, evolution and regulation. Eur. J. Biochem. **262(2)**: 247–257.

Liu, L., Zhu, K., Yang, Y., Wu, J., Chen, F. and Yu, D. 2008. Molecular cloning, expression profiling and trans–activation property studies of a DREB2–like gene from chrysanthemum (*Dendranthema vestitum*). J. Plant Res. **121(2)**: 215–226. doi 10.1007/s10265–007–0140–x.

Liu, Q., Kasuga, M., Sakuma, Y., Abe, H., Miura, S., Yamaguchi–Shinozaki, K. and Shinozaki, K. 1998. Two transcription factors, DREB1 and DREB2, with an EREBP/AP2 DNA binding domain separate two cellular signal transduction pathways in drought– and low–temperature–responsive gene expression, respectively, in *Arabidopsis*. Plant Cell **10(8)**: 1391–1406.

Liu, Q., Zhao, N., Yamaguch–Shinozaki, K. and Shinozaki, K. 2000. Regulatory role of DREB transcription factors in plant drought, salt and cold tolerance. Chinese Sci. Bull. **45(11)**: 970–975. doi 10.1007/bf02884972.

Liu, Q.L., Xu, K.D., Zhao, L.J., Pan, Y.Z., Jiang, B.B., Zhang, H.Q. and Liu, G.L. 2011c. Overexpression of a novel chrysanthemum NAC transcription factor gene enhances salt tolerance in tobacco. Biotechnol Lett **33(10)**: 2073–2082. doi 10.1007/s10529–011–0659–8.

Liu, X., Bai, X., Wang, X. and Chu, C. 2007. OsWRKY71, a rice transcription factor, is involved in rice defense response. J. Plant Physiol. **164(8)**: 969–979. doi S0176–1617(06)00199–4 [pii]10.1016/j.jplph.2006.07.006.

Liu, X., Hong, L., Li, X.Y., Yao, Y., Hu, B. and Li, L. 2011d. Improved drought and salt tolerance in transgenic *Arabidopsis* overexpressing a NAC transcriptional factor from *Arachis hypogaea*. Biosci. Biotechnol. Biochem. **75(3)**: 443–450.

Lorenzo, O., Chico, J.M., Sanchez–Serrano, J.J. and Solano, R. 2004. JASMONATE–INSENSITIVE1 encodes a MYC transcription factor essential to discriminate between different jasmonate–regulated defense responses in *Arabidopsis*. Plant Cell **16 (7)**: 1938–1950.

Lu, P.L., Chen, N.Z., An, R., Su, Z., Qi, B.S., Ren, F., Chen, J. and Wang, X.C. 2007. A novel drought–inducible gene, ATAF1, encodes a NAC family protein that negatively regulates the expression of stress–responsive genes in *Arabidopsis*. Plant Mol. Biol. **63(2)**: 289–305. doi 10.1007/s11103–006–9089–8.

Luo, H., Song, F. and Zheng, Z. 2005 Overexpression in transgenic tobacco reveals different roles for the rice homeodomain gene OsBIHD1 in biotic and abiotic stress responses. J. Exp. Bot. **56(420)**: 2673–2682. doi 10.1093/jxb/eri260.

Ma, Q., Dai, X., Xu, Y., Guo, J., Liu, Y., Chen, N., Xiao, J., Zhang, D., Xu, Z., Zhang, X. and Chong, K. 2009. Enhanced tolerance to chilling stress in OsMYB3R–2 transgenic rice is mediated by alteration in cell cycle and ectopic expression of stress genes. Plant Physiol. **150**: 244–256.

Mallikarjuna, G., Mallikarjuna, K., Reddy, M.K. and Kaul, T. 2011. Expression of OsDREB2A transcription factor confers enhanced dehydration and salt stress tolerance in rice (*Oryza sativa* L.). Biotechnol. Lett. **33(8)**: 1689–1697.

Mare, C., Mazzucotelli, E., Crosatti, C., Francia, E., Stanca, A.M. and Cattivelli, L. 2004. Hv–WRKY38: A new transcription factor involved in cold– and drought–response in barley. Plant Mol. Biol. **55**: 399–416.

Martin, C. and Paz–Ares, J. 1997. MYB transcription factors in plants. Trends Genet. **13(2)**: 67–73.

Martin, L., Leblanc–Fournier, N., Azri, W., Lenne, C., Henry, C., Coutand, C. and Julien, J.L. 2009. Characterization and expression analysis under bending and other abiotic factors of PtaZFP2, a poplar gene encoding a Cys2/His2 zinc finger protein. Tree Physiol. **29**: 125–136.

Masucci, J.D. and Schiefelbein, J.W. 1996. Hormones act downstream of TTG and GL2 to promote root hair outgrowth during epidermis development in the *Arabidopsis* root. Plant Cell **8 (9)**: 1505–1517.

Matsukura, S., Mizoi, J., Yoshida, T., Todaka, D., Ito, Y., Maruyama, K., Shinozaki, K. and Yamaguchi–Shinozaki, K. 2010. Comprehensive analysis of rice DREB2–type genes that encode transcription factors involved in the expression of abiotic stress–responsive genes. Mol. Genet. Genomics **283(2)**: 185–96.

Mattana, M., Biazzi, E., Consonni, R., Locatelli, F., Vannini, C., Provera, S. and Coraggio, I. 2005. Overexpression of Osmyb4 enhances compatible solute accumulation and increases stress tolerance of *Arabidopsis thaliana*. Physiol. Plantarum **125(2)**: 212–223.

McConnell, J.R., Emery, J., Eshed, Y., Bao, N., Bowman, J. and Barton, M.K. 2001. Role of *PHABULOSA* and *PHAVOLUTA* in determining radial patterning in shoots. Nature **411**: 709–713.

Medina, J., Bargues, M., Terol, J., Perez–Alonso, M. and Salinas, J. 1999. The *Arabidopsis* CBF gene family is composed of three genes encoding AP2 domain–containing proteins whose expression is regulated by low temperature but not by abscisic acid or dehydration. Plant Physiol. **119(2)**: 463–470.

Meissner, R.C., Jin, H., Cominelli, E., Denekamp, M., Fuertes, A., Greco, R., Kranz, H.D., Penfield, S., Petroni, K., Urzainqui, A., Martin, C., Paz–Ares, J., Smeekens,

S., Tonelli, C., Weisshaar, B., Baumann, E., Klimyuk, V., Marillonnet, S., Patel, K., Speulman, E., Tissier, A.F., Bouchez, D., Jones, J.J., Pereira, A., Wisman, E., *et al.*, 1999. Function search in a large transcription factor gene family in *Arabidopsis*: assessing the potential of reverse genetics to identify insertional mutations in R2R3 MYB genes. Plant Cell **11(10)**: 1827–1840.

Meng, C., Cai, C., Zhang, T. and Guo, W. 2009. Characterization of six novel NAC genes and their responses to abiotic stresses in *Gossypium hirsutum* L. Plant Sci. **176(3)**: 352–359.

Mengiste, T., Chen, X., Salmeron, J. and Dietrich, R. 2003. The BOTRYTIS SUSCEPTIBLE1 gene encodes an R2R3MYB transcription factor protein that is required for biotic and abiotic stress responses in *Arabidopsis*. Plant Cell. **15**: 2551–2565. doi: 10.1105/tpc.014167tpc.014167 [pii].

Mittler, R. 2006. Abiotic stress, the field environment and stress combination. Trends Plant Sci. **11(1)**: 15–19.

Mochida, K., Yoshida, T., Sakurai, T., Ogihara, Y.and Shinozaki, K. 2009. TriFLDB: a database of clustered full–length coding sequences from Triticeae with applications to comparative grass genomics. Plant Physiol. **150(3)**: 1135–1146. doi pp.109.138214 [pii]10.1104/pp.109.138214.

Morran, S., Eini, O., Pyvovarenko, T., Parent, B., Singh, R., Ismagul, A., Eliby, S., Shirley, N., Langridge, P. and Lopato, S. 2011. Improvement of stress tolerance of wheat and barley by modulation of expression of DREB/CBF factors. Plant Biotechnol. J **9**: 230–49.

Nakashima, K., Shinwari, Z.K., Sakuma, Y., Seki, M., Miura, S., Shinozaki, K. and Yamaguchi–Shinozaki, K. 2000. Organization and expression of two *Arabidopsis* DREB2 genes encoding DRE–binding proteins involved in dehydration– and high–salinity–responsive gene expression. Plant Mol. Biol. **42(4)**: 657–665.

Nakashima, K., Fujita, Y., Katsura, K., Maruyama, K., Narusaka, Y., Seki, M., Shinozaki, K. and Yamaguchi–Shinozaki, K. 2006. Transcriptional regulation of ABI3– and ABA–responsive genes including RD29B and RD29A in seeds, germinating embryos, and seedlings of *Arabidopsis*. Plant Mol. Biol. **60(1)**: 51–68.

Nakashima, K., Tran, L.S., Van Nguyen, D., Fujita, M., Maruyama, K., Todaka, D., Ito, Y., Hayashi, N., Shinozaki, K. and Yamaguchi–Shinozaki, K. 2007. Functional analysis of a NAC–type transcription factor OsNAC6 involved in abiotic and biotic stress–responsive gene expression in rice. Plant J. **51(4)**: 617–630. doi TPJ3168 [pii]10.1111/j.1365–313X.2007.03168.x.

Narusaka, Y., Nakashima, K., Shinwari, Z.K., Sakuma, Y., Furihata, T., Abe, H., Narusaka, M., Shinozaki, K. and Yamaguchi–Shinozaki, K. 2003. Interaction between two cis–acting elements, ABRE and DRE, in ABA–dependent expression of *Arabidopsis* rd29A gene in response to dehydration and high–salinity stresses. Plant J. **34(2)**: 137–148. doi 1708 [pii].

Navarro, M., Ayax, C., Martinez, Y., Laur, J., El Kayal, W., Marque, C. and Teulieres, C. 2011 Two EguCBF1 genes overexpressed in *Eucalyptus* display a different impact on stress tolerance and plant development. Plant Biotechnol. J. **9(1)**: 50–63. doi PBI530 [pii]10.1111/j.1467–7652.2010.00530.x.

Nelson, D.E., Repetti, P.P., Adams, T.R., Creelman, R.A., Wu, J., Warner, D.C., Anstrom, D.C., Bensen, R.J., Castiglioni, P.P., Donnarummo, M.G., Hinchey, B.S., Kumimoto, R.W., Maszle, D.R., Canales, R.D., Krolikowski, K.A., Dotson, S.B., Gutterson, N., Ratcliffe, O.J. and Heard, J.E. 2007. Plant nuclear factor Y (NF–Y) B subunits confer drought tolerance and lead to improved corn yields on water–limited acres. Proc. Natl. Acad. Sci. U S A **104(42)**: 16450–16455. doi 0707193104 [pii]10.1073/pnas.0707193104.

Niu, Y., Hu, T., Zhou, Y. and Hasi, A. 2010. Isolation and characterization of two *Medicago falcata* AP2/EREBP family transcription factor cDNA, MfDREB1 and MfDREB1s. Plant Physiol. Biochem. **48**: 971–976.

Nogueira, F.T.S., Schlögl, P.S., Camargo, S.R., Fernandez, J.H., De Rosa, Jr. V.E., Pompermayer, P. and Arruda, P. 2005. SsNAC23, a member of the NAC domain protein family, is associated with cold, herbivory and water stress in sugarcane. Plant Sci. **169(1)**: 93–106.

Nuruzzaman, M., Manimekalai, R., Sharoni, A.M., Satoh, K., Kondoh, H., Ooka, H. and Kikuchi, S. 2010. Genome–wide analysis of NAC transcription factor family in rice. Gene **465(1–2)**: 30–44. doi S0378–1119(10)00257–X [pii]10.1016/j.gene.2010.06.008.

Oh, J.E., Kwon, Y., Kim, J.H., Noh, H., Hong, S.W. and Lee, H. 2011. A dual role for MYB60 in stomatal regulation and root growth of *Arabidopsis thaliana* under drought stress. Plant Mol. Biol. **77(1–2)**: 91–103. doi 10.1007/s11103–011–9796–7.

Oh, S.J., Song, S.I., Kim, Y.S., Jang, H.J., Kim, S.Y., Kim, M., Kim, Y.K., Nahm, B.H. and Kim, J.K. 2005. *Arabidopsis* CBF3/DREB1A and ABF3 in transgenic rice increased tolerance to abiotic stress without stunting growth. Plant Physiol. **138(1)**: 341–351. doi pp.104.059147 [pii]10.1104/pp.104.059147.

Ohnishi, T., Sugahara, S., Yamada, T., Kikuchi, K., Yoshiba, Y., Hirano, H.Y. and Tsutsumi, N. 2005. OsNAC6, a member of the NAC gene family, is induced by various stresses in rice. Genes Genet. Syst. **80(2)**: 135–139. doi JST.JSTAGE/ggs/80.135 [pii].

Olsen, A.N., Ernst, H.A., Leggio, L.L. and Skriver, K. 2005. NAC transcription factors: structurally distinct, functionally diverse. Trends Plant Sci. **10(2)**: 79–87.

Olsson, A.S., Engstrom, P. and Soderman, E. 2004. The homeobox genes ATHB12 and ATHB7 encode potential regulators of growth in response to water deficit in *Arabidopsis*. Plant Mol. Biol. **55 (5)**: 663–677.

Ooka, H., Satoh, K., Doi, K., Nagata, T., Otomo, Y., Murakami, K., Matsubara, K., Osato, N., Kawai, J., Carninci, P., Hayashizaki, Y., Suzuki, K., Kojima, K., Takahara, Y., Yamamoto, K. and Kikuchi, S. 2003. Comprehensive analysis of NAC family genes in *Oryza sativa* and *Arabidopsis thaliana*. DNA Res. **10(6)**: 239–247.

Orellana, S., Yañez, M., Espinoza, A., Verdugo, I., González, E., Ruiz–Lara, S. and Casaretto, J.A. 2010. The transcription factor SlAREB1 confers drought, salt stress tolerance and regulates biotic and abiotic stress–related genes in tomato. Plant Cell Environ. **33**: 2191–208.

Otting, G., Qian, Y.Q., Billeter, M., Muller, M., Affolte, M., Gehring, W.J. and Wuthrich, K. 1990 Protein–DNA contacts in the structure of a homeodomain–DNA complex determined by nuclear magnetic resonance spectroscopy in solution. EMBO J. **9 (10)**: 3085–3092.

Pabo, C.O., Peisach, E. and Grant, R.A. 2001. Design and selection of novel Cys2His2 zinc finger proteins. Annu. Rev. Biochem. **70**: 313–340. doi 70/1/313 [pii]10.1146/annurev.biochem.70.1.313.

Park, M.R., Yun, K.Y., Mohanty, B., Herath, V., Xu, F., Wijaya, E., Bajic, V.B., Yun, S.J. and De Los Reyes, B.G. 2010. Supra–optimal expression of the cold–regulated OsMyb4 transcription factor in transgenic rice changes the complexity of transcriptional network with major effects on stress tolerance and panicle development. Plant Cell Environ. **33(12)**: 2209–2230. doi PCE2221 [pii]10.1111/j.1365–3040.2010.02221.x.

Pellegrineschi, A., Reynolds, M., Pacheco, M,, Brito, R.M., Almeraya, R., Yamaguchi–Shinozaki, K. and Hoisington, D. 2004. Stress–induced expression in wheat of the *Arabidopsis thaliana* DREB1A gene delays water stress symptoms under greenhouse conditions. Genome **47(3)**: 493–500.

Peng, H., Cheng, H.Y., Chen, C., Yua, X.W., Yang, J.N., Gao, W.R., Shi, Q.C., Zhang, H., Li, J.G. and Ma, H. 2009. A NAC transcription facto gene of Chickpea (*Cicer arietinum*), CarNAC3, is involved in drought stress response and various developmental processes. J. Plant Physiol. **166**: 1934–1945.

Perisic, O. and Lam, E. 1992. A tobacco DNA–binding protein that interacts with a light–responsive box II element. Plant Cell **4**: 831–838.

Pinheiro, G.L., Marques, C.S., Costa, M.D., Reis, P.A., Alves,M.S., Carvalho, C.M., Fietto, L.G. and Fontes, E.P. 2009. Complete inventory of soybean NAC transcription factors: sequence conservation and expression analysis uncover their distinct roles in stress response. Gene **444(1–2)**: 10–23.

Puranik, S., Bahadur, R.P., Srivastava, P.S. and Prasad, M. 2011. Molecular cloning and characterization of a membrane associated NAC family gene, SiNAC from foxtail millet [*Setaria italica* (L.) P. Beauv]. Mol. Biotechnol. **49(2)**: 138–150. doi 10.1007/s12033–011–9385–7.

Qin, F., Sakuma, Y., Li, J., Liu, Q., Li, Y.Q., Shinozaki, K. and Yamaguchi–Shinozaki, K. 2004. Cloning and functional analysis of a novel DREB1/CBF transcription factor involved in cold–responsive gene expression in *Zea mays* L. Plant Cell Physiol. **45(8)**: 1042–1052. doi 10.1093/pcp/pch11845/8/1042 [pii].

Qiu, Y. and Yu, D. 2009. Over–expression of the stress–induced OsWRKY45 enhances disease resistance and drought tolerance in *Arabidopsis*. Environ. Exp. Bot. **65(1)**: 35–47.

Rabinowicz, P.D., Braun, E.L., Wolfe, A.D., Bowen, B. and Grotewold, E. 1999. Maize R2R3 Myb genes: Sequence analysis reveals amplification in the higher plants. Genetics, **153 (1)**: 427–444.

Rahaie, M., Xue, G.P., Naghavi, M.R., Alizadeh, H. and Schenk, P.M. 2010. A MYB gene from wheat (*Triticum aestivum* L.) is up–regulated during salt and drought stresses and differentially regulated between salt–tolerant and sensitive genotypes. Plant Cell Rep. **29**: 835–844.

Ramamoorthy, R., Jiang, S.Y., Kumar, N., Venkatesh, P.N. and Ramachandran, S. 2008. A comprehensive transcriptional profiling of the WRKY gene family in rice under various abiotic and phytohormone treatments. Plant Cell Physiol. **49**: 865–879.

Ramanjulu, S. and Bartels, D. 2002. Drought– and desiccation–induced modulation of gene expression in plants. Plant Cell Environ. **25(2)**: 141–151.

Romero, I., Fuertes, A., Benito, M.J., Malpica, J.M., Leyva, A. and Paz–Ares, J. 1998. More than 80R2R3–MYB regulatory genes in the genome of *Arabidopsis thaliana*. Plant J. **14 (3)**: 273–284.

Ren, X., Chen, Z., Liu, Y., Zhang, H., Zhang, M., Liu, Q., Hong, X., Zhu, J.K. and Gong, Z. 2010. ABO3, a WRKY transcription factor, mediates plant responses to abscisic acid and drought tolerance in *Arabidopsis*. Plant J. doi TPJ4248 [pii]10.1111/j.1365–313X.2010.04248.x.

Rizhsky, L., Liang, H. and Mittler, R. 2002. The combined effect of drought stress and heat shock on gene expression in tobacco. Plant Physiol. **130**: 1143–1151.

Riano–Pachon, D.M., Ruzicic, S., Dreyer, I. and Mueller–Roeber, B. 2007 PlnTFDB: an integrative plant transcription factor database. BMC Bioinformatics **8**: 42. doi 1471–2105–8–42 [pii]10.1186/1471–2105–8–42.

Riechmann, J.L., Heard, J., Martin, G., Reuber, L., Jiang, C.Z., *et al.,* 2000. *Arabidopsis* transcription factors: genome–wide comparative analysis among eukaryotes. Science **290**: 2105–2110.

Rizhsky, L., Liang, H. and Mittler, R. 2002. The combined effect of drought stress and heat shock on gene expression in tobacco. Plant Physiol. **130(3)**: 1143–1151. doi 10.1104/pp.006858.

Ross, C.A., Liu, Y. and Shen, Q.J. 2007. The WRKY gene family in rice (*Oryza sativa*). J. Int. Plant Biol. **49**: 827–842.

Rushton, P.J., Macdonald, H., Huttly, A.K., Lazarus, C.M. and Hooley, R. 1995. Members of a new family of DNA–binding proteins bind to a conserved *cis*-element in the promoters of *alpha–Amy2* genes. Plant Mol. Biol. **2**: 691–702.

Rushton, P.J., Bokowiec, M.T., Han, S., Zhang, H., Brannock, J.F., Chen, X., Laudeman, T.W. and Timko, M.P. 2008 Tobacco transcription factors: novel insights into transcriptional regulation in the Solanaceae. Plant Physiol. **147(1)**: 280–295.

Sakamoto, H., Araki, T., Meshi, T. and Iwabuchi, M. 2000. Expression of a subset of the *Arabidopsis* Cys(2)/His(2)–type zinc–finger protein gene family under water stress. Gene **248 (1–2)**: 23–32.

Saad, R.B., Fabre, D., Mieulet, D., Meynard, D., Dingkuhn, M., Al–Doss, A., Guiderdoni, E. and Hassairi, A. 2011. Expression of the *Aeluropus littoralis* AlSAP gene in rice confers broad tolerance to abiotic stresses through maintenance of photosynthesis. Plant Cell Environ. doi 10.1111/j.1365–3040.2011.02441.x.

Schenk, P.M., Kazan, K., Manners, J.M., Anderson, J.P., Simpson, R.S., Wilson, I.W., Somerville, S.C. and Maclean, D.J. 2003. Systemic gene expression in *Arabidopsis* during an incompatible interaction with *Alternaria brassicicola*. Plant Physiol. **132(2)**: 999–1010. doi 10.1104/pp.103.021683pp.103.021683 [pii].

Seki, M., Narusaka, M., Ishida, J., Nanjo, T., Fujita, M., Oono, Y., Kamiya, A., Nakajima, M., Enju, A., Sakurai, T., Satou, M., Akiyama, K., Taji, T., Yamaguchi–Shinozaki, K., Carninci, P., Kawai, J., Hayashizaki, Y. and Shinozaki, K.2002. Monitoring the expression profiles of 7000 *Arabidopsis* genes under drought, cold and high–salinity stresses using a full–length cDNA microarray. Plant J. **31(3)**: 279–292.

Seo, P.J., Xiang, F., Qiao, M., Park, J.Y., Lee, Y.N., Kim, S.G., Lee, Y.H., Park, W.J. and Park, C.M. 2009. The MYB96 transcription factor mediates abscisic acid signaling during drought stress response in *Arabidopsis.* Plant Physiol. **151(1)**: 275–289.

Seong, E.S. and Wang, M.H. 2008. A novel CaAbsi1 gene induced by early–abiotic stresses in pepper. BMB Rep. **41(1)**: 86–91.

Shen, Y.G., Zhang, W.K., He, S.J., Zhang, J.S., Liu, Q. and Chen, S.Y. 2003a. An EREBP/AP2–type protein in *Triticum aestivum* was a DRE–binding transcription factor induced by cold, dehydration and ABA stress. Theor. Appl. Genet. **106(5)**: 923–930. doi 10.1007/s00122–002–1131–x.

Shen, Y.G., Zhang, W.K., Yan, D.Q., Du, B.X., Zhang, J.S., Liu, Q. and Chen, S.Y. 2003b. Characterization of a DRE–binding transcription factor from a halophyte *Atriplex hortensis*. Theor. Appl. Genet. **107(1)**: 155–161. doi 10.1007/s00122–003–1226–z.

Shen, Q. and Ho, T.H. 1995. Functional dissection of an abscisic acid (ABA)–inducible gene reveals two independent ABA–responsive complexes each containing a G–box and a novel cis–acting element. Plant Cell. **7 (3)**: 295–307.

Shin, D., Moon, S.J., Han, S., Kim, B.G., Park, S.R., Lee, S.K., Yoon, H.J., Lee, H.E., Kwon, H.B., Baek, D., Yi, B.Y. and Byun, M.O. 2011. Expression of StMYB1R–1, a novel potato single MYB–like domain transcription factor, increases drought tolerance. Plant Physiol. **155(1)**: 421–432. doi pp.110.163634 [pii]10.1104/pp.110.163634.

Shinozaki, K. and Yamaguchi–Shinozaki, K. 1996. Molecular responses to drought and cold stress. Curr. Opin. Biotechnol. **7(2)**: 161–167. doi S0958–1669(96)80007–3 [pii].

Shinozaki, K. and Yamaguchi–Shinozaki, K. 1997. Gene expression and signal transduction in water–stress response. Plant Physiol. **115(2)**: 327–334. doi 115/2/327 [pii].

Shinozaki, K. and Yamaguchi–Shinozaki, K. 2000. Molecular responses to dehydration and low temperature: differences and cross–talk between two stress signaling pathways. Curr. Opin. Plant Biol. **3(3)**: 217–223. doi S1369–5266(00)00067–4 [pii].

Shinozaki, K., Yamaguchi–Shinozaki, K. and Seki, M. 2003. Regulatory network of gene expression in the drought and cold stress responses. Curr Opin Plant Biol **6(5)**: 410–417. doi S136952660300092X [pii].

Shinwari, Z.K., Nakashima, K., Miura, S., Kasuga, M., Seki, M., Yamaguchi–Shinozaki, K. and Shinozaki, K. 1998. An *Arabidopsis* gene family encoding DRE/CRT binding proteins involved in low–temperature–responsive gene expression. Biochem. Biophys. Res. Commun. **250(1)**: 161–170. doi S0006–291X(98)99267–1 [pii]10.1006/bbrc.1998.9267.

Siddiqua, M. and Nassuth, A. 2011. Vitis CBF1 and Vitis CBF4 differ in their effect on *Arabidopsis* abiotic stress tolerance, development and gene expression. Plant Cell Environ. **34(8)**: 1345–1359. doi 10.1111/j.1365–3040.2011.02334.x.

Siefers, N., Dang, K.K., Kumimoto, R.W., Bynum, W.Et., Tayrose, G. and Holt, B.F., 3rd 2009. Tissue–specific expression patterns of *Arabidopsis* NF–Y transcription factors suggest potential for extensive combinatorial complexity. Plant Physiol. **149(2)**: 625–641. doi pp.108.130591 [pii]10.1104/pp.108.130591.

Singh, K., Foley, R.C. and Onate–Sanchez, L. 2002. Transcription factors in plant defense and stress responses. Curr. Opin. Plant Biol. **5(5)**: 430–436.

Smalle, J., Kurepa, J., Haegman, M., Gielen, J., Van Montagu, M. and Van Der Straeten, D. 1998. The trihelix DNA–binding motif in higher plants is not restricted to the transcription factors GT–1 and GT–2. Proc. Natl. Acad. Sci. USA **95**: 3318–3322.

Soderman, E., Mattsson, J. and Engstrom, P. 1996. The *Arabidopsis* homeobox gene ATHB–7 is induced by water deficit and by abscisic acid. Plant J. **10(2)**: 375–381.

Soderman, E., Hjellstrom, M., Fahleson, J. and Engstrom, P. 1999. The HD–Zip gene ATHB6 in *Arabidopsis* is expressed in developing leaves, roots and carpels and up–regulated by water deficit conditions. Plant Mol. Biol. **40**: 1073–1083.

Song, Y., Ai, C.R., Jing, S.J. and Yu, D.Q. 2010. Research progress on functional analysis of rice WRKY genes. Rice Sci. **17(1)**: 60–72.

Song, Y., Jing, S.J. and Yu, D.Q. 2009. Overexpression of the stress–induced OsWRKY08 improves osmotic stress tolerance in *Arabidopsis.* Chinese Sci. Bull. **54(24)**: 4671–4678. doi 10.1007/s11434–009–0710–5.

Stephenson, T.J., McIntyre, C.L., Collet, C. and Xue, G.P. 2007. Genome–wide identification and expression analysis of the NF–Y family of transcription factors in *Triticum aestivum.* Plant Mol. Biol. **65(1–2)**: 77–92. doi 10.1007/s11103–007–9200–9.

Steponkus, P.L., Uemura, M., Joseph, R.A., Gilmour, S.J. and Thomashow, M.F. 1998. Mode of action of the COR15a gene on the freezing tolerance of *Arabidopsis thaliana.* Proc. Natl. Acad. Sci. U S A **95(24)**: 14570–14575.

Stockinger, E.J., Gilmour, S.J. and Thomashow, M.F. 1997. *Arabidopsis thaliana* CBF1 encodes an AP2 domain–containing transcriptional activator that binds to the

C–repeat/DRE, a cis–acting DNA regulatory element that stimulates transcription in response to low temperature and water deficit. Proc. Natl. Acad. Sci. U S A **94(3)**: 1035–1040.

Stracke, R., Werber, M. and Weisshaar, B. 2001. The R2R3–MYB gene family in *Arabidopsis thaliana*. Curr. Opin. Plant Biol. **4 (5)**: 447–456.

Su, C.F., Wang, Y.C., Hsieh, T.H., Lu, C.A., Tseng, T.H. and Yu, S.M. 2010 A novel MYBS3–dependent pathway confers cold tolerance in rice. Plant Physiol. **153(1)**: 145–158. doi pp.110.153015 [pii]10.1104/pp.110.153015.

Sun, J., Jiang, H., Xu, Y., Li, H., Wu, X., Xie, Q. and Li, C. 2007. The CCCH–type zinc finger proteins AtSZF1 and AtSZF2 regulate salt stress responses in *Arabidopsis*. Plant Cell Physiol. **48(8)**: 1148–1158.

Sugano, S., Kaminaka, H., Rybka, Z., Catala, R., Salinas, J., Matsui, K., Ohme–Takagi, M. and Takatsuji, H. 2003. Stress–responsive zinc finger gene ZPT2–3 plays a role in drought tolerance in petunia. Plant J. **36 (6)**: 830–841.

Sun, S., Yu, J.P., Chen, F., Zhao, T.J., Fang, X.H., Li, Y.Q. and Sui, S.F. 2008. TINY, a dehydration–responsive element (DRE)–binding protein–like transcription factor connecting the DRE– and ethylene–responsive element–mediated signaling pathways in *Arabidopsis*. J. Biol. Chem. **283(10)**: 6261–6271. doi M706800200 [pii]10.1074/jbc.M706800200.

Sun, S.J., Guo, S.Q., Yang, X., Bao, Y.M., Tang, H.J., Sun, H,, Huang, J. and Zhang, H.S. 2010. Functional analysis of a novel Cys2/His2–type zinc finger protein involved in salt tolerance in rice. J. Exp. Bot. **61(10)**: 2807–2818. doi erq120 [pii]10.1093/jxb/erq120.

Taji, T., Ohsumi, C., Iuchi, S., Seki, M., Kasuga, M., Kobayashi, M., Yamaguchi–Shinozaki, K. and Shinozaki, K. 2002. Important roles of drought– and cold–inducible genes for galactinol synthase in stress tolerance in *Arabidopsis thaliana*. Plant J. **29**: 417–426.

Takasaki, H., Maruyama, K., Kidokoro, S., Ito, Y., Fujita, Y., Shinozaki, K., Yamaguchi–Shinozaki, K. and Nakashima, K. 2010. The abiotic stress–responsive NAC–type transcription factor OsNAC5 regulates stress–inducible genes and stress tolerance in rice. Mol. Genet. Genomics **284(3)**: 173–183. doi 10.1007/s00438-010-0557-0.

Takatsuji, H. 1998. Zinc–finger transcription factors in plants. Cell Mol. Life Sci. **54(6)**: 582–596.

Takatsuji, H., Mori, M., Benfey, P.N., Ren, L. and Chua, N.H. 1992. Characterization of a zinc finger DNA–binding protein expressed specifically in *Petunia* petals and seedlings. EMBO J. **11 (1)**: 241–249.

Takeda, S. and Matsuoka, M. 2008. Genetic approaches to crop improvement: responding to environmental and population changes. Nat. Rev. Genet. **9(6)**: 444–457. doi nrg2342 [pii]10.1038/nrg2342.

Tamaoki, M., Kusaba, S., Kano–Murakami, Y. and Matsuoka, M. 1997. Ectopic expression of a tobacco homeobox gene, NTH15, dramatically alters leaf morphology and hormone levels in transgenic tobacco. Plant Cell Physiol.**38 (8)**: 917–927.

Tang, M., Liu, X., Deng, H. and Shen, S. 2011. Over–expression of JcDREB, a putative AP2/EREBP domain–containing transcription factor gene in woody biodiesel plant *Jatropha curcas*, enhances salt and freezing tolerance in transgenic *Arabidopsis thaliana*. Plant Sci. **181(6)**: 623–631. doi S0168–9452(11)00257–3 [pii]10.1016/j.plantsci.2011.06.014.

The Arabidopsis Genome Initiative 2000. Analysis of the genome sequence of the flowering plant *Arabidopsis thaliana*. Nature **408(6814)**: 796–815. doi 10.1038/35048692.

Thomashow, M.F. 1999. Plant cold acclimation: Freezing tolerance genes and regulatory mechanisms. Annu. Rev. Plant. Physiol. Plant. Mol. Biol. **50**: 571–599. doi 10.1146/annurev.arplant.50.1.571.

Thomashow, M.F. 2001. So what's new in the field of plant cold acclimation? lots! Plant Physiol. **125(1)**: 89–93.

Tian, A.G., Wang, J., Cui, P., Han, Y.J., Xu, H., Cong, L.J., Huang, X.G., Wang, X.L., Jiao, Y.Z., Wang, B.J., Wang, Y.J., Zhang, J.S. and Chen, S.Y. 2004. Characterization of soybean genomic features by analysis of its expressed sequence tags. Theor Appl. Genet. **108(5)**: 903–913. doi 10.1007/s00122–003–1499–2.

Tian, X.H., Li, X.P., Zhou, H.L., Zhang, J.S., Gong, Z.Z. and Chen, S.Y. 2005. OsDREB4 genes in rice encode AP2–containing proteins that bind specifically to the dehydration–responsive element. J. Integ. Plant Biol. **47(4)**: 467–476. doi doi: 10.1111/j.1744–7909.2005.00028.x.

Tian, Z.D., Zhang, Y., Liu, J. and Xie, C.H. 2010. Novel potato C2H2–type zinc finger protein gene, StZFP1, which responds to biotic and abiotic stress, plays a role in salt tolerance. Plant Biol. (Stuttg) **12(5)**: 689–697. doi PLB276 [pii]10.1111/j.1438–8677.2009.00276.x.

Toledo–Ortiz, G., Huq, E. and Quail, P.H. 2003. The *Arabidopsis* basic/helix–loop–helix transcription factor family. Plant Cell **15(8)**: 1749–1770.

Tong, Z., Hong, B., Yang, Y., Li, Q., Ma, N., Ma, C. and Gao, J. 2009. Overexpression of two chrysanthemum DgDREB1 group genes causing delayed flowering or dwarfism in *Arabidopsis*. Plant Mol. Biol. **71(1–2)**: 115–129. doi 10.1007/s11103–009–9513–y.

Tran, L.S., Nakashima, K., Sakuma, Y., Osakabe, Y., Qin, F., Simpson, S.D., Maruyama, K., Fujita, Y., Shinozaki, K. and Yamaguchi–Shinozaki, K. 2007. Co–expression of the stress–inducible zinc finger homeodomain ZFHD1 and NAC transcription factors enhances expression of the ERD1 gene in *Arabidopsis*. Plant J. **49(1)**: 46–63. doi TPJ2932 [pii]10.1111/j.1365–313X.2006.02932.x.

Tran, L.S., Nakashima, K., Sakuma, Y., Simpson, S.D., Fujita, Y., Maruyama, K., Fujita, M., Seki, M., Shinozaki, K. and Yamaguchi–Shinozaki, K. 2004. Isolation and functional analysis of *Arabidopsis* stress–inducible NAC transcription factors that bind to a drought–responsive cis–element in the early responsive to dehydration stress 1 promoter. Plant Cell **16(9)**: 2481–2498. doi 10.1105/ tpc.104.022699tpc.104.022699 [pii].

Tuteja, N. 2007. Abscisic Acid and abiotic stress signaling. Plant Signal Behav. **2(3)**: 135–138.

Ulker, B. and Somssich, *I.E.* 2004. WRKY transcription factors: from DNA binding towards biological function. Curr. Opin. Plant Biol.**7**: 491–498.

Uno, Y., Furihata, T., Abe, H., Yoshida, R., Shinozaki, K. and Yamaguchi–Shinozaki, K. 2000. *Arabidopsis* basic leucine zipper transcription factors involved in an abscisic acid–dependent signal transduction pathway under drought and high–salinity conditions. Proc. Natl. Acad. Sci. U S A **97(21)**: 11632–7.

Vannini, C., Locatelli, F., Bracale, M., Magnani, E., Marsoni, M., Osnato, M., Mattana, M., Baldoni, E. and Coraggio, I. 2004. Overexpression of the rice Osmyb4 gene increases chilling and freezing tolerance of *Arabidopsis thaliana* plants. Plant J. **37 (1)**: 115–127.

Vanjildorj, E., Bae, T.W., Riu, K.Z., Kim, S.Y. and Lee, H.Y. 2005. Overexpression of *Arabidopsis* ABF3 gene enhances tolerance to droughtand cold in transgenic lettuce (*Lactuca sativa*). Plant Cell Tissue Organ Cult. **83(1)**: 41–50.

Villalobos, M.A., Bartels, D. and Iturriaga, G. 2004. Stress tolerance and glucose insensitive phenotypes in *Arabidopsis* overexpressing the CpMYB10 transcription factor gene. Plant Physiol. **135 (1)**: 309–324.

Wang, F., Hou, X., Tang, J., Wang, Z., Wang, S., Jiang, F. and Li, Y. 2011. A novel cold–inducible gene from Pak–choi (*Brassica campestris* ssp. *chinensis*), BcWRKY46, enhances the cold, salt and dehydration stress tolerance in transgenic tobacco. Mol. Biol. Rep. doi 10.1007/s11033–011–1245–9.

Wang, Q., Guan, Y., Wu, Y., Chen, H., Chen, F. and Chu, C. 2008. Overexpression of a rice OsDREB1F gene increases salt, drought, and low temperature tolerance in both *Arabidopsis* and rice. Plant Mol. Biol. **67(6)**: 589–602. doi 10.1007/s11103–008–9340–6.

Wang, H. and Cutler, A.J. 1995. Promoters from kin1 and cor6.6, two *Arabidopsis thaliana* low–temperature– and ABA–inducible genes, direct strong beta–glucuronidase expression in guard cells, pollen and young developing seeds. Plant Mol. Biol **28**: 619–634.

Wang, Z., Zhu, Y., Wang, L., Liu, X., Liu, Y., Phillips, J. and Deng, X. 2009. A WRKY transcription factor participates in dehydration tolerance in *Boea hygrometrica* by binding to the W–box elements of the galactinol synthase (BhGolS1) promoter. Planta **230(6)**: 1155–1166. doi 10.1007/s00425–009–1014–3.

Wei, W., Zhang, Y., Han, L., Guan, Z. and Chai, T. 2008. A novel WRKY transcriptional factor from *Thlaspi caerulescens* negatively regulates the osmotic stress tolerance of transgenic tobacco. Plant Cell Rep. **27(4)**: 795–803. doi 10.1007/s00299–007–0499–0.

Wu, X., Shiroto, Y., Kishitani, S., Ito, Y. and Toriyama, K. 2009. Enhanced heat and drought tolerance in transgenic rice seedlings overexpressing OsWRKY11 under the control of HSP101 promoter. Plant Cell Rep. **28(1)**: 21–30.

Xia, N., Zhang, G., Liu, X.Y., Deng, L., Cai, G.L., Zhang, Y., Wang, X.J., Zhao, J., Huang, L.L. and Kang, Z.S. 2010. Characterization of a novel wheat NAC transcription factor gene involved in defense response against stripe rust pathogen infection and abiotic stresses. Mol. Biol. Rep. **37(8)**: 3703–3712. doi 10.1007/s11033–010–0023–4.

Xianjun, P., Xingyong, M., Weihong, F., Man, S., Liqin, C., Alam, I., Lee, B.H., Dongmei, Q., Shihua, S. and Gongshe, L. 2011. Improved drought and salt tolerance of *Arabidopsis thaliana* by transgenic expression of a novel DREB gene from *Leymus chinensis*. Plant Cell Rep. **30(8)**: 1493–1502. doi 10.1007/s00299–011–1058–2.

Xie, Q., Frugis, G., Colgan, D. and Chua, N.H. 2000. *Arabidopsis* NAC1 transduces auxin signal downstream of TIR1 to promote lateral root development. Genes Dev. **14(23)**: 3024–3036.

Xie, Z., Zhang, Z.L., Zou, X., Huang, J., Ruas, P., Thompson, D. and Shen, Q.J. 2005. Annotations and functional analyses of the rice WRKY gene super family reveal positive and negative regulators of abscisic acid signaling in aleurone cells. Plant Physiol. **137**: 176–189.

Xie, Z.M., Zou, H.F., Lei, G., Wei, W., Zhou, Q.Y., Niu, C.F., Liao, Y., Tian, A.G., Ma, B., Zhang, W.K., Zhang, J.S. and Chen, S.Y. 2009. Soybean Trihelix transcription factors GmGT–2A and GmGT–2B improve plant tolerance to abiotic stresses in transgenic *Arabidopsis*. PLoS One **4(9)**: e6898. doi 10.1371/journal.pone.0006898.

Xiong, L., Schumaker, K.S. and Zhu, J.K. 2002. Cell signaling during cold, drought, and salt stress. Plant Cell **14** Suppl: S165–183.

Xiong, X., James, V., Zhang, H. and Altpeter, F. 2010. Constitutive expression of the barley HvWRKY38 transcription factor enhances drought tolerance in turf and forage grass (*Paspalum notatum* Flugge). Mol. Breeding **25(3)**: 419–432. doi 10.1007/s11032–009–9341–4.

Xiong, Y. and Fei, S.Z. 2006. Functional and phylogenetic analysis of a DREB/CBF–like gene in perennial ryegrass (*Lolium perenne* L.). Planta **224(4)**: 878–888. doi 10.1007/s00425–006–0273–5.

Xu, D.Q., Huang, J., Guo, S.Q., Yang, X., Bao, Y.M., Tang, H.J. and Zhang, H.S. 2008 Overexpression of a TFIIIA–type zinc finger protein gene ZFP252 enhances drought and salt tolerance in rice (*Oryza sativa* L.). FEBS Lett. **582(7)**: 1037–1043. doi S0014–5793(08)00167–1 [pii]10.1016/j.febslet.2008.02.052.

Xu, S., Wang, X. and Chen, J. 2007. Zinc finger protein 1 (ThZF1) from salt cress (*Thellungiella halophila*) is a Cys–2/His–2–type transcription factor involved in drought and salt stress. Plant Cell Rep. **26 (4)**: 497–506.

Xu, D.Q., Huang, J., Guo, S.Q., Yang, X., Bao, Y.M., Tang, H.J. and Zhang, H.S. 2008. Overexpression of a TFIIIA–type zinc finger protein gene ZFP252 enhances drought and salt tolerance in rice (*Oryza sativa* L.). FEBS Lett. **582 (7)**: 1037–1043.

Xu, Z.S., Ni, Z.Y., Li, Z.Y., Li, L.C., Chen, M., Gao, D.Y., Yu, X.D., Liu, P. and Ma, Y.Z. 2009. Isolation and functional characterization of HvDREB1–a gene encoding a dehydration–responsive element binding protein in *Hordeum vulgare*. J. Plant Res. **122(1)**: 121–130. doi 10.1007/s10265–008–0195–3.

Xue, G.P., Way, H.M., Richardson, T., Drenth, J., Joyce, P.A. and McIntyre, C.L. 2011. Overexpression of TaNAC69 leads to enhanced transcript levels of stress up–regulated genes and dehydration tolerance in bread wheat. Mol. Plant. **4(4)**: 697–712.

Yadav, V., Mallappa, C., Gangappa, S.N., Bhatia, S. and Chattopadhyay, S. 2005. A basic helix–loop–helix transcription factor in *Arabidopsis*, MYC2, acts as a repressor of blue light–mediated photomorphogenic growth. Plant Cell **17 (7)**: 1953–1966.

Yamaguchi–Shinozaki, K. and Shinozaki, K. 1994. A novel cis–acting element in an *Arabidopsis* gene is involved in responsiveness to drought, low–temperature, or high–salt stress. Plant Cell **6(2)**: 251–264.

Yamasaki, K., Kigawa,T., Inoue, M., Tateno, M., Yamasaki, T., Yabuki, T., Aoki, M., Seki, E., Matsuda, T., Tomo, Y., Hayami, N., Terada, T., Shirouzu, M., Tanaka, A., Seki, M., Shinozaki, K. and Yokoyama, S. 2005. Solution structure of an *Arabidopsis* WRKY DNA binding domain. Plant Cell **17**: 944–956.

Yang, X., Sun, C., Hu, Y. and Lin, Z. 2008. Molecular cloning and characterization of a gene encoding RING zinc finger ankyrin protein from drought–tolerant *Artemisia desertorum*. J. Biosci. **33(1)**: 103–112.

Yang, R., Deng, C., Ouyang, B. and Ye, Z. 2011a. Molecular analysis of two salt–responsive NAC–family genes and their expression analysis in tomato. Mol. Biol. Rep. **38(2)**: 857–863. doi 10.1007/s11033–010–0177–0.

Yang, S., Tang, X.F., Ma, N.N., Wang, L.Y. and Meng, Q.W. 2011b. Heterology expression of the sweet pepper CBF3 gene confers elevated tolerance to chilling stress in transgenic tobacco. J. Plant Physiol. **168(15)**: 1804–1812. doi S0176–1617(11)00258–6 [pii]10.1016/j.jplph.2011.05.017.

Yang, W., Liu, X.D., Chi, X.J., Wu, C.A., Li, Y.Z., Song, L.L., Liu, X.M., Wang, Y.F., Wang, F.W., Zhang, C., Liu, Y., Zong, J.M. and Li, H.Y. 2011c Dwarf apple MbDREB1 enhances plant tolerance to low temperature, drought, and salt stress via both ABA–dependent and ABA–independent pathways. Planta **233**: 219–229.

Yoo, C.Y., Pence, H.E., Jin, J.B., Miura, K., Gosney, M.J., Hasegawa, P.M., Mickelbart, M.V. 2010 The *Arabidopsis* GTL1 transcription factor regulates water use efficiency and drought tolerance by modulating stomatal density via transrepression of SDD1. Plant Cell **22(12)**: 4128–4141. doi tpc.110.078691 [pii]10.1105/tpc.110.078691.

Yoshida, T., Fujita, Y., Sayama, H., Kidokoro, S., Maruyama, K., Mizoi, J., Shinozaki, K. and Yamaguchi–Shinozaki, K. 2010. AREB1, AREB2, and ABF3 are master transcription factors that cooperatively regulate ABRE–dependent ABA signaling involved in drought stress tolerance and require ABA for full activation. Plant J. **61**: 672–685.

Yu, H., Chen, X., Hong, Y.Y., Wang, Y., Xu, P., Ke, S.D., Liu, H.Y., Zhu, J.K., Oliver, D.J.and Xiang, C.B. 2008. Activated expression of an *Arabidopsis* HD–START protein confers drought tolerance with improved root system and reduced stomatal density. Plant Cell **20(4)**: 1134–1151. doi tpc.108.058263 [pii]10.1105/tpc.108.058263.

Zhang, J., Klueva, N., Wang, Z., Wu, R., Ho, T.H. and Nguyen, H. 2000. Genetic engineering for abiotic stress resistance in crop plants. *In Vitro* Cell Develop Biol – Plant **36(2)**: 108–114.

Zhang, J., Peng, Y. and Guo, Z. 2008. Constitutive expression of pathogen–inducible OsWRKY31 enhances disease resistance and affects root growth and auxin response in transgenic rice plants. Cell Res.**18(4)**: 508–521. doi cr2007104 [pii]10.1038/cr.2007.104.

Zhang, J.Z. 2003. Overexpression analysis of plant transcription factors. Curr. Opin. Plant Biol. **6(5)**: 430–440. doi S1369526603000815 [pii].

Zhang, J.Z., Creelman, R.A. and Zhu, J.K. 2004. From laboratory to field. Using information from *Arabidopsis* to engineer salt, cold, and drought tolerance in crops. Plant Physiol. **135(2)**: 615–621.

Zhang, Y., Chen, C., Jin, X.F., Xiong, A.S., Peng, R.H., Hong, Y.H., Yao, Q.H. and Chen, J.M. 2009. Expression of a rice DREB1 gene, OsDREB1D, enhances cold and high–salt tolerance in transgenic *Arabidopsis*. BMB Rep. **42**: 486–492.

Zhang, X., Wang, L., Meng, H., Wen, H., Fan, Y. and Zhao, J. (2011a) Maize ABP9 enhances tolerance to multiple stresses in transgenic *Arabidopsis* by modulating ABA signaling and cellular levels of reactive oxygen species. Plant Mol. Biol. **75(4–5)**: 365–78.

Zhang, Y.J., Yang, J.S., Guo, S.J., Meng, J.J., Zhang, Y.L., Wan, S.B., He, Q.W. and Li, X.G. 2011b. Over–expression of the *Arabidopsis* CBF1 gene improves resistance of tomato leaves to low temperature under low irradiance. Plant Biol. (Stuttg)**13(2)**: 362–367.

Zhan, W., Ruan, J., Ho, T.H., You, Y., Yu, T. and Quatrano, R.S. 2005. Cis–regulatory element based targeted gene finding: genome–wide identification of abscisic acid– and abiotic stress–responsive genes in *Arabidopsis thaliana*. Bioinformatics **21 (14)**: 3074–3081.

Zhao, T.J., Sun, S., Liu, Y., Liu, J.M., Liu, Q., Yan, Y.B. and Zhou, H.M. 2006. Regulating the drought–responsive element (DRE)–mediated signaling pathway by synergic functions of trans–active and trans–inactive DRE binding factors in *Brassica napus*. J. Biol. Chem. **281**: 10752–10759. doi: M510535200 [pii]10.1074/jbc.M510535200.

Zhao, L., Hu, Y., Chong, K. and Wang, T. 2010. ARAG1, an ABA–responsive DREB gene, plays a role in seed germination and drought tolerance of rice. Ann. Bot. **105**: 401–409.

Zheng, X., Chen, B., Lu, G. and Han, B. 2009. Overexpression of a NAC transcription factor enhances rice drought and salt tolerance. Biochem. Biophys. Res. Commun. **379(4)**: 985–989. doi S0006–291X(09)00006–0 [pii]10.1016/j.bbrc.2008.12.163.

Zhou, Q.Y., Tian, A.G., Zou, H.F., Xie, Z.M., Lei, G., Huang, J., Wang, C.M., Wang, H.W., Zhang, J.S. and Chen, S.Y. 2008. Soybean WRKY–type transcription factor genes, GmWRKY13, GmWRKY21, and GmWRKY54, confer differential tolerance to abiotic stresses in transgenic *Arabidopsis* plants. Plant Biotechnol. J. **6(5)**: 486–503. doi PBI336 [pii]10.1111/j.1467–7652.2008.00336.x.

Zou, X., Shen, Q.J. and Neuman, D. 2007. An ABA inducible WRKY gene integrates responses of creosote bush (*Larrea tridentata*) to elevated CO_2 and abiotic stresses. Plant Sci. **172(5)**: 997–1004.

Zhu, J., Shi, H., Lee, B.H., Damsz, B., Cheng, S., Stirm, V., Zhu, J.K., Hasegawa, P.M. and Bressan, R.A. 2004. An *Arabidopsis* homeodomain transcription factor gene, HOS9, mediates cold tolerance through a CBF–independent pathway. Proc. Natl. Acad. Sci. U S A **101(26)**: 9873–9878.

2013, Abiotic Stress and Biotechnology
Editor: T. Pullaiah
Published by: REGENCY PUBLICATIONS, NEW DELHI

Pages 127–135

Chapter 6

In vitro Propagation of Talinum cuneifolium (Vahl) Willd. through Leaf Culture

N. Savithramma, P. Venkateswarlu, A. Sasikala and Beena Prabha

Department of Botany, S.V. University, Tirupati – 517 502, A.P, India

ABSTRACT

The present study is on a reproducible and efficient protocol for mass propagation of *Talinum cuneifolium* a multipurpose tuberous shrub used in traditional medicinal system. This green leafy vegetable has been successfully *in-vitro* propagated through leaf as explant on MS (Murashige and Skoog) nutrient medium. Auxins and cytokinins were examined for their effects on callus and multiple shoot induction. Morphogenic callus induced on MS medium supplemented with BA (Benzyladenine) and various combinations and concentrations of auxins (NAA, IBA, IAA and 2,4-D). Multiple shoots were induced from callus on MS medium supplemented with 1 and 2 mg/l BA in combination of 1 mg/l Kn (Kinetin) and 0.1 mg/l IAA (Indole 3-acetic acid). Maximum number of multiple shoots (68.31 per cent) was induced on MS medium fortified with 1 mg/l Kn in combination of 2 mg/l BA. The rooted plantlets were gradually acclimatized to lab and greenhouse environmental conditions. Hardened plantlets were transplanted in 1:1 (sand :soil) with ¼ MS medium for highest survival rate. An attempt was also made to determine the extent of clonal purity of the *in vitro* regenerated plants at the biochemical level by employing peroxidase isozyme as marker, in order to get an insight into the impact of somaclonal variations in the course of their regeneration.

Keywords: In vitro propagation, Talinum cuneifolium, Leaf explant.

Introduction

Talinum cuneifolium (Vahl) Willd., [Syn.: *Talinum portulacifolium* (Forsk.) Asch. ex Schw.] – an erect shrub with subterranean tuber belongs to the family Portulacaceae. The leaves of *Talinum cuneifolium* commonly known as Ceylone bachalli are eaten as a cooked vegetable or raw as a salad, alone or with young stem parts. It is cultivated in Africa (like spinach), and is used as a green leafy vegetable due to its rich vitamin A and mineral content (Anon, 2004). The leaves can also be stored dry for later use. The plant is a palatable fodder for cattle and goats. It is also an important medicinal plant in the local system of medicine. Indian system of medicine (ISM) refers that the leaves and roots are medicinally important parts. The supplementation of the leaves of this plant is reported to be a better diet for strengthening the body. The powdered leaves are used in treatment of diabetic, mouth ulcer, and aphrodisiac; roots are used for cough, gastritis, diarrhoea and pulmonary tuberculosis. In Ethiopia the leaves are applied medicinally against eye diseases and the root against cough and gonorrhoea. This valuable plant has markedly depleted to satisfy the local food needs and no attempts for its replenishment. The growth was very slow and takes long time. One of the constraints associated with the conventional propagation was very short span of seed viability. A low survival rate by stem cuttings in *Talinum cuneifolium* restricts its mass propagation via conventional methods. No alternative mode of multiplication was available to propagate and to conserve genetic stock of this plant. *In vitro* multiple shoot regeneration may give higher rate of propagation within very short time and space.

Isoenzymes can be considered to be the direct expression of the gene function of cells during differentiation and their variations were often associated with somaclonal variations. A detailed analysis of their changing patterns during development may lead to some understanding of the basic mechanism of cellular differentiation to obtain efficient plant regeneration *in vitro*. Peroxidases and esterases were widely distributed among higher plants. The application of isozymes as markers in morphological and regeneration studies has been reported by several workers (Bhatt *et al.*, 1992; Feuser *et al.*, 2003) to detect clonal fidelity. The current problem facing the regenerating system in plant tissue culture was the occurrence of uncontrollable somaclonal variations that were undesirable in any clonal propagation and conservation programme. Propagation by all methods of indirect organogenesis carries a risk that the regenerated plants will differ genetically from each other and from the mother plant (George, 1993). There have been no reports of a regeneration system for *T. cuneifolium*. Therefore, the present study was undertaken with an aim of establishing an efficient protocol for *in vitro* plant regeneration from leaf explant.

Materials and Methods

Leaves of *Talinum cuneifolium* (Vahl) Willd. were collected from S.V. University Botanical gardens, Tirupati, A.P. Explants were initially washed under running tap water with Teepol solution (5 per cent v/v) for 15 min. followed by 4-5 washings with distilled water. Disinfestations of these explants was then made under laminar air flow chamber by keeping them in 70 per cent alcohol for 60 sec. followed by rinsing for 3 times in sterile distilled water. Finally the explants were immersed in 0.1 per cent $HgCl_2$ (Mercuric chloride) for 3 min. The surface sterilization was followed by 5-6 rinses in sterile distilled water. The surface sterilized explants were cultured on MS basal medium (Murashige and Skoog, 1962) containing 3 per cent (w/v) sucrose and 0.8 per cent (w/v) agar. Explants were implanted in different combinations and concentrations of growth regulators (BA, IAA, IBA, NAA and 2,4-D) singly as well as in combinations for shoot proliferation. The pH of the medium was adjusted to 5.8 using 0.1 N HCl or 0.1 N NaOH (Sodium hydroxide) solutions before autoclaving. All cultures were incubated in a culture room at $25 \pm 2^\circ C$ with a relative humidity of 50-60 per cent and 16 h photoperiod at a photon flux density of 15-20 mE m^2/s^{-1} from white cool fluorescent tubes. For each treatment 12 replicates were used and each experiment was repeated at least thrice. The cultures were examined periodically.

The aseptic leaf explant was used for morphogenic callus induction on MS medium supplemented with different auxins (IAA, NAA, IBA and 2,4-D) in combination with BA. Callus developed on the above mentioned optimal medium was used for the studies to evaluate the effect of various Cytokinins (BA, Kn) at various concentrations and combinations on shoot regeneration. These *in vitro* elongated multiple shoots were excised and transferred on MS medium supplemented with IBA, NAA and IAA in different concentrations and combinations separately for root induction. These *in vitro* raised plantlets with well developed roots were taken from the culture tubes and washed thoroughly to remove the traces of agar. They were then planted in plastic cups containing mixture soil rite and soil (1:1) and maintained in the hardening with controlled temperature, light and relative humidity. The acclimatized complete plantlets were then transferred to the field and its survival frequency was recorded. Non-SDS-PAGE (Poly Acrylamide Gel Electrophoresis) of peroxidase isozyme was carried out to test the clonal purity as per the method of Van Eldic *et al.* (1980).

Results and Discussion

In the present study among the different types of sterilents used 5 per cent teepol, H_2O_2 (Hydrogen peroxide), NaOCl (Sodium hypochloride), 70 per cent alcohol for 60 seconds and 0.1 per cent $HgCl_2$ (Mercuric chloride) for 10 minutes were identified as an appropriate surface sterilizing agents for producing aseptic shoots form leaf explant. The decision on using type of media for the metabolic needs for the cultured cells and tissues was a major factor of success in plant regeneration process. Each plant requires different quantities of inorganic and organic nutrients for its morphogenic response, so no single medium will give satisfactory results with all tissues used. Selection of appropriate nutrient medium was also essential for the success of all experimental systems in plant tissue culture. Bhojwani and Razdan (1983) suggested that in order to formulate a suitable medium for a new system, it was best to start with a well known basal medium such as MS. Accordingly MS medium was initially used in the present study. Medium composition greatly affected the callus and shoot regeneration of *T. cuneifolium*.

Leaf explant produced fast growing light green coloured compact nodular calli with patches of pink coloured on MS medium supplemented with 0.2 mg/l BA along with 0.5 mg/l 2,4-D (Table 6.1) (Plate 6.1A). MS medium fortified with 0.2 mg/l BA and 0.5 mg/l IAA produced light whitish green coloured with patches pink friable loose nodular calli from leaf explant (Table 6.1) (Plate 6.1B). Light green coloured glossy loose friable calli was produced from leaf explant when inoculated on MS medium supplemented with 0.2 mg/l BA and 0.5 mg/l NAA (Table 6.1) (Plate 6.1C). The formation of pink colour may be due to the formation of Anthocyanins in the callus tissue.

Table 6.1: Effect of different plant growth regulators on callus induction from leaf explant of *Talinum cuneifolium.*

Plant Growth Growth Regulators (mg l[-1])				Nature of response and Morphogenic ability	
BA	NAA	IAA	2,4-D		
0.2			0.5	Light green coloured compact nodular calli with patches of pink coloured calli	+
0.2		0.5		Light whitish green coloured with patches of pink friable loose nodular calli	+
0.2	0.5			Light green coloured glossy loose friable calli	+

+: Represents organogenetic ability of callus.

Table 6.2: Effect of different PGR on indirect shoot regeneration from the callus derived from leaf explant of *Talinum cuneifolium.*

Plant Growth Regulators (mg l[-1])			Frequency of Shoot Regeneration	Mean No. of Shoots/Explant	Mean Length of the Shoot (cm)
BA	Kn	IAA			
1			48.8 ± 0.05^e	3.55 ± 0.16^d	0.78 ± 0.02^b
2			59.6 ± 0.16^i	5.27 ± 0.10^g	0.67 ± 0.02^{ab}
2	1		68.31 ± 0.09^l	7.18 ± 0.08^i	0.75 ± 0.01^b
2		0.1	45.1 ± 0.06^c	2.43 ± 0.03^b	0.53 ± 0.02^a

Values represented above are the means of 12 replicates. '±' indicates the standard error. Observations after 6 weeks of culture. Mean values having the same letter in each column don't differ significantly at Pd"0.05 (Duncans Test)

Auxins and Cytokinins play a significant role in *in-vitro* culture of plants (Evans *et al.,* 1981; Vasil *et al.,* 1994). Callus developed on the above mentioned optimal medium was used for further studies to evaluate the effect of BA and Kn at various concentrations on Indirect shoot regeneration. In general high concentrations of cytokinin and low concentration of auxin promotes the induction of shoot organogenesis. There were also numerous cases, where the observed type of morphogenesis was opposite that which was expected of the phytohormone balance to which the explants were exposed (Thorpe, 1980). MS medium supplemented with 1mg/l BA produced 3.55 mean numbers of shoots from the callus induced from leaf (Table 602) (Plate 6.1D). 2 mg/l BA produced 5.27 mean numbers of shoots from the

Plate 6.1: Induction of callus from leaf of *Talinum cuneioflium* and *invitro* plant regeneration through one stage process. MS medium supplemented with (A) 0.2mg/l BA + 0.5 mg/l 2,4-D (1cm Bar = 5.55 mm), (B) 0.2 mg/l BA + 0.5 mg/l IAA (1cm Bar = 3.84 mm), (C) 0.2 mg/l BA + 0.5 mg/l NAA (1cm Bar = 5.95 mm).

In vitro shoot regeneration from different callus tissues of *Talinum cuneifolium* through two stage process. Multiple shoot regeneration from leaf callus on MS medium containing D) 1 mg/l BA (1 cm Bar = 6.75 mm), E) 2 mg/l BA (1 cm Bar = 3.96 mm), F) 2 mg/l BA + 1 mg/l Kn (1 cm Bar = 8.06 mm).

callus induced from leaf (Plate 6.1E). 2 mg/l BA and 1 mg/l Kn in MS medium produced maximum number of multiple shoots (7.18) from the leaf callus with 68.31 per cent frequency of shoot regeneration (Table 6.2) (Plate 6.1F). 2 mg/l BA and 0.1 mg/l IAA in MS medium produced 2.43 mean numbers of shoots from the callus induced from leaf (Table 6.2).

Efficient rooting of *in vitro* regenerated plants and subsequent field establishment was the last and crucial stage of rapid clonal propagation. An *in vitro* rooting experiment reiterates the importance of auxins on root induction. About 3-4 cms long microshoots were isolated from proliferating bud cultures growing on MS medium and used for *in vitro* rooting. Among the three auxins used, IBA alone was found to be most effective when compared with IAA and NAA for root induction. The IBA improved rooting efficiency. Microshoots when subjected to rooting on ½ strength MS medium supplemented with 0.5 mg/l IBA also exhibited elongation at the earlier stages of inoculation while at the later stages produced a bunch of small thin roots with lateral roots (8.64) with out basal callus within two weeks. High frequency (85.7 per cent) of root regeneration and maximum mean number (12.46) of thin slender lengthy roots were also induced from the base of the shoots with out interference of callus on ½ strength MS medium supplemented with 1.0 mg/l IBA. The superiority of IBA in rhizogenesis was seen in *Lonicera tatarica* (Palacios *et al.*, 2001), *Artemisia judaica* (Liu *et al.*, 2002), apricot (Tornero and Burgos, 2000), *Piper longum* (Soniya and Das, 2002), mung bean (Tivarekar and Eapen, 2001) and *Cunila galioides* (Fracaro and Echeverrigaray, 2001). Efficiency of IBA at lower concentration *in vitro* rooting has been reported by Das and Rout (2002), Soniya and Das (2002), Martin *et. al.* (2003) and Prasad (2004). Heloir *et al.* (1997) reported that IBA provided a suitable auxin for *in vitro* rooting of *Vitis vinifera*. The observations coincide with the present study as the higher number of roots obtained in the MS medium supplemented with higher concentrations of IBA than NAA.

Table 6.3: Effect of different auxins on the root induction of *in vitro* raised shoots of *Talinum cuneifolium* on half strength MS medium.

Plant Growth Regulators (mg l⁻¹)			Frequency of Regeneration	Mean No. of Roots	Mean Length of the Root
NAA	IBA	IAA			
	0.5		78.81 ± 0.09[k]	8.64 ± 0.01[j]	3.50 ± 0.04[h]
	1.0		85.70 ± 0.07[l]	12.46 ± 0.02[k]	4.60 ± 0.01[i]
		0.5	46.70 ± 0.02[b]	2.18 ± 0.01[a]	1.28 ± 0.03[b]
		1.0	54.36 ± 0.04[e]	3.36 ± 0.02[c]	1.50 ± 0.03[c]
0.5			58.21 ± 0.05[g]	4.62 ± 0.08[f]	2.15 ± 0.02[d]
1.0			67.41 ± 0.03[i]	7.90 ± 0.05[i]	3.26 ± 0.02[g]
0.5	0.1		74.64 ± 0.02[j]	5.31 ± 0.01[g]	3.03 ± 0.10[f]
1.0	0.1		63.92 ± 0.01[h]	7.60 ± 0.03[h]	3.30 ± 0.03[g]
	0.5	0.1	52.63 ± 0.02[d]	4.40 ± 0.01[e]	1.60 ± 0.04[c]
	1.0	0.1	55.05 ± 0.03[f]	5.32 ± 0.06[g]	2.55 ± 0.06[e]
0.1		0.5	44.71 ± 0.01[a]	2.83 ± 0.09[b]	1.02 ± 0.02[a]
0.1		1.0	52.23 ± 0.02[c]	3.60 ± 0.04[d]	1.18 ± 0.01[b]

Values represented above are the means of 12 replicates. '±' indicates the standard error. Observations after 4 weeks of culture. Mean values having the same letter in each column don't differ significantly at Pd"0.05 (Duncans Test)

The primary target of a micropropagation system was the best acclimatization and field establishment of regenerated plants. Acclimatization of regenerated plants to the external environment was very important and depends on different plant and environmental factors as suggested by Donelly and Tindall (1993). Acclimatization of the regenerated plants under ambient environmental condition was the last stage of micrpropagation and its success depends upon different factors as suggested by George and Sherrington (1984). The high humidity does not allow synthesis of cuticle on the epidermis of leaves of regenerated plants and *in vitro* cultivated plants lack the necessary anatomical features to withstand variations in the natural environment (Thakur *et al.,* 1998). Consequently when such plants were transferred to the natural conditions, they undergo desiccation and death. To overcome these problems, it was necessary to increase the temperature and decrease relative humidity. Accordingly rooted plants were gradually acclimatized with an increase in temperature from 25-28°C and decrease in relative humidity form 80-50 per cent for a period of 15-20 days. Therefore, in the present study, the paper cups containing *in vitro* derived plantlets were kept in the culture room temperature and cups were covered with polythene bags to maintain high humidity and kept in mist chamber covered with coir mat. Survival of *in vitro* planting was largely dependent on the components of the potting media. In the present study, garden soil and sand (1:1) was found suitable for highest survival of plants. These plants were irrigated with ¼ strength MS salts and exposed gradually to external environment.

The observations obtained with peroxidase isoenzyme pattern in the current investigation substantiated the uniformity of the multiple shoots derived from leaf explant which was most desirable in any micropropagation system. These findings were also supporting the view that biochemical traits such as isozyme provides an evidence to study the extent of somaclonal variations in a manner analogous to their use in elucidating genetic variation in natural population (Bhaskaran *et al.,* 1987; Ramamurthy and Savithramma, 2003). Plants derived from tissue culture as having superior field performance to those derived from stem cuttings in terms of survival rate, fruit yield, rhizome production and total plant weight. After two weeks, by which time a fresh leaf appeared from the potted plants, were transferred to green house and then transferred to field condition. 70-75 per cent of regenerated plants were successfully acclimatized to natural environment. The direct regeneration of plants from leaf buds results in the maintenance of genotypic stability without the risk of somaclonal variations normally associated with adventitious regeneration via callus (Levieille and Wilson, 2002). *In vitro* derived plantlets were morphologically similar to *in vivo* plants. The present study has successfully established a high frequency, mass propagation system for *Talinum cuneifolium* a valuable leafy vegetable having known medicinal benefits. This protocol provides a successful and rapid technique that can be used for *ex situ* conservation.

References

Anon, 2004. The Wealth of India, A dictionary of Indian raw materials and industrial products, first supplement series. NISCAIR. **5**: 184.

Bhaskaran, S., Smith, R.H., Paliwal, S. and Schert. K.F. 1987. Somaclonal variation from *Sorghum bicolor* (L.) Moench cell culture. Plant Cell Tiss. Org. Cult. **9**: 189–196.

Bhatt, S.R., Kackar A. and Chandel, K.P.S. 1992. Plant regeneration from callus cultures of *Piper longum* L. by organogenesis. Plant Cell Rep. **11**: 5252–5254.

Bhojwani, S.S. and Razdan, M.K. 1983. Plant tissue culture: Theory and practice. Elsevier Science, Amsterdam.

Das, G. and Rout, G.R. 2002. Direct plant regeneration from leaf explants of *Plumbago* species. Plant Cell Tiss. Org. Cult. **68**: 311–314.

Donelly, D.J. and Tindall, L. 1993. Acclimatization Strategies for micropropagated Plants, In: Micropropagation of Woody Plants, Ahuja, M.R. (ed.), Kluwer Academic Publishers, Dordrecht, London, UK, pp.153–166.

Evans, D.A., Sharp, W.R. and Flick, C.E. 1981. Growth and behaviour in cell cultures – Embryogenesis and Organogenesis. In plant cell culture: Methods and Applications in Agriculture, Thorpe, T.A. (ed.), Academic Press, New York, pp.45–113.

Feuser, S., Meler, K., Daqunita, M., Guerra M.P. and Nodari, R.O. 2003. Genotypic fidelity of micro propagated pineapple (*Ananas csomosus*) plantlets assessed by isozyme and RAPD markers. Plant Cell Tiss. Org. Cult., **73**: 221–227.

Fracaro, F. and Echeverrigaray, S. 2001. Micropropagation of *Cunila galioides*, a popular medicinal plant of South Brazil. Plant Cell Tiss. Org. Cult. **64**: 1–4.

George, E.F. and Sherrington, P.D. 1984. Plant Propagation by Tissue Culture, In: Handbook and Directory of Commercial Laboratory, Eastern Press, Great Britain.

George, E.F. 1993. Plant Propagation by Tissue Culture Part I, 2nd edn., Exegetics Ltd., Edington, England.

Heloir, M.C., Fournioux, J.C., Oziol, L. and Bessis, R. 1997. An improved procedure for the propagation *in vitro* of grapevine (*Vitis vinifera* cv. Pinot noir) using auxiliary–bud microcuttings. Plant Cell Tiss. Org. Cult. **49**: 223–225.

Levieille, G. and Wilson, G. 2002. *In vitro* propagation and iridoid analysis of the medicinal species *Harpagophytum procumbens* and *H. zeyheri.* Plant Cell Rep. **21**: 220–225.

Liu, C.Z., Murch, S.J., Demerdash, M. and Saxena, P.K. 2002. Regeneration of the Egyptian medicinal plant *Artemisia judaica* L. Plant Cell Rep. **21**: 525–530.

Martin, K.P., Beena, P.K. and Domini Joseph. 2003. High frequently auxiliary bud multiplication and *ex vitro* rooting of *Wedelia chinensis* (Osbeck) Merr. – A medicinal plant. Indian J. Exp. Biol. **41**: 262–266.

Murashige, P. and Skoog, F. (1962). A revised medium for rapid growth and bioassay with tobacco tissue. Physiol. Plant.. **15**: 473–497.

Palacios, N., Christou P. and Leech. M.J. 2001. Regeneration of *Lonicera tatarica* plants via adventitious organogenesis from cultured stem explants Plant Cell Rep. 20: 808–813.

Prasad, P.J.N. 2004. *In vitro* studies of *Cryptolepis buchannini* Roem. & Schult. and *Sarcostemma intermedium* Dcne. (Asclepiadaceae). Ph.D. Thesis submitted to S.K. University, Anantapur, Andhra Pradesh, India.

Ramamurthy, N. and Savithramma, N. 2003. Shoot bud regeneration from leaf explants of *Albizzia amara* Boiv. Indian J. Plant Physiol. **8**(4): 372–376.

Soniya, E.V. and Das, M.R. 2002, *In vitro* micropropagation of *Piper longum* – an important medicinal plant. Plant Cell Tiss. Org. Cult. **70**: 325–327.

Thakur, R., Rao P.S. and Bapat, V.A. 1998. *In vitro* plant regeneration in *Melia azedarach* L. Plant Cell Rep. **18**: 127–131.

Thorpe, T.A. 1980. Organogenesis in *in vitro* structural physiological and biochemical aspects, In : Perspectives in Plant Cell Tissue, Vasil, I.K.(ed.), Academic Press, New York. pp. 71–111.

Tivarekar, S. and Eapen, S. 2001. High frequency plant regeneration from immature cotyledons of mungbean. *Plant Cell Tiss. Org. Cult.,* **66**: 227–230.

Tornero, O.P. and Burgos. L. 2000. Different media requirements for micropropagation of apricot cultivars. Plant Cell Tiss. Org. Cult. **63**: 133–141.

Van Eldic, L. J., Grossman, A.R., Iversion D.B. and Watterson, D.M. 1980. Isolation and characterization of calmodulin from spinach leaves and *in vitro* translation mixtures. Proc.Natl.Acad.Sci. USA. **77**: 1912–1916.

Vasil, I.K. and Thrope, T.A. 1994. Plant cell and Tissue Culture, Kluwer Academic Publishers, Dordrecht.

2013, Abiotic Stress and Biotechnology
Editor: T. Pullaiah
Published by: REGENCY PUBLICATIONS, NEW DELHI

Pages 137–168

Chapter 7

Contribution of Marine Actinobacteria to Human Health

Mangamuri Usha Kiranmayi[1], Sudhakar Poda[2], Chitta Padmavethi[2] and Muvva Vijayalakshmi[1] *

[1]Department of Botany and Microbiology,
[2]Department of Biotechnology,
Acharya Nagarjuna University, Nagarjunanagar,
Guntur – 522 510, Andhra Pradesh, India

ABSTRACT

Most of the drugs that are in use today are the secondary metabolites gifted by nature. Actinobacteria are the most efficient group of secondary metabolite producers. Several species have been isolated and screened for the production of antimicrobial metabolites from terrestrial habitats in the past decades. Consequently the chance of isolating novel compounds from the terrestrial habitats have been reduced and forced the scientists to explore the unexplored marine habitats, thereby increasing the chance of isolating novel actinobacteria that produce novel carbon skeletons that have new biological activities. The revolutionary demand for the search of new secondary metabolites is to combat the problem of multi drug resistance of pathogens that are no longer susceptible to the currently used drugs. The diversity and novelty among actinobacteria exist

[1] Corresponding Author: E mail: muuuoul@yahoo.co.in

in the marine environments. These actinobacteria produce different types of metabolites that have tremendous potential to be developed into therapeutic agents of different categories. Marine actinobacteria are a novel prolific, but are unexplored source for the discovery of novel secondary metabolites.

Keywords: *Actinobacterial diversity, Carbon skeletons, Natural products, Therapeutic agents, Marine habitats.*

Introduction

The most economically and biotechnologically valuable prokaryotes are actinobacteria as they are well known for their ability to produce biologically active substances (Lazzarini *et al.,* 2000). They are responsible for production of half of the bioactive metabolites that are discovered (Berdy, 2005) including notable antitumor agents (Cragg *et al.,* 2005), immunosuppressive agents (Mann, 2001), enzymes (Pecznska-Czoch *et al.,* 1988) and antibiotics (Strohl, 2004). Actinobacteria have an excellent fruiting track record of their ability in this regard. For successful drug screening programmes, efforts have been focused on the successful isolation of the novel actinobacteria from the terrestrial sources during the past five decades (Bruna *et al.,* 1973; Omura, 1986; Malkova *et al.,* 1991; Wu *et al.,* 2000; Caffrey *et al.,* 2001; Narayana *et al.,* 2004; Narayana *et al.,* 2008a; 2008b; Kavitha *et al.,* 2009a; 2009b; 2009c). Methods for drug screening have been focused towards marine habitats because the rate of discovery of the novel compounds from terrestrial actinobacteria has been decreased and the rate of re-isolation of the known compounds has increased (Kin, 2006; Fenical *et al.,* 1999; Fenical and Jensen, 2006). This crucial situation diverts the attention of biotechnologists to focus on the unexplored habitats for the discovery of the novel secondary metabolites.

No doubt that the diversity of the terrestrial environment is great but the greatest diversity is in marine ecosystem (Donia and Hamann, 2003) that has virtually important untapped source of novel actinobacterial diversity (Stach *et al.,* 2003) and also for novel metabolites (Fiedler *et al.,* 2005). The focus on actinobacteria in the marine ecosystem is largely neglected, because of the elusive assumption that very little scope to isolate strains for the discovery of novel metabolites (Prauser, 1964). However the current research suggests that marine actinobacteria are the prime source for the search and discovery of the novel natural products. The environmental conditions of the marine ecosystem are extremely different from that of the terrestrial ones and the marine actinobacteria have different characteristics as compared to their terrestrial counterparts and likely to produce different types of bioactive compounds (Fiedler *et al.,* 2005; Bull *et al.,* 2007). In addition the environmental conditions have a profound effect on the genetic and metabolic diversity on the marine actinobacteria that remain unknown (Cross, 1989).

The actinobacteria have a high commercial value as producers of antibiotics and other therapeutically useful compounds and exhibit a range of life cycles unique among the prokaryotes and play a major role in cycling of organic matter in the soil ecosystem (Veiga *et al.,* 1983). They hold a prominent position due to their diversity and their capability to produce new compounds (Surajit *et al.,* 2008). Actinobacterial population has been initially considered as one of the major groups of soil microorganisms; however it has also been increasingly isolated from various marine samples. The frequent breakdown of the epidemic diseases, immediate emergence of drug resistant pathogens and the magnitude of their dissemination among people demand the production of effective antibiotics to be continued (Imada *et al.,* 2007). The need to search for the novel therapeutic agents by the pharma industry is to combat the increasing incidence of infection due to antibiotic resistant pathogens linked with the search for novel anti-tumour and anti-viral compounds (Gontang *et al.,* 2007). Natural products have been the major source of numerous therapeutic agents of various categories (Hayakawa, 2003). More than half of the drugs that are in use today are the natural product therapeutic agents (Berdy, 2005).

Search for Novel Drugs and Novel Actinobacteria from Unexplored Habitats

The natural products that are of microbial origin are the most promising source for future medicine that society expects to be developed. These microbes are the origin for most of the antibiotics in the market today (Kaltenpoth, 2009; Thumar *et al.,* 2010). The search for new antibiotics from novel strains of microorganisms producing bioactive metabolites continues to be of utmost importance in research programmes due to increase in drug resistant pathogens and extreme side effects produced by some existing antibiotics. There is a great demand and urgent need for new antibiotics to halt and reverse the spread of the drug resistant pathogens that are responsible for life threatening infections and risk that became a hurdle and nightmare to the health care system (Talbot *et al.,* 2006). In addition to drug resistance problem, patient's sensitivity and inability to control certain infectious diseases have given necessity for continuous search of new antibiotics all over the world. The multi drug resistant organisms can only be combated by introduction of antimicrobial compounds or the antibiotics essential from new sources (Saha *et al.,* 2010). Very few new antibiotics have been developed despite the critical need for new antibiotics required to treat drug–resistant infections and other infectious diseases (Jain and Pundir, 2011). Pharma industry in response to microbial resistance has produced a remarkable range of antibiotics (Luzhetskyy *et al.,* 2007). The two parallel problems that are experienced are 1) to find new drugs to treat known diseases due to the emergence of resistant bacterial strains 2) to find new drugs to fight new diseases.

Taking into consideration of the above facts there is a need to discover antibiotics at war front to treat patients infected with multi-drug resistant infections (Luzhetskyy *et al.,* 2007). Over the years several hundreds of drugs have been isolated from microbes, but clinical applications are attributed to very few of them (Thomashow *et al.,* 2008). The reason is that only compounds with selective toxicity can be used clinically. Effective antimicrobial drug discovery and development with novel mechanisms, new targets and new modes of action have become an urgent task for infectious disease research programmes (Jain and Pundir, 2011). Novel families of new drugs must enter the market at regular intervals if modern medicine is to continue in its present form. However aggressive screening programmes for the selection of novel natural and modified chemical compounds will be necessary to produce novel antibiotics against resistant bacteria in the coming years.

The outstanding role of marine actinobacteria as source for the antibiotic manufacturing unit is prominent. To the present date only 1 per cent of the microbial world has been explored. With the advent of modern and sensitive techniques for studying the microbial world and extraction of nucleic acids from soils and marine habitats develop access to a vast uncaged reservoir of genetic and metabolic diversity (Sanchez and Olson, 2005). The ability of the microorganisms to produce important unusual metabolites rests on their capability to grow in extreme habitats as a means of chemical defence. Many biosynthetic pathways are involved in the production of anti-microbial metabolites (peptides, polyketides, isoprenes, oligosaccharides, aromatic compounds and β-Lactum rings) and understanding the underlying knowledge of the enzyme activity in the pathways are still unknown (Paradkar *et al.,* 2003). The enzymatic pathways occurring with individual proteins are free or complexed or as large multifunctional polypeptides carry a multi step process in the biosynthesis of these different kinds of antibiotics (Demain, 1998). The genes that encode these enzymes are either organized on the chromosome or on the plasmid. But in prokaryotes these genes are clustered and not necessarily as single operons. The antibiotic encoding gene clusters in actinomycetes are of different sizes (Piepersberg and Distler, 1997; Marti´nez-Bueno *et al.,* 1990; Brown *et al.,* 1996; Hopwood, 1999). The ideal growth phase is the stationary phase for the formation of antimicrobial compounds. Biosynthesis of these compounds is usually influenced by several features like exhaustion of nutrients, addition of inducers/precursors/ inhibitors etc (Sanchez and Demain, 2002; Bibb, 2005).

Marine Actinobacteria as a Reservoir for Novel Antibiotics

Marine actinobacteria are fascinating resources due to their ability of producing novel natural products with antimicrobial activities (Xiangyang *et al.,* 2010). The seemingly unlimited capacity to produce secondary metabolites with diverse chemical structures and biological activities have been especially useful to the pharmaceutical industry. These species are considered to be an important valuable treasure to the human health due to their capability to produce secondary metabolites that may have pharmaceutical and biotechnological application (Valan arasu *et al.,* 2008; Solanki *et al.,* 2008). Marine actinobacteria have proven to be efficient producers of novel secondary metabolites as shown in Table 7.1. Some of the actinobacterial genera identified by cultural and molecular techniques from different marine ecological niches include *Actinomadura, Actinosynnema, Amycolatopsis, Arthrobacter, Blastococcus, Brachybacterium, Corynebacterium, Dietzia, Frankia, Frigoribacterium, Geodermatophilus, Gordonia, Janibacter, Kitasatospora, Marinispora, Micromonospora, Micrococcus, Microbacterium, Mycobacterium, Nocardioides, Nocardiopsis, Nonomurea, Pseudonocardia, Rhodococcus, Saccharopolyspora, Salinispora, Serinicoccus, Solwaraspora, Streptomyces, Streptosporangium, Thermoactinomyces, Tsukamurella, Turicella, Verrucosispora* and *Williamsia* (Stach *et al.,* 2004; Jensen *et al.,* 2005; Ward and Bora 2006; Das *et al.,* 2006). Among them *Streptomyces, Saccharopolyspora, Amycolatopsis, Micromonospora* and *Actinoplanes* are the major producers of commercially important bio molecules. Over the past decade, information on the diversity of actinobacteria in marine habitats has grown considerably, and also their ability to produce secondary metabolites has continued quite strongly (Nonomura, 1974; Stackebrandt *et al.,* 1997).

Table 7.1: Important *novel* metabolites produced by marine actinobacteria.

Sl.No.	Compound	Source	Activity
1.	Abyssomicin C (Bister *et al.,* 2004)	*Verrucosispora*	Antibacterial
2.	Actinofuranones A and B (Cho *et al.,* 2006)	*Streptomyces* spp.	Cytotoxic
3.	Aureoverticillactum (Mitchell *et al.,* 2004)	*Streptomyces aureoverticillatus*	Anticancer
4.	Bonactin (Schumacher *et al.,* 2003)	*Streptomyces* spp.	Antibacterial; Antifungal
5.	Butenolide (Li *et al.,* 2006)	*Streptoverticillium luteoverticillatum*	Anticancer
6.	Caboxamycin (Hohmann *et al.,* 2009)	*Streptomyces spp. NTK 937*	Antibacterial activity
7.	Caprolactones (Stritzke *et al.,* 2004)	*Streptomyces* spp.	Anticancer
8.	Chandrananimycins (Maskey *et al.,* 2003)	*Actinomadura* spp.	Antialgal, Antibacterial; Anticancer; Antifungal
9.	Chinikomycins	*Streptomyces* spp.	Anticancer
10.	Cyclotyrosyl prolyl (Rofiq *et al.,* 2010)	*Streptomyces* spp. *A11*	Antibacterial activity
11.	CyclomarinA (Renner *et al.,* 1999)	*Streptomyces* spp.	Antiinflammatory, antiviral
12.	Daryamides (Asolkar *et al.,* 2004)	*Streptomyces* spp. *CNQ-085*	Anticancer, antifungal
13.	Dehydroxynocardamine and Desmethylenylnocardamine (Lee *et al.,* 2005)	*Streptomyces* spp.	Enzyme sortase B inhibitor
14.	Diazepinomycin (Charan *et al.,* 2004)	*Micromonospora* spp.	Antibacterial, anticancer, Antiinflammatory
15.	Frigocyclinone (Brutner *et al.,* 2005)	*Streptomyces griseus*	Antibacterial
16.	Glaciapyrroles A (Cho *et al.,* 2006)	*Streptomyces* spp. *NPS008187*	Antibacterial
17.	Helquinolone (Asolkar *et al.,* 2004)	*Janibacter limosus*	Antibacterial
18.	Himalomycins (Maskey *et al.,* 2003)	*Streptomyces* spp.	Antibacterial
19.	Komodoquinone A (Itoh *et al.,* 2003)	*Streptomyces* spp.	Neutrigenic activity

Contd...

Table 7.1–*Contd...*

Sl.No. Compound	Source	Activity
20. Lazollamycin (Manam *et al.*, 2005)	*Streptomyces nodosus*	Antibacterial
21. Lynamicins A-E (Katherine *et al.*, 2008)	*Marinispora* spp.	Antibacterial activity Methicillin-resistant *Staphylococcus aureus* and Vancomycin-resistant *Enterococcus faecium*
22. Marinomycins (Kwon *et al.*, 2006)	*Marinispora*	Antibacterial; Anticancer
23. Mechercharmycins (Kanoh *et al.*, 2005)	*Thermoactinomyces* spp.	Antitumor
24. Neomarinone (Macherla *et al.*, 2005)	*Strain CNH-099*	Cytotoxic
25. Piericidins C7 and C8 (Hayakawa *et al.*, 2007)	*Streptomyces*	Anticancer
26. Piperazimycins (Miller *et al.*, 2007)	*Streptomyces* spp.	Anticancer
27. Pyridinium salt (Ravi *et al.*, 2011)	*Amycolatopsis alba*	Antibacterial and cytotoxic activity
28. Requinomycin (Bavya *et al.*, 2011)	*Streptomyces filamentosus (R1)*	Antiphage activity
29. Resistomycin (Shiono *et al.*, 2002)	*Streptomyces corchorusii AUBN(1)/7*	Antiviral
30. Salinamides A and B (Moore *et al.*, 1999)	*Streptomyces* spp.	Antibacterial; Anti-inflammatory
31. Saliniketal A, Saliniketal B (William *et al.*, 2007; Jensen *et al.*, 2007)	*Salinispora arenicola*	Anticancer
32. Salinosporamide A (Feling *et al.*, 2003)	*Salinispora tropica*	Anticancer
33. Sporolides (Buchanan *et al.*, 2005)	*Salinispora tropica*	Unknown biological activity
34. Thiocoraline (Romero *et al.*, 1996)	*Micromonospora*	Anticancer, antibacterial
35. Thiopeptide antibiotic TP-1161(Kerstin *et al.*, 2010)	*Nocardiopsis* spp.	Antibacterial and Antifungal activity
36. Urukthapelstatin (Matsuo *et al.*, 2007)	*Mechercharimyces asporophorigenes YM11-542*	Anticancer

Bioactive compounds from marine actinomycetes possess distinct chemical structures that may form the basis for synthesis of new drugs that could be used to combat resistant pathogens (Solanki *et al.*, 2008). It has been emphasized that

that could potentially yield a broad spectrum of secondary metabolites (Maskey *et al.,* 2003). Lot of scepticism on the existence of the marine actinobacteria raised from the fact that terrestrial bacteria produce resistant spores that are known to be transported from land to sea, where they remain dormant for many years (Bull *et al.,* 2000; Cross, 1981; Goodfellow and Haynes, 1984; Kin, 2006).

Factors that Govern the Yield of the Marine Natural Products

Marine bioactive secondary metabolites possess interesting chemical structures but bringing these chemical structures as new drugs at industry is always not fruitful because of low yield. That's why there is a limitation of natural product at the industry level. These limitations can be de-hurdled when there is a possibility of organic synthesis that is an alternate to the chemical synthesis which is highly expensive. But optimized production and extraction of the microbial natural products can be targeted by determining the bench culturing conditions and scaling up to reach adequate levels of metabolite (Gogoi *et al.,* 2008). The condition that needs to be scaled up is designing the seed and fermentation media which is important in production of secondary metabolites (Gao *et al.,* 2009). The success relies on the knowledge and experience in developing the suitable medium that plays an important role in optimization. In addition the success of the product formation depends on various environmental factors including the nutrients (nitrogen, phosphorous and carbon source), growth rate, feedback control, enzyme inactivation and variable conditions (oxygen supply, temperature, light and pH) (Sánchez *et al.,* 2010). In addition the metabolite production by the actinomycetes differs qualitatively and quantitatively depending on the strains used in fermentation. The inoculum age and volume of the organism and fermentation medium have profound effect on product formation directly or indirectly. Concentrations of the media composition also influence the yield of the product. Low concentration of nutrients may lead to poor growth and poor yields because of under nutrition. High concentration may have deleterious effect on the growth. Catabolite repression and toxicity may also be observed. So the medium should contain a judicious combination of various nutrients at concentration that favours the growth of organism and product formation (Venkata *et al.,* 2011).

Marine Natural Products

Actinobacteria are virtually an unlimited source for the production of novel natural compounds that are secondary metabolites with many therapeutic values. Marine actinobacteria hold a prominent position due to their ability to produce natural products. These natural products played a key role in treating and preventing human diseases (Jones *et al.,* 2006). The value of natural products in this regard can be assessed using 3 criteria: (1) The rate of introduction of new chemical entities of wide structural diversity, including serving as templates for semi synthetic and total synthetic modification (2) The number of diseases treated or prevented by these substances and (3) Their frequency of use in the treatment of disease. Large proportion of the natural products in drug discovery has stemmed from diverse structures and the complicated carbon skeletons of natural products. These secondary metabolites from natural sources that are associated with the living systems show more likeliness and biological friendliness than totally synthetic molecules (Koehn and Carter, 2005) making them good candidates for further drug development (Balunas and Kinghorn, 2005).

The different kinds of natural products of marine actinobacteria are:

Peptides

These are short protein molecules containing short polymers of amino acids. Peptides obtained from *Streptomyces* contain further rare structural elements such as chromophores or uncommon amino acids. Some of the notable peptides of marine origin include:

Cyclomarins A-C

These are cyclic in nature and produced by *Streptomyces* spp. They show anti-inflammatory and antiviral activities (Renner *et al.,* 1999).

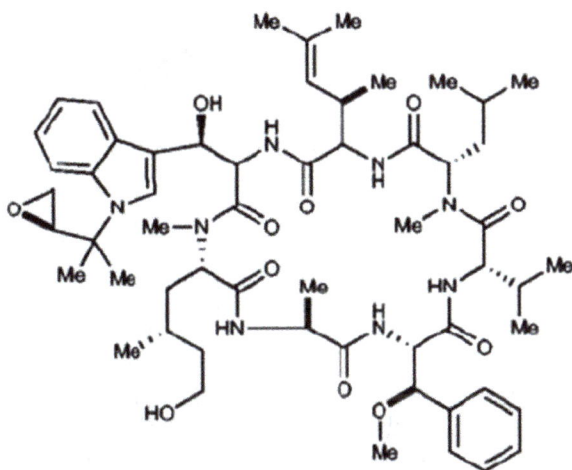

Piperazimycins A-C

These are cytotoxic hexadepsipeptides that are obtained from *Streptomyces* spp. and show cytotoxic and anti-viral activity (Miller *et al.,* 2007: Shirling and Gottlieb, 1966).

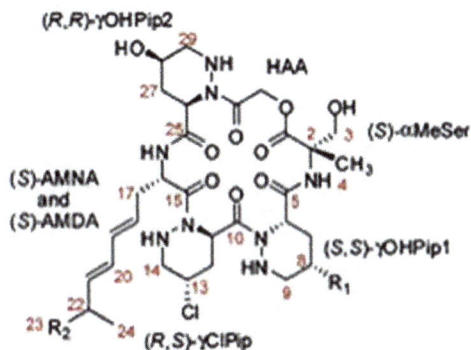

Piperazimycin A R_1=OH, R_2=CH$_3$
Piperazimycin B R_1=H, R_2=CH$_3$
Piperazimycin C R_1=OH, R_2=CH$_2$CH$_3$

Mechercharmycins A and B

These are novel antimicrobial metabolites obtained from marine isolate *Thermoactinomyces* spp. Of the two compounds only the Mechercharmycins A exhibit potential anti-tumour activity (Kanoh *et al.*, 2005).

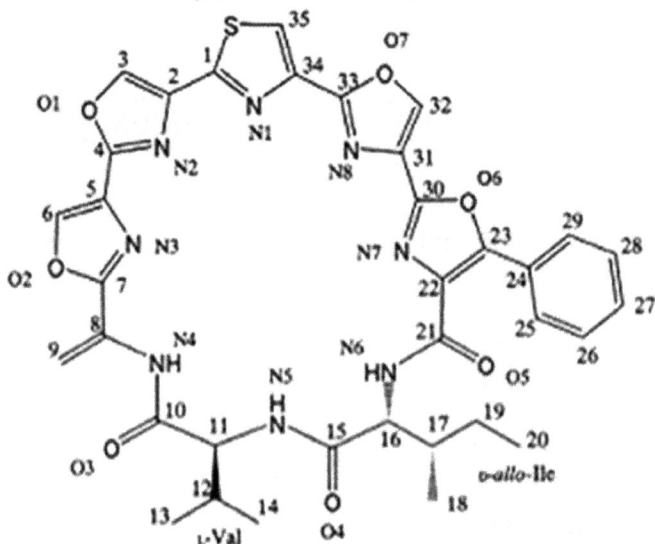

Mechercharmycin A

Lucentamycins A and B

These are four new 1-methyl-4-ethylideneproline containing peptides designated as Lucentamycins A-D from the cultured broth of *Nocardiopsis lucentensis.* Of the four compounds only the A and B compounds exhibited *in vitro* cytotoxicity against human colon cancer (Cho *et al.*, 2007).

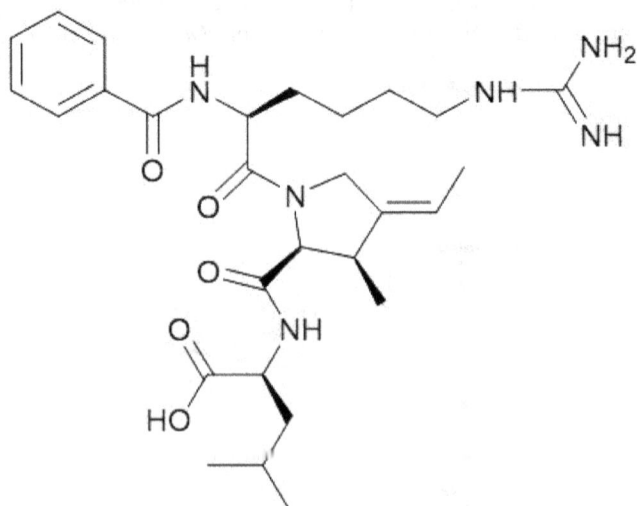

Quinones

Quinones are conjugated cyclicdione structures that are the secondary metabolites with high biological activity obtained from marine actinobacteria. Some of the valuable quinones of marine origin are:

Himalomycins A and B

These are two new quinone antibiotics from *Streptomyces* spp. They exhibit potent antibacterial activity (Maskey *et al.*, 2003).

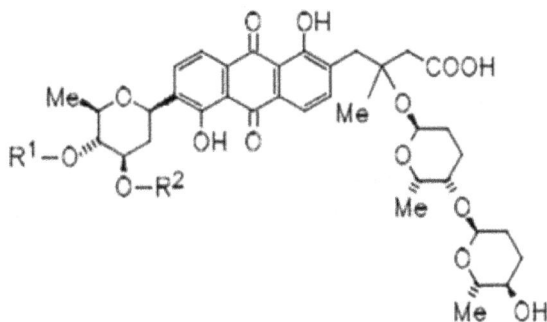

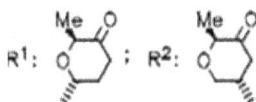

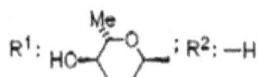

Resistoflavine

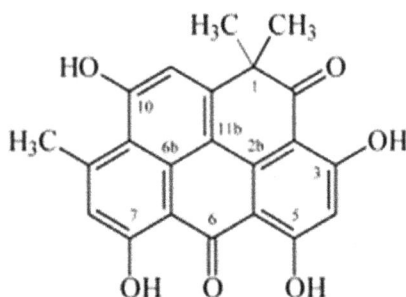

This drug is produced by *Streptomyces chibaensis*. It shows cytotoxic and antibacterial activity (Kock *et al.*, 2005; Gorajana *et al.*, 2006).

Resistomycin

This is an antibiotic related to quinones, is produced by *Streptomyces corchorusii*, an inhibitor of HIV-1 protease (Shiono *et al.*, 2002).

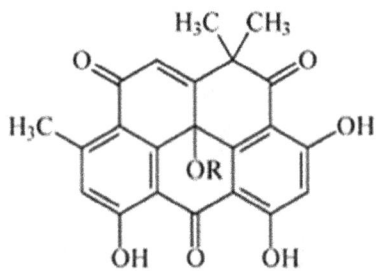

Resistoflavine

Resistomycin

Mansouramycin A-D

These are isoquinoline quinone derivatives isolated from the extracts of *Streptomyces* spp. Mansouramycin A-D known as 3-methyl-7-5-8-isoquinolinedione is active against 36 different cell lines indicated significant cytotoxicity with great degree of selectivity for lung cancer, breast cancer, melanoma and prostate cancer cells (Hawas *et al.,* 2009).

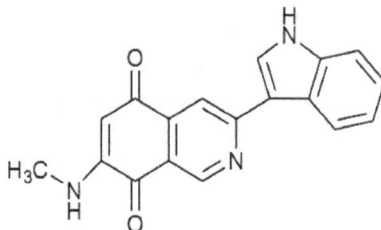

Macrolides

A group of drugs that are obtained from the actinobacteria containing a marocyclic lactone ring. Presence of a macrolide ring is responsible for its biological activity. These are protein synthesis inhibitors. Some of the important macrolides of marine origin include:

Sporolide A and Sporolide B

Salinispora tropica strain produces two macrolide compounds that are called Sporolide A and B. These two compounds appear to be synthesised from polyketides containing the oxidised carbons. These structures demonstrate the tremendous potential of marine actinomycetes for the production of novel secondary metabolites.

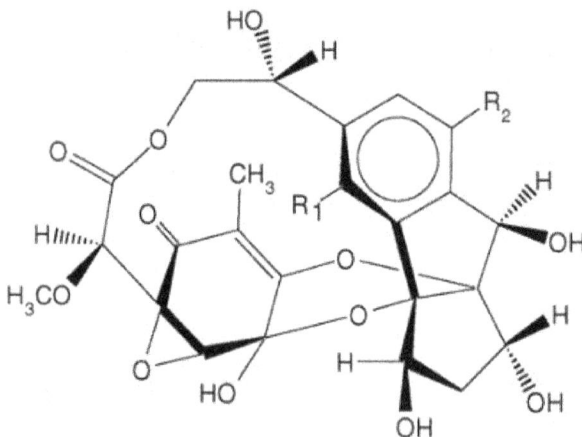

Sporolide A: R_1=Cl; R_2=H; Sporolide B: R_1=H; R_2=Cl

Chalcomycin A

A macrolide antibiotic produced by a marine *Streptomyces* spp. M491 and some terpenes (Wu *et al.,* 2007).

Arenicolide

A macrolide antibiotic produced by some strains of *Salinispora arenicola*, exhibits antibacterial activity against drug resistant bacteria (Jensen *et al.,* 2007; Williams *et al.,* 2007).

$R_1 = Cl, R_2 = R_3 = OH$

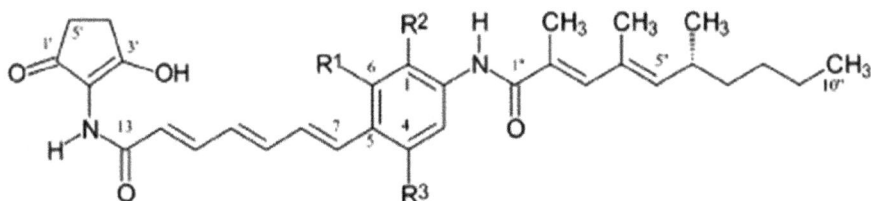

Chalcomycin A

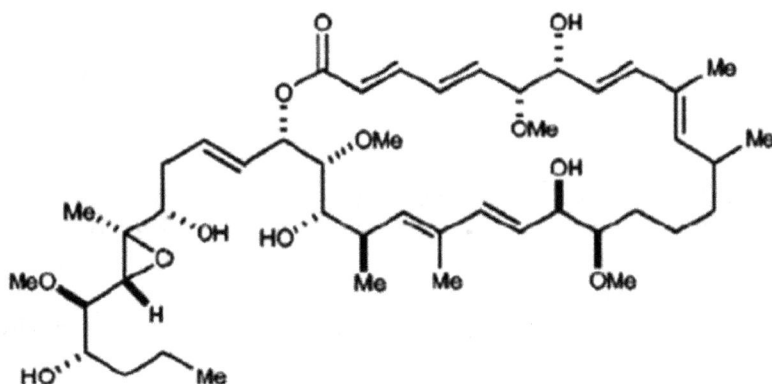

Arenicolide

Marinomycins

These are polyene like macrolides produced by *Marinispora* with potent antitumor activity. Marinomycin A inhibits the growth of methicillin resistant *Staphylococcus aureus* and vancomycin resistant *Enterococcus faecium* (Kwon *et al.,* 2006).

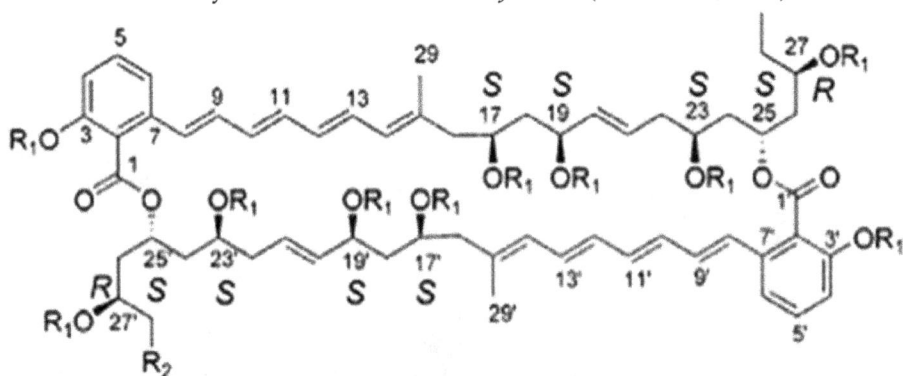

$$R_1 = H, \quad R_2 = H$$

Tartrolon D

This is a macrolide isolated from the *Streptomyces* spp. MDG-04-17-069 exhibiting a strong cytotoxic activity against lung tumour and breast cancer (Pérez *et al.,* 2009).

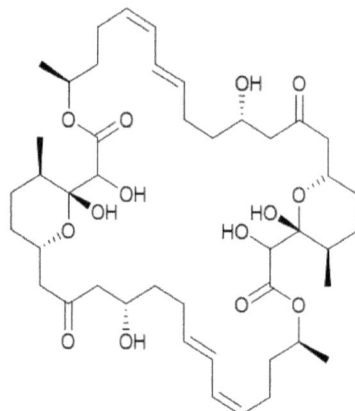

Terpenes

These are the major biosynthetic building blocks in all the living creatures with a large varied class of hydrocarbons that are derived from isoprene. Novobiocin was isolated as the first antibiotic with a terpenoid side chain from *Streptomyces niveus*. These compounds are classified sequentially by size as hemiterpenes, monoterpenes, sesquiterpenes, diterpenes, sesterterpenes, triterpenes and tetraterpenes. Some of the notable terpenes of marine origin include:

Neomarinone

This is a novel sesquiterpene metabolite produced by a novel marine actinomycete (strain CNH-099), exhibits moderate cytotoxic activity against human cancer cells (Hardt *et al.*, 2000).

Amorphane Sesquiterpenes

These are produced by *Streptomyces* spp. M491 (Wu *et al.*, 2007).

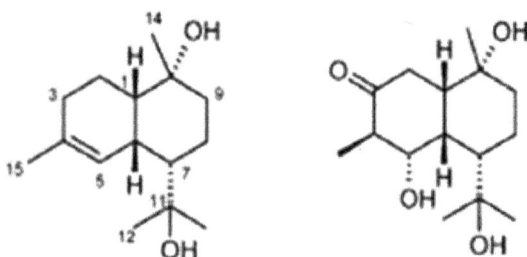

Azamerone

This is a meroterpenoid by a novel marine bacterium belongs to the genus *Streptomyces*, with a phthalazione ring (Cho *et al.*, 2006).

T-Muurolol Sesquiterpenes

These are isolated from *Streptomyces* strain M491 derived from marine sediment Qingdao coast, China. These sesquiterpenes of the T-muurolol family were tested for

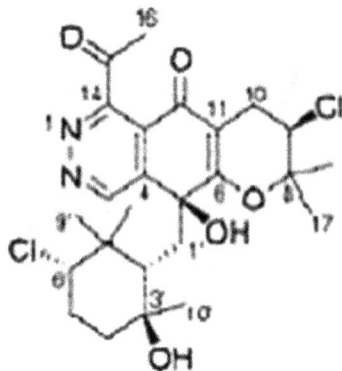

Azamerone

their cytotoxicity against 37 human tumour cell lines, but only 15-hydroxy-T-muurolol showed moderate cytotoxic activity (Ding *et al.*, 2009).

Polyketides

These are the compounds derived from marine actinobacteria that are synthesized by multifunctional, mono or bi- functional enzymes called polyketide synthases. These compounds have a great demand in medicine. The compounds like doramectin, epirubicin, aurantimycin, chartreusin, concanamycin, kirromycin, lysolipin, polyketomycin were used for medical treatments (Weber *et al.*, 2003). Some of the polyketides of industrial importance include:

Abyssomicin C

A novel polycyclic polyketide antibiotic that was obtained from the marine *Verrucosispora* strain (Riedlinger *et al.*, 2004; Bister *et al.*, 2004). It shows strong antibacterial activity even against the vancomycin resistant *Staphylococcus aureus* and Gram positive bacteria. This drug or its analogue has the potential to be developed as an antibiotic against the drug resistant pathogens (Rath *et al.*, 2005).

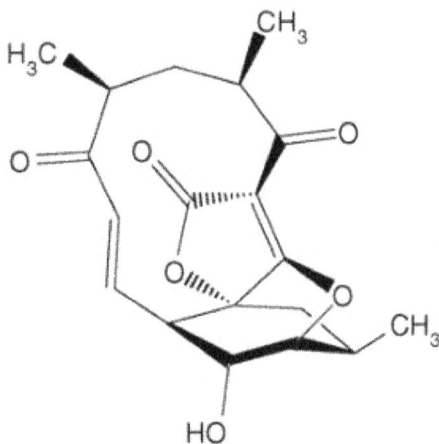

Daryamides

These are cytotoxic polyketides isolated from *Streptomyces* strain CNQ-085. These compounds show moderate cytotoxic activity against human colon carcinoma cell line HCT-116 (Asolkar *et al.*, 2006).

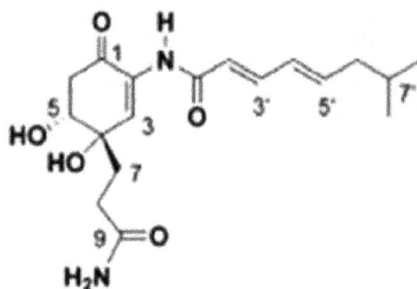

Daryamide A

Actinofuranones A and B

These are isolated from marine actinobacteria belong to genus *Streptomyces*. These bioactive compounds exhibit cytotoxic activity (Cho *et al.*, 2006).

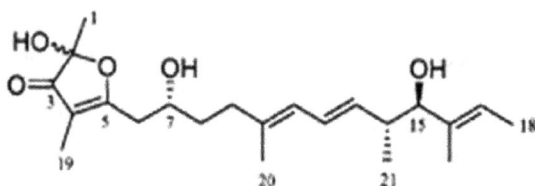

Actinofuranone A

Lactams

Salinisporamide A and Aureoverticillactum are the lactams from marine actinobacteria which are different from β-lactam compounds. These two compounds exhibit cytotoxic activity against cancer cells.

Salinosporamide A

The chemical structure comprises of β-lactam and γ-lactam. It is isolated from the fermentation broth of the novel actinomycetes *Salinispora tropica* (Maldonado *et al.*, 2005; Feling *et al.*, 2003). It is yet under clinical trials and its mode of action is interesting that it induces the apoptosis in multiple myeloma cells whose mechanism is distinct from the known anti-cancer drug Bortezomib (Chauhan *et al.*, 2005).

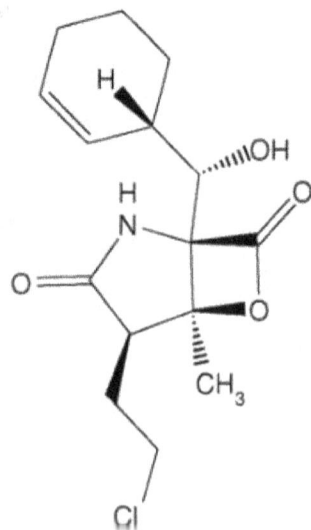

Salinosporamide A

Aureoverticillactum

This is a new 22 atom macrocyclic lactam isolated from *Streptomyces aureoverticillaris*. It exhibits cytotoxic activity against various tumor cell lines.

Tetramic Acid Derivatives

Tirandamycin C and D

A novel dienoyl tetramic acid derivatives that have been isolated from *Streptomyces* spp. 307-9, showed activity against vancomycin resistant *Enterococcus faecalis*. These compounds are similar in structure with already identified compounds such as Tirandamycins A and B with a slight variation in the pattern of oxygenation on the bicyclic ketal system (Carlson *et al.*, 2009).

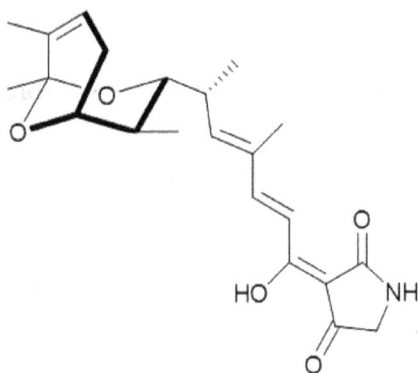

Trioxacarcins

Maskey *et al.* (2008) isolated them from the marine actinomycetes. The derivatives designated as Trioxacarcins D, E and F are analogues to previously isolated Trioxacarcins A, B and C. All the compounds showed antibacterial activity, in addition some of them showed very high anti-tumour and anti-malarial activity (Maskey *et al.*, 2008).

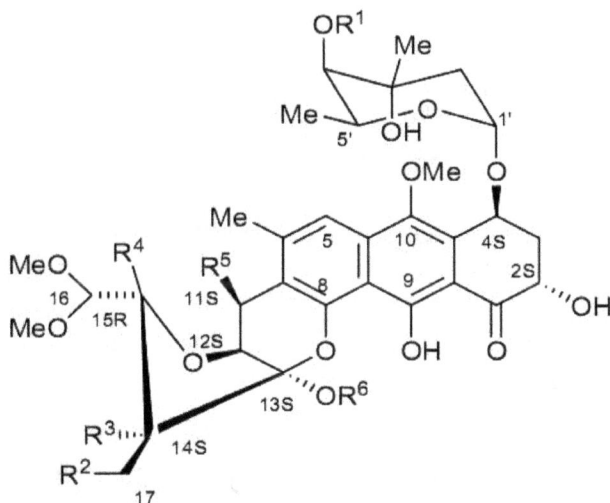

In addition the other important drugs that the marine actinobacteria produce are indole compounds, Piericidins, Methylpyridine, Marinopyrroles, Sisomicin, Triazolopyrimidine, Esters, Caprolactones, Manumycin derivatives, Streptochlorin, Chartreusin, Chinikomycins, Butenolides, Alkaloids, Flavonoids and Polycyclic xanthones.

Enzyme Inhibitors

Enzyme inhibitors are the molecules that bind to enzyme and decrease the activity. The binding of an inhibitor can stop a substrate to enter into the active site of enzymes or hinder the enzyme from catalysing the reaction. The activity of the inhibitor may be reversible or irreversible. Some of the commercially important enzyme inhibitors of marine origin include:

Pyrostatins A and B

These are inhibitors of n-acetyl beta-glucosaminidase produced by *Streptomyces* spp.SA-3501 (Aoyama *et al.,* 1995).

Alpha Amylase Inhibitor

This is isolated from novel *Streptomyces corchorusii* (Imada and Simidu, 1988).

Pyrizinostatin

This is an inhibitor of pyroglutamyl peptidase, isolated from *Streptomyces* spp. SA-2289 (Aoyagi *et al.,* 1992).

Enzymes

Some of the commercially valuable enzymes that are obtained from marine actinomyctes are L-asparaginase, alkaline protease and chitinases.

Alkaline Proteases

These are isolated from marine *S. clavuligerus* strain derived from the west coast of India (Thumar and Singh, 2009). Thermostable alkaline protease was characterised from marine *S. fungicidicus* MML1614 (Ramesh *et al., 2009).*

Chitinases

These are isolated from marine *Streptomyces* spp. DA11 associated with sponge *Craniella australiensis* from South China, which showed antifungal activities against *Aspergillus niger* and *Candida albicans* (Han *et al.,* 2009).

L-Asparaginases

These are isolated from marine *Streptomyces* strains (PDK2, PDK7) of South Indian coast that exhibited cytotoxic effect on JURKAT cells (Acute T cell leukemia) and K562 cells (chronic myelogenous leukemia) (Dhevagi and Poorani, 2006).

Summarized chemical structures of secondary metabolites produced by marine actinobacteria are represented in Table 7.2.

Role of Bioinformatics for Marine Drug Discovery

The bioinformatics created a paradigm shift in our approach to discover novel natural products. The taxonomic databases help the predictive road maps to natural

Table 7.2: Secondary metabolites produced by marine actinobacteria.

Chemical Group	Compound	Source	Activity	References
Peptide	Mechercharmycins	*Thermoactinomyces* spp.	Antitumor	Kanoh *et al.*, 2005
Peptide	Thiocoraline	*Micromonospora* spp.	Anticancer, antibacterial	Romero *et al.*, 1997
Peptide	Cyclomarin A	*Streptomyces* spp.	Antiviral, Anti-inammatory	Renner *et al.*, 1999
Peptide	Piperazimycins	*Streptomyces* spp.	Anticancer	Miller *et al.*, 2007
Peptide	Salinamides A and B	*Streptomyces* spp.	Antibacterial, antiinflammatory	Moore *et al.*, 1999
Quinone	Resistomycin	*Streptomyces corchorusii*	Antiviral	Shiono *et al.*, 2002
Quinone	Resistoflavine	*Streptomyces chibaensis* AUBN(1)/7	Anticancer, antibacterial	Kock *et al.*, 2005; Gorajana *et al.*, 2006
Quinone	Komodoquinone A	*Streptomyces* spp. K53	Neuritogenic activity	Itoh *et al.*, 2003
Quinone	Himalomycins A and B	*Streptomyces* spp. B6921	Antibacterial	Maskey *et al.*, 2003
Quinone	Helquinoline	*Janibacter limosus*	Antibacterial	Asolkar *et al.*, 2004
Macrolide	Macrolide Chalcomycin A	*Streptomyces* spp. M491	None	Wu *et al.*, 2007
Macrolide	Arenicolide A	*Salinispora arenicola*	Antibacterial	Jensen *et al.*, 2007; Williams *et al.*, 2007
Macrolide	Marinomycins	*Marinispora*	Antibacterial, Anticancer	Kwon *et al.*, 2006
Meroterpenoid	Azamerone	*Streptomyces* spp.	None	Cho *et al.*, 2006
Pyrrolosesquiterpenes	Glaciapyrroles A, B and C	*Streptomyces* spp. NPS008187	Antibacterial	Macherla *et al.*, 2005
Amorphane sesquiterpenes	10α, 15-dihydroxyamorph-4-en-3-one, 10α, 11-dihydroxyamorph-4-ene and5α, 10α, 11-trihydroxyamor-phan-3-one	*Streptomyces* spp. M491	None	Wu *et al.*, 2007
Sesquiterpene	Neomarinone	Strain CNH-099	Cytotoxic	Hardt *et al.*, 2000

Contd...

Table 2–*Contd...*

Chemical Group	Compound	Source	Activity	References
Polyketide	Saliniketal A, saliniketal B	*Salinispora arenicola*	Anticancer	William *et al.*, 2007; Jensen *et al.*, 2007
Polyketide	Abyssomicin C	*Verrucosispora*	Antibacterial	Bister *et al.*, 2004
Polyketide	Daryamides	*Streptomyces* spp. CNQ085	Anticancer, antifungal	Asolkar *et al.*, 2006
Polyketide	Actinofuranones A and B	*Streptomyces* spp.	Cytotoxic	Cho *et al.*, 2006
Gamma lactam beta lactone	Salinosporamide A	*Salinispora tropica*	Anticancer	Feling *et al.*, 2003
Macro–cyclic lactam	Aureoverticillactam	*Streptomyces aureoverticillaris*	Anticancer	Mitchell *et al.*, 2004
Enzyme inhibitor	Alpha-amylase inhibitor	*Streptomyces corchorusii* subsp. *rhodomarinus*	Enzyme Inhibition	Imada and Simidu, 1988
Enzyme inhibitor	Pyrostatins A and B	*Streptomyces* spp. SA-3501	N-acetyl-betaglucos aminidase inhibition	Aoyama *et al.*, 1995
Enzyme inhibitor	Pyrizinostatin	*Streptomyces* spp. SA2289	Pyroglutamyl peptidase inhibition	Aoyagi *et al.*, 1992
Enzyme	Alkaline proteases	*S. clavuligerus*	-	Thumar and Singh, 2009
Enzyme	Chitinases	*Streptomyces* sp. DA11	Antifungal	Han *et al.*, 2009
Enzyme	L-Asparaginases	*Streptomyces* strains (PDK2, PDK7)	Anti Cancer	Dhevagi and Poorani, 2006
Dienoyl tetramic acid	Tirandamycin C and D	*Streptomyces* spp.307-9	Vancomycin resistant	Carlson *et al.*, 2009
Alkylating agents	Trioxacarcins D,E, F	*Streptomyces* spp. B8652	Anti-tumor	Maskey *et al.*, 2008
Caprolactone	R-10-methyl-6-undecanolide (6R 10S)10-methyl-6-dodeconolide	*Streptomyces* spp. B6007	Phytotoxic, anticancer	Stritzke *et al.*, 2004
Butenolide	Butenolide	*Streptoverticillium luteoverticillatum*	Anticancer	Li *et al.*, 2006

Contd...

Table 7.2–*Contd...*

Chemical Group	Compound	Source	Activity	References
Caprolactone	R-10-methyl-6-undecanolide (6R 10S)10-methyl-6-dodeconolide	*Streptomyces* spp. B6007	Phytotoxic, anticancer	Stritzke *et al.*, 2004
Polycyclic xanthone	IB 00208	*Actinomadura*	Anticancer, antibacterial	Malet *et al.*, 2003
Piericidin	Piericidins C7 and C8	*Streptomyces*	Anticancer	Hayakawa *et al.*, 2007
Alkaloid	K252c and arcyriaavin A	Z (2)0392	Anticancer	Liu *et al.*, 2007
Ester	Bonactin	*Streptomyces* spp. BD21	Antibacterial, antifungal	Schumacher *et al.*, 2003
Manumycin derivatives	Chinikomycins A and B	*Streptomyces* spp. M045	Anticancer	Li *et al.*, 2005
Complex compounds	Trioxacarcins	*Streptomyces ochraceus* and *Streptomyces bottropensis*	Anticancer, antimalarial	Maskey *et al.*, 2004
Methylpyridine	Streptokordin	*Streptomyces* spp. KORDI-3238	Anticancer	Jeong *et al.*, 2006
Gamma lactam beta lactone	Salinosporamide A	*Salinispora tropica*	Anticancer	Feling *et al.*, 2003

product discovery. The data base also provides the information related to the identification of the species based on the 16S rRNA gene sequence. Analysis of these sequences and calculating the phylogenic distances will provide the information of the unexplored actinobacterial taxonomy (Ward and Bora, 2006).

Metabolite Database

The profiling of the different metabolites from marine actinobacteria in the past was carried out by using High pressure Liquid chromatography with diode array detection (HPLC-DAD) (Fiedler *et al.*, 2005) but for interrogating the microbial metabolism need technologies with greater resolving power. The profiling technologies of choice include Nuclear magnetic resonance (NMR), differential analysis of two dimensional (2D) NMR spectra arrays (Schroeder *et al.*, 2007).

Genomics

These genomic technologies help to eliminate the misconception that marine organisms are exhausted of natural products. Genomics study also help to know the gene cluster that produce the secondary metabolites (Bentley *et al.*, 2002). Automated screening and Robotic platform provides the technology to genome mining. Automated colony picking and isolation coupled to high-throughput genomic dereplication will enable large libraries of taxonomically unique actinomycetes to be identified. Such libraries when combined with metadata (environment descriptions, taxonomy, screening hits, genome scanning) can promote rationally guided bio discovery. The application of microarray technology enables rapid screening of millions of strains for gene of interest (Zhang *et al.*, 2004).

Conclusion

Since the greatest biological diversity exists in marine habitats, it is recognized as the persistant reservoir of untapped chemical entities. Due to the extreme conditions of the dwelling habitats of the marine actinomycetes, they have their own genomic and metabolic maps. Effects should be riddled at exploring the novel actinomycetes as source for novel secondary metabolites. Marine natural products have acknowledged the developments in organic chemistry leading to the synthetic methods to make the analogues of the original lead compounds. Technical drawbacks associated with natural product research have been reduced, and there are better opportunities to explore the biological activity of previously inaccessible sources of marine natural products. With the increasing acceptance that the chemical diversity of marine natural products is well suited to provide the core scaffolds for future drugs, there will be further developments in the use of novel natural products and chemical libraries based on natural products in drug discovery campaigns.

Acknowledgements

One of the authors (MUK) is thankful to CSIR for providing financial assistance.

References

Aadinarayana, G., Venkateshan, M.R., Bapiraju, V.V., Sujatha, P., Premkumar, J., Ellaiah, P and Zeeck, A. 2006. Cytotoxic compounds from the marine actinobacterium. Bio. Org. Chem. **32**: 328–334.

Aoyagi, T., Hastu, M., Imada, C., Naganawa, H., Okami, Y. and Takeuchi, T.1992. Pyrizinostatin: A new inhibitor of pyroglutamyl peptidase. J. Antibiot. **45**: 1795–1796.

Aoyama, T., Kojima, F., Imada, C., Muraoka, Y., Naqanawa, H., Okami, Y., Takeuchi, T. and Aoyaqi, T. 1995. Pyrostatins A and B, new inhibitors of N–acetyl–beta–D–glucosamidase, produced by *Streptomyces* sp. SA3501. J. Enzyme Inhib. **8**: 223–232.

Asolkar, R.N., Jensen, P.R., Kauffman, C.A. and Fenical, W. 2006. Daryamides A–C weakly cytotoxic polyketides from a marine derived actinomycete of the genus *Streptomyces* strain CNQ–085. J. Nat. Prod. **69**: 1756–1759.

Asolkar, R.N., Schroder, D., Heckmann, R., Lang, S., Wagner–Dobler, I. and Laatsch, H. 2004. Helquinoline, a new tetrahydroquinoline antibiotic from *Janibacter limosus* Hel 1. J. Antibiot. **57**: 17–23.

Balunas, M.J. and Kinghorn, A.D. 2005. Drug discovery from medicinal plants. Life Sci. **78**: 431 – 44.

Bavya, M., Mohanapriya, P., Pazhanimurugan, R. and Balagurunathan, R. 2011. Potential bioactive compound form marine actinomyctes against biofouling bacteria. Indian J. Geo–Marine Sci. **40**: 578–582.

Bentley, S.D., Chater, K.F., Cerdeño–Tárraga, K.M., Challis, G.L., Thomson, N.R., James, K.D., Harris, D.E., Quail, M.A., Kieser, H., Harper, D., Bateman, A., Brown, S., Chandra, G., Chen, C.W., Collins, M., Cronin, A., Fraser, A., Goble, A., Hidalgo, J., Hornsby, T., Howarth, S., Huang, C.H., Kieser, T., Larke, L., Murphy, L., Oliver, K., O'Neil, S., Rabbinowitsch, E., Rajandream, m.A., Rutherford, K., Rutter, S., Seeger, K., Saunders, D., Sharp, S., Squares, S., Taylor, K., Warren, T., Wietzorrek, A., Woodward, J., Barrell, B., Parkhill, J. and D.A. Hopwood. 2002. Complete genome sequence of the model actinomycetes *Streptomyces coelicolor*. Nature **417**: 141–147.

Berdy, J. 2005. Bioactive microbial metabolites. J. Antibiot. **58**: 1–26.

Bibb, M.J. 2005. Regulation of secondary metabolism in *Streptomycetes*. Curr. Opin. Microbiol. **8**: 208–215.

Bister, B., Bischoff, D., Strobele, M., Riedlinger, J., Reicke, A., Wolter, F., Bull, A.T., Zahner, H., Fiedler, H.P. and Sussmuth, R.D. 2004. Abyssomicin C–a polycyclic antibiotic from a marine *Verrucosispora* strain as an inhibitor of the p–aminobenzoic acid/tetrahydrofolate biosynthesis pathway. Angew Chem. Int. Ed. Engl. **43**: 2574–2576.

Bister, B., Bischoff, D., Strobele, M., Riedlinger, J., Reicke, A., Wolter, F., Bull, A.T., Zahner, H., Fiedler, H.P. and Sussmuth, R.D. 2004. Abyssomicin C a polycyclic antibiotic from a marine *Verrucosispora* strain as an inhibitor of the p–aminobenzoic acid/tetrahydrofolate biosynthesis pathway. Chem. Int. Ed. **43**: 2574–2576.

Brown, D.W., Yu, J.H., Kelkar, H.S., Fernandes, M., Nesbitt, T.C., Keller, N.P., Adams, T.H. and Leonard, T.J. 1996. Twenty–ve coregulated transcripts dene a sterigmatocystin gene cluster in *Aspergillus nidulans*. Proc. Natl. Acad. Sci., USA **93**: 1418–1422.

Bruntner, C., Binder, T., Pathom–aree, W., Goodfellow, M., Bull, A.T., Potterat, O., Puder, C., Horer, S., Schmid, A., Bolek, W., Klaus, W., Gerhard, M. and Hans–Peter, F. 2005. Frigocyclinone, a novel angucyclinone antibiotic produced by a *Streptomyces griseus* strain from Antarctica. J. Antibiot. 58: 346–349.

Buchanan, G.O., Williams, P.G., Feling, R.H., Kauffman, C.A., Jensen, P.R., Fenical, W. 2005. Sporolides A and B: structurally unprecedented halogenated macrolides from the marine actinomycete *Salinispora tropica*. Org. Lett. **7**: 2731–2734.

Bull, A.T and Stach, J.E.M. 2007. Marine actinobacteria: new opportunities for natural product search and discovery. Trends in Microbiol. **15**: 491–498.

Bull, A.T., Ward, A.C. and Goodfellow, M. 2000. Search and discovery strategies for biotechnology: the paradigm shift. Microbiol. Mol. Biol. Rev. **64**: 573–606.

Caffrey, P., Lynch, S., Flood, E., Finnan, S. and Oliynyk, M. 2001. Amphotericin biosynthesis in *Streptomyces nodosus*: deductions from analysis of polyketide synthase and late genes. Chem Biol. **8**: 71–723.

Charan, R.D., Schlingmann, G., Janso, J., Bernan, V., Feng, X., Carter, G.T. 2004. Diazepinomicin, a new antimicrobial alkaloid from marine *Micromonospora* sp. J. Nat. Prod. **67**: 1431–1433.

Cho, J.Y., Kwon, H.C., Williams, P.G., Kauffman, C.A., Jensen, P.R. and Fenical, W. 2006. Actinofuranones A and B, polyketides from a marine derived bacterium related to the genus *Streptomyces* (Actinomycetales). J. Nat. Prod. **69**: 425–428.

Cho, J.Y., Kwon, H.C., Williamsm, P.G., Jensenm, P.R. and Fenicalm, W. 2006. Azamerone, a terpenoid phthalazinone from a marine derived bacterium related to the genus *Streptomyces* (Actinomycetales). Org. Lett. **8**: 2471–2474.

Cho,J.Y., Williams, P.G., Kwon, H.C., Jensen, P.R. and Fenical, W. 2007. Lucentamycins A–D, cytotoxic peptides from the marine–derived actinomycete *Nocardiopsis lucentensis*. J. Nat. Prod. **70**: 1321–1328.

Cragg, G.M., Kingston, D.G.I. and Newman, D.J. (Eds): 2005. Anticancer Agents from Natural Products. Taylor and Francis.

Cross, T. 1981. Aquatic actinomycetes: a critical survey of the occurrence, growth and role of actinomycetes in aquatic habitats. J. Appl. Bacteriol. **50**: 397–423.

Cross, T. 1989. Growth and examination of Actinomycetes Some Guidelines. In Bergey's Manual of Systematic Bacteriology, **4**: 2340–2343.

Das, S., Lyla, P.S. and Ajmal Khan, S. 2006. Marine microbial diversity and ecology: importance and future perspectives. Curr. Sci. **25**: 1325–1335.

Demain, A.L.1998. Induction of microbial secondary metabolism. Int. Microbiol. **1**: 259–264.

Demain, A.L.1999. Pharmaceutically active secondary metabolites of microoganisms. Appl. Microbiol. Biotechnol. **52**: 455–463.

Dhevagi, P. and Poorani, E. 2006. Isolation and characterization of L–asparaginase from marine actinomycetes. Indian J. Biotech. **5**: 514–520.

Ding, L., Pfoh, R., Ruhl, S., Qin, S. and Laatsch, H. 2009. T–muurolol sesquiterpens from the marine *Streptomyces* sp. M491 and revision of the configuration of previously reported amorphanes. J. Nat. Prod. **72**: 99–101.

Donia, M. and Hamann, M.T. 2003. Marine natural products and their potential applications as anti–infective agents. Lancet Infect Dis. **3**: 338–348.

Engelhardt, K., Degnes, K.F., Kemmler, M., Bredholt, H., Fjærvik, E., Klinkenberg, G. Sletta, H., Ellingsen, T.E. and Zotchev, S.B. 2010. Production of a New Thiopeptide Antibiotic, TP–1161, by a Marine *Nocardiopsis* Species. Applied and Environmental Microbiol. **76**: 4969–4976.

Feling, R.H., Buchanan, G.O., Mincer, T.J., Kauffman, C.A., Jensen, P.R. and Fenical, W. 2003. Salinosporamide A: a highly cytotoxic proteasome inhibitor from a novel microbial source, a marine bacterium of the new genus *Salinospora*. Angew Chem. Int. Ed. Engl. **42**: 355–357.

Fenical, W. and Jensen, P.R. 2006. Developing a new resource for drug discovery: marine actinomycete bacteria. Nat. Chem. Biol. **2**: 666–673.

Fenical, W., Baden, D., Burg, M., Goyet, C.V., Grimes, J.D., Katz, M., Marcus, N.H., Pomponi, S., Rhines, P., Tester, P. and Vena, J. 1999. Marine–derived pharmaceuticals and related bioactive compounds. In: From monsoons to microbes: understanding the ocean's role in human health. National Academies Press, Washington, pp 71–86.

Fiedler, P.H., Bruntner, C., Bull, A.T., Ward, A.C., Goodfellow, M., Potterat,O., Puder, C. and Mihm, G. 2005. Marine actinomycetes as a source of novel secondary metabolites. Antonie Van Leeuwenhoek. **87**: 37–42.

Gallo, M.L., Seldes, A.M. and Cabrera, G.M. 2004. Antibiotic long–chain α–unsaturated aldehydes from the culture of the marine fungus *Cladosporium* sp. Biochem. Systemat. Ecol. 32: 554–551.

Gao, H., Liu, M., Liu, J., Dai, H., Zhou, X., Liu, X., Zhuo, Y., Zhang, W. and Zhang, L. 2009. Medium optimization for the production of avermectin B1a by *Streptomyces avermitilis* 14–12A using response surface methodology. Bioresour. Technol. **100**: 4012–4016.

Gogoi, D.K., Boruah, H.P.D., Saikia, R. and Bora, T.C. 2008. Optimization of process parameters for improved production of bioactive metabolite by a novel endophytic fungus *Fusarium* sp. DF2 isolated from *Taxus wallichiana* of north east India. World J. Microbiol. Biotechnol. **1**: 79–87.

Gontang, EA., Fenical, W. and Jensen, P.R. 2007. Phylogenetic diversity of Gram–positive bacteria cultured from marine sediments. Appl. Environ. Microb. **73**: 3272–3282.

Goodfellow, M. and Haynes, J.A. 1984. Actinomycetes in marine sediments. In Biological, Biochemical, and Biomedical Aspects of Actinomycetes. Edited by Ortiz–Ortiz L, Bojalil, L.F., Yakoleff, V. New York: Academic Press. pp 453–472.

Gorajana, A.M.V., Vinjamuri, S., Kurada, B.V., Peela, S., Jangam, P., Poluri, E. and Zeeck, A. 2006. Resistoavine cytotoxic compound from a marine actinomycete, *Streptomyces chibaensis* AUBN(1)/7.Microbiol Res. 29.

Gorajana, A.M.V., Vinjamuri, S., Kurada, B.V., Peela, S., Jangam, P., Poluri, E. and Zeeck, A. 2007. Resistoflavine cytotoxic compound from a marine actinomycete, *Streptomyces chibaensis* AUBN(1)/7. J. Antibiot. **162**: 526–529.

Han, Y., Yang, B., Zhang, F., Miao, X. and Li, Z. 2009. Characterization of antifungal chitinase from marine *Streptomyces* sp. DA11 associated with South China Sea Sponge *Craniella australiensis*. Mar. Biotechnol. **11**: 132–140.

Hanada, M., Kaneta, K., Nishiyama, Y., Hoshino, Y., Konishi, M. and Oki, T. 1991. Hydramycin: a new antitumor antibiotic. Taxonomy, isolation, physico–chemical properties, structure and biological activity. J. Antibiot. **44**: 824–831.

Hardt, I.H., Jensen, P.R. and William, F. 2000. Neomarinone and new cytotoxic marinone derivatives, produced by a marine lamentous bacterium (Actinomycetales). Science. **41**: 2073–2076.

Hayakawa, M. 2003. Selective isolation of rare actinomycete genera using pretreatment techniques. In *Selective Isolation of Rare Actinomycetes*; Kurtböke, D.I., Ed.; Queensland Complete Printing Services: Nambour, Australia., pp. 56–81.

Hayakawa, Y., Shirasaki, S., Shiba, S., Kawasaki, T., Matsuo, Y., Adachi, K. and Shizuri, Y. 2007. Piericidins C7 and C8, new cytotoxic antibiotics produced by a marine *Streptomyces* sp. J Antibiot., 60: 196–200.

Hohmann, C., Schneider, K., Bruntner, C., Irran, E., Nicholson, G., Bull, A.T., Jones, A.L., Brown, R., Stach, J.E., Goodfellow, M., Beil, W., Krämer, M., Imhoff, J.F., Süssmuth, R.D. and Fiedler, H.P. 2009. Caboxamycin, a new antibiotic of the benzoxazole family produced by the deep–sea strain *Streptomyces* sp. NTK 937. J. Antibiot. **62**: 99–104.

Hopwood, D. A. 1999. Forty years of genetics with *Streptomyces*: from *in vivo* through *in vitro* to *in silico*. Microbiology **145**: 2183–2202.

Ikeda, H., Nonomiya, T., Usami, M., Ohta, T. and Omura, S. 1999. Organization of the biosynthetic gene cluster for the polyketide anthelmintic macrolide avermectin in *Streptomyces avermitilis*. Biochem. **96**: 9509–9514.

Imada, C. and Simidu, U. 1988. Isolation and characterization of an alpha amylase inhibitor producing actinomycete from marine environment. Nippon Suisan Gakkaishi.,54: 1839–1845.

Imada, C., Koseki, N., Kamata, M., Kobayashi, T. and Hamada–Sato, N. 2007. Isolation and characterization of antimicrobial substances produced by marine actinomycetes in the presence of seawater. Actinomycetologica 21: 27–31.

Itoh, T., Kinoshita, M., Aoki, S. and Kobayashi, M. 2003. Komodoquinone A, a novel neuritogenic anthracycline from marine *Streptomyces* sp. KS3. J. Nat. Prod. **66**: 1373 1377.

Jain, P. and Pundir, R.K. 2011. Effect of fermentation medium, pH and temperature variations on antibacterial soil fungal metabolite production. J. Agricultural Technol. **7**: 247–269.

Jensen, P.R., Williams, P.G., Oh, D.C., Zeigler, L. and Fenical, W. 2007. Species–specific secondary metabolite production in marine actinomycetes of the genus *Salinispora*. Appl. Environ. Microbiol.**73**: 1146–1152.

Jeong, S.Y., Shin, H.J., Kim, T.S., Lee, H.S., Park, S.K. and Kim, H.M. 2006. Streptokordin a new cytotoxic compound of the methylpyridine class from a marine derived *Streptomyces* sp. KORDI–3238. J. Antibiot. **59**: 234–240.

Johnson, LE. and Dietz, A. 1968. Kalafungin, a new antibiotic produced by *Streptomyces tanashiensis* strain kala. Appl. Microbiol. **16**: 1815–1821.

Jones, W.P., Chin, Y–W. and Kinghorn, A.D. 2006. The role of pharmacognosy in modern medicine and pharmacy. Curr. Drug Targets **7**: 247 – 264.

Kaltenpoth, M. 2009. Actinobacteria as mutualists: general healthcare for insects? Trends Microbiol. **17**: 529–535.

Kanoh, K., Matsuo, Y., Adachi, K., Imagawa, H., Nishizawa, M. and Shizuri, Y. 2005. Mechercharmycins A and B cytotoxic substances from marine derived *Thermoactinomyces* sp. YM 3–251. J. Antibiot. **58**: 289–292.

Katherine, A. McArthur, Scott S. Mitchell., Ginger Tsueng, Arnold Rheingold, Donald J. White, Jennifer Grodberg, Kin S. Lam and Barbara C. M. Potts. 2008. Lynamicins A–E, Chlorinated Bisindole Pyrrole Antibiotics from a Novel Marine Actinomycete. J. Nat. Prod. **71**: 1732–1737.

Kavitha, A., Prabhakar, P., Narasimhulu, M., Vijayalakshmi, M., Venkateswarlu, Y., Venkateswara Rao, K. and Raju, V.B.S. 2010. Isolation, characterization and biological evaluation of bioactive metabolites from *Nocardia levis* MK–VL_113. Microbiol. Res., Elsevier **165 (3)**: 199–210.

Kavitha, A., Prabhakar, P., Vijayalakshmi, M., Venkateswarlu, Y., 2009. Production of Bioactive metabolites by *Nocardia levis* MK–VL_113. Letters in Appl.Microbiol. **49 (4)**: 484–490.

Kavitha, A., Prabhakar, P., Vijayalakshmi, M., Venkateswarlu, Y., 2010. Purification and biological evaluation of the metabolites produced by *Streptomyces* sp. TK_VL 333. Res. Microbiol., Elsevier **161**: 335–345.

Kin, S. Lam. 2006. Discovery of novel metabolites from marine actinomycetes. Current Opinion in Microbiol. **9**: 245–25.

Kock, I., Maskey, R.P., Biabani, M.A.F., Helmke, E. and Laatsch, H. 2005. 1–hydroxy–1–norresistomycin and resistoavine methyl ether new antibiotics from marine derived Streptomycetes. J. Antibiot. **58**: 530–534.

Koehn, F.E. and Carter, G.T. 2005. The evolving role of natural products in drug discovery. Nat. Rev. Drug Discov. **4**: 206 – 220.

Krishna, P.S.M., Venkateshwarlu, G. and Rao, L.Y. 1998. Studies on fermentative production of rifamycin using *Amycolatopsis mediterranei*. J. Microbiol. Biotechnol. **14**: 689–691.

Kwon, H.C., Kauffman, C.A., Jensen, P.R. and Fenical, W. 2006. Marinomycins a–d, antitumor antibiotics of a new structure class from a marine actinomycete of the recently discovered genus '*Marinispora*'. J. Am. Chem. Soc. **128**: 1622–1632.

Laurence Catley., Guilan Li., Klaus Podar., Teru Hideshima., Mugdha Velankar., Constantine Mitsiades., Nicolas Mitsiades., Hiroshi Yasui., Anthony Letai., Huib Ovaa., Celia Berkers., Benjamin Nicholson., Ta–Hsiang Chao., Saskia T.C. Neuteboom., Paul Richardson., Michael A. Palladino., Kenneth C. Anderson. 2005. A novel orally active proteasome inhibitor induces apoptosis in multiple myeloma cells with mechanisms distinct from Bortezomib. Cancer Cell **8**: 407–419.

Lazzarini, A., Caveletti, L., Toppo, G. and Marinelli, F. 2000. Rare genera of actinomycetes as potential of new antibiotics. Antonie van Leuwenhoek **78**: 399–405.

Lee, H.S., Shin, H.J., Jang, K.H., Kim, T.S., Oh, K.B. and Shin, J. 2005. Cyclic peptides of the Nocardamine class from a marine derived bacterium of the genus *Streptomyces*. J. Nat. Prod. **68**: 623–625.

Li, F., Maskey, R.P., Qin, S., Sattler, I., Fiebig, H.H., Maier, A., Zeeck, A., Laatsch, H. 2005. Chinikomycins A and B: isolation, structure elucidation, and biological activity of novel antibiotics from a marine *Streptomyces* sp. Isolate M045. J Nat Prod. **68**: 349–353.

Li, D.H., Zhu, T.J., Liu, H.B., Fanq, Y.C., Gu, O.O. and Zhu, W.M. 2006. Four butenolides are novel cytotoxic compounds isolated from the marine derived bacterium, *Streptoverticillium luteoverticillatum* 11014. Arch. Pharm. Res. **29**: 624–626.

Liu, X., Ashforth, E., Ren, B., Song, F., Dai, H., Liu, M., Wang, J., Xie, Q. and Zhang, L. 2010. Bioprospecting microbial natural product libraries from the marine environment for drug discovery. The Journal of Antibiotics **63**: 415–422.

Luzhetskyy, A., Pelzer, S. and Bechthold, A. 2007. The future of natural products as a source of new antibiotics. Current Opinion in Investigational Drugs **8**: 608–613.

Macherla, V.R., Liu, J., Bellows, C., Teisan, S., Nicholson, B., Lam, K.S. and Potts, B.C.M. 2005. Glaciapyrroles A, B and C pyrrolosesquiterpenes from a *Streptomyces* sp. isolated from an Alaskan marine sediment. J. Nat. Prod. **68**: 780–783.

Maldonado, L.A., Fenical, W., Jensen, P. R., Kauffman, C.A., Mincer, T.J., Ward, A.C., Bull, A.T. and Goodfellow, M. 2005. *Salinispora arenicola* gen. nov., sp. nov. and *Salinispora tropica* sp. nov., obligate marine actinomycetes belonging to the family *Micromonosporaceae*. Int. J. Syst. Evol. Microbial. **55**: 1759–1766.

Malet Cascon, L., Romero, F., Espliego Vazquez, F., Gravalos, D. and Fernandez Puentes, JL. 2003. IB00208, a new cytotoxic polycyclic xanthone produced by a marine derived *Actinomadura*. Isolation of the strain, taxonomy and biological activities. J Antibiot. **56**: 219–225.

Manam, R.R., Teisan, S., White, D.J., Nicholson, B., Grodberg, J., Neutchoom, S.T.C. Lam, K.S., Mosca, D.A., Lloyd, G.K. and Potts, B.C.M. 2005. Lajollamycin, a nitro–

tetraene spiro–b–lactone–g–lactam antibiotic from the marine actinomycete *Streptomyces nodosus*. J. Nat. Prod. **68**: 240–243.

Mann, J. 2001. Natural products as immunosuppressive agents. Nat. Prod. Rep. **18**: 417–430.

Mao, Y., Varoglu, M. and Sherman, D.H. 1999. Molecular characterization and analysis of the biosynthetic gene cluster for the antitumor antibiotic Mitomycin C from *Streptomyces lavendulae* NRRL 2564. Chem. Biol. **6**: 251–263.

Marti´nez–Bueno, M., Ga´lvez, A., Valdivia, E. and Maqueda, M. A. 1990. A transferable plasmid associated with AS–48 production in *Enterococcus faecalis*. J. Bacteriol., **172**: 2817–2818.

Maskey, R.P., Helmke, E. and Laatsch, H. 2003. Himalomycin A and B isolation and structure elucidation of new fridamycin type antibiotics from a marine *Streptomyces* isolate. J.Antibiot. **56**: 942–949.

Maskey, R.P., Li, F.C., Quin, S., Feibig, H.H.H. and Laatsch, H. 2003. Chandrananimycins A approximately C: production of a novel anticancer antibiotics from marine actinomadura sp. Isolate M048 by variation of medium composition and growth conditions. J. Antibiot. **56**: 622–629.

Maskey, RP., Helmke, E. and Laatsch, H. 2003. Himalomycin A and B: isolation and structure elucidation of new fridamycin type antibiotics from a marine *Streptomyces* isolate. J. Antibiot. **56**: 942–949.

Maskey, R.P., Li, F.C., Qin, S., Fiebig, H.H., Laatsch, H. 2003. Chandrananimycins A! C: production of novel anticancer antibiotics from a marine *Actinomadura* sp. isolate M048 by variation of medium composition and growth conditions. J. Antibiot., **56**: 622–629.

Miller, E.D., Kauffman, C.A., Jensen, P.R. and Fenical, W. 2007. Piperazimycins cytotoxic hexadepsipeptides from a marine derived bacterium of the genus *Streptomyces*. J. Org. Chem. **72**: 323–330.

Miller, E.D., Kauffman, C.A., Jensen, P.R. and Fenical, W. 2007. Piperazimycins cytotoxic hexadepsipeptides from a marine derived bacterium of the genus *Streptomyces*. J Org Chem. **72**: 323–330.

Mitchell, S.S., Nicholson, B., Teisan, S., Lam, K.S. and Potts, B.C. 2004. Aureoverticillactam, a novel 22–atom macrocyclic lactam from the marine actinomycete *Streptomyces aureoverticillatus*. J. Nat. Prod. **67**: 1400–1402.

Moore, B.S., Trischman, J.A., Seng, D., Kho, D., Jensen, P.R. and Fenical, W. 1999. Salinamides, anti-inflammatory depsipeptides from a marine *Streptomycete*. J. Org. Chem. **64**: 1145–1150.

Narayana, K.J.P., Prabhakar, P., Vijayalakshmi, M., Venkateswarlu, Y. and Krishna, P.S.J. 2008a. Study on bioactive compounds from *Streptomyces* sp. ANU 6277. Polish. J. Microbiol., **57**: 35–39.

Narayana, K.J.P., Prabhakar, P., Vijayalakshmi, M., Venkateswarlu, Y. and Krishna, P.S.J. 2008b. Biological activity of phenyl propionic acid from a terrestrial *Streptomyces*. Polish. J. Microbiol.**56**: 191–197.

Narayana, K.J.P., Ravikiran, D., Vijayalakshmi, M. 2004. Production of antibiotics from *Streptomyces* sp. isolated from virgin soils. Indian J. Microbiol. **44**: 147–148.

Nonomura, H. 1974. Key for classification and identification of 458 species of the *Streptomycetes* included in ISP. J. Ferment. Technol. **52**: 78–92.

Paradkar, A., Trefzer, A., Chakraburty, R. and Stassi, D. 2003. *Streptomyces* genetics: a genomic perspective. Crit. Rev. Biotechnol., 23: 1–27.

Parenti, F., Beretta, G., Berti, M. and Arioli, V. 1978. Teichomycins, new antibiotics from *Actinoplanes techomyceticus* nov. sp. I. Description of the producer strain, fermentation studies and biological properties. J. Antibiot. (Tokyo) **31**: 276–283.

Parenti, F., Pagani, H. and Beretta, G. 1975. Lipiarmycin, a new antibiotic from *Actinoplanes*. I. Description of the producer strain and fermentation studies. J. Antibot. **4**: 247–252.

Pecznska–Czoch, W. and Mordarski, M. 1988. Actinomycete enzymes. In: Goodfellow M, Williams S.T, Mordarski M. (eds) Actinomycetes in Biotechnology. London: Academic Press. pp 219–283.

Piepersberg, W. and Distler, J. 1997. Aminoglycosides and sugar components in other secondary metabolites in products of secondary metabolism. (eds Rehm, H.J., Reed, G.) 397–488. VCH–Verlagsgesells–chaft, Weinheim.

Prauser, H. 1964. Aptness and application of color for exact description of colors of *Streptomyces*. Z. Allg. Mikrobiol. **4**: 95–98.

Pridham, TG. and Gottlieb, D. 1948. The utilization of carbon compounds by some actinomycetes as an aid for species determination. J. Bacteriol. **56**: 107–114.

Ramesh, S., Rajesh, M. and Mathivanan, N. 2009. Characterization of a thermostable alkaline protease produced by marine *Streptomyces fungicidicus* MML1614. Bioprocess Biosyst Eng. **32**: 791–800.

Ravi Kumar, V. R. D., Murali Krishna Kumar, M., Murali Yugandhar, N., Sri Rami Reddy, D. 2011. Novel Pyridinium compound from marine actinomycete, *Amycolatopsis alba* var. nov. DVR D4 showing antimicrobial and cytotoxic activities *in vitro*. Microbiological Research. *http*: //dx.doi.org/10.1016/j.micres.2011.12.003.

Renner, M.K., Shen, Y.C., Cheng, X.C., Jensenm P.R, Frankmoelle, W., Kauffman, C.A., Fenical, W., Lobkovsky, E. and Cladry, J. 1999. Cyclomarins A–C, new anti-inammatory cyclic peptides produced by a marine bacterium (*Streptomyces* sp.). J. Am. Chem. Soc. **121**: 11273–11276.

Renner, M.K., Shen, Y.C., Cheng, X.C., Jensen, P.R., Frankmoelle, W., Kauffman, C.A., Fenical, W., Lobkovsky, E. and Cladry, J. 1999. Cyclomarins A–C, new anti inammatory cyclic peptides produced by a marine bacterium (*Streptomyces* sp.). J. Am. Chem. Soc., 121: 11273–11276.

Riedlinger, J., Reicke, A., Zahner, H., Krismer, B., Bull, A.T., Maldonado, L. A , Ward, A.C., Goodfellow, M., Bister, B., Bischoff, D., Sussmuth, R.D. and Fiedler, H.P.

2004. Abyssomicins, inhibitors of the para–aminobenzoic acid pathway produced by the marine *Verrucosispora* strain AB–18032. J. Antibiot. **57**: 271–279.

Rofiq, S., Bambang, M., Tun, T.I., Zainal, A. M. and Liesbetini, H. 2010. Antibiotic compound from marine actinomycetes (*Streptomyces* sp a11): Isolation and structure elucidation. Indo. J. Chem. **10**: 226 – 232.

Romero, F., Espliego, F., Perez, B.J., Garcia de, Q.T, Gravalos, D., de la Calle, F. and Fernandez, P.J.L. 1997. Thiocoraline a new depsipeptide with antitumor activity produced by a marine *Micromonospora*. Taxonomy, fermentation, isolation and biological activities. J. Antibiot. **50**: 734–737.

Ross, A. and Schugerl, K. 2005. Tetracycline production by *Streptomyces aureofaciens*: the time lag of production. Appl. Microbiol. Biotechnol. **29**: 174–180.

Saha, M.R., Ripa, F.A., Islam, M.Z. and Khondkar, P. 2010. Optimization of conditions and *in vitro* antimicrobial activity of secondary metabolite isolated from *Streptomyces* sp. MNK7. J. Appl. Sci. Res. **6**: 453–459.

Sanchez, S. and Demain, A. L. 2002. Regulation of fermentation processes. Enzyme Microb. Technol. **31**: 895–906.

Sanchez, S. and Olson, B. 2005. The bright and promising future of microbial manufacturing. Curr. Opin. Microbiol. **8**: 229–233.

Sánchez, S., Chavez, A., Forero, A., arcýa–Huante, Y., Romero, A., Sánchez, M., Rocha, D., Sánchez, B., Avalos, M., Guzman–Trampe, S., Rodrýguez–Sanoja, R., Langley, E. and Ruiz, B. 2010. Carbon source regulation of antibiotic production. J. Antibiot. **63**: 442–459.

Schroeder,F.C., Gibson, DM., Churchill, A.C.L., Sojikul, P., Wursthorn, EJ., Krasnoff, SB., and Clard, J. 2007. Differential analysis of 2D NMR spectra: new natural products from a pilot–scale fungal extract library. Angew. Chem. Int. Ed. **46**: 901–904.

Schumacher, R.W., Talmage, S.C., Miller, S.A., Sarris, K.E., Davidson, B.S., Goldberg, A. 2003. Isolation and structure determination of an antimicrobial ester from a marine–derived bacterium. J Nat Prod. **66**: 1291–1293.

Schumacher, R.W., Talmage, S.C., Miller, S.A., Sarris, K.E., Davidson, B.S. and Goldberg, A. 2003. Isolation and structure determination of an antimicrobial ester from a marine sediment derived bacterium. J Nat Prod. **66**: 1291–1293.

Shiono, Y., Shiono, N., Seo, S., Oka, S. and Yamazaki, Y. 2002. Effects of polyphenolic anthrone derivatives resistomycin and hypericin on apoptois in human megakaryoblastic leukemia CMK–7cell2. Natuforsch. **57**: 923–929.

Shiono, Y., Shiono, N., Seo, S., Oka, S. and Yamazaki, Y. 2002. Effects of polyphenolic anthrone derivatives resistomycin and hypericin on apoptois in human megakaryoblastic leukemia CMK–7cell2. Natuforsch. **57**: 923–929.

Shirling, E.B. and Gottlieb, D. 1966. Methods for characterization of *Streptomyces* species. Int. J. Syst. Bacteriol. **16**: 313–340.

Solanki, R., Khanna, M. and Lal, R. 2008. Bioactive compounds from marine actinomycetes. Indian J. Microbiol. **48**: 410–431.

Stach, J. E. M., Maldonado, L. A., Ward, A.C., Goodfellow, M. and Bull, A. T. 2003. New primers for the class Actinobacteria: application to marine and terrestrial environments. Environ Microbiol. **5**: 828–84.

Stach, J.E.M., Maldonado, L.A., Ward, A.C., Bull, A.T. and Goodfellow, M. 2004. *Williamsia maris* sp. nov., a novel actinomycete isolated from the Sea of Japan. Int. J. Syst. Evol. Microbiol. **54**: 191–194.

Stackebrandt, E., Rainey, F.A. and Ward–Raine, N.L. 1997. Proposal for a new hierarchic classication system, Actinobacteria classis nov. Int. J. Syst. Bacteriol. **47**: 479–491.

Stritzke, K., Schulz, S., Laatsch, H., Helmke, E. and Beil, W. 2004. Novel caprolactones from a marine *Streptomycete*. J. Nat. Prod. **67**: 395–401.

Sujatha, P., Bapi Raju, KV. and Ramana, T. 2005. Studies on a new marine *Streptomycete* BT 408 producing polyketide antibiotic SBR–22 effective against methicillin resistant *Staphylococcus aureus*. Microbiol. Res. **160**: 119–126.

Surajit, D.A.S., Lyla, P.S. and Ajmal Khan, S. 2008. Distribution and generic composition of culturable marine actinomycetes from the sediments of Indian continental slope of Bay of Bengal. Chinese Journal of Oceanology and Limnology **26**: 166–177.

Talbot, G. H., Bradley, J., Edwards, J.E. Jr., Gilbert, D., Scheld, M. and Bartlett, J. G. 2006. Bad Bugs Need Drugs: An update on the development pipeline from the antimicrobial availability task force of the infectious diseases society of America. Clin. Infect. Dis., **42**: 657–668.

Thomashow, L.S., Bonsall, R.F. and David, M. 2008. Detection of antibiotics produced by soil and rhizosphere microbes *in situ*. In: Karlovsky, P. (Ed.). Secondary Metabolites in Soil Ecology., Springer. Berlin Heidelberg.

Thumar, J.T. and Singh, S.P. 2007. Secretion of an alkaline protease from a salt tolerant and alkaliphilic, *Streptomyces clavuligerus* strain MIT–1. Brazil. J. Microbiol. **38**: 766–772.

Thumar, J.T., Dhulia, K. and Singh, S.P. 2010. Isolation and partial purification of an antimicrobial agent from halotolerant alkaliphilic *Streptomyces aburaviensis* strain Kut–8. World J. Microbiol. Biotechnol. **26**: 2081–2087.

Valan Arasu, M., Duraipandiyan, V., Agastian, P. and Ignacimuthu, S. 2008. Antimicrobial activity of *Streptomyces* sp. ERI–26 recovered from Western Ghats of Tamil Nadu. J. Medical Mycol. **18**: 147–153.

Veiga, M., Esparis, A. and Fabregas, J. 1983. Isolation of cellulolytic actinomycetes from marine sediments. Appl. Environ. Microbiol. **46**: 286–287.

Venkata R.R.K.D., Murali, Y.N. and Sri Rami Reddy, D. 2011. Screening of Antagonistic Marine Actinomycetes: Optimization of Process Parameters for the Production of Novel Antibiotic by *Amycolatopsis alba* var. nov. DVR D4. J. Microbial. Biochem. Technol. **3**: 092–098.

Waksman, S.A., 1961. The Actinomycetes: Classification, identification and descriptions of genera and species. The Williams and Wilkins Company, Baltimore, **2**: 61–292.

Ward, A.C. and Bora, N. 2006. Diversity and biogeography of marine actinobacteria. Curr. Opin. Microbiol. **9**: 279–286.

Warnick–Pickle, D.J., Byrne, K.M., Pandey, R.C. and White, R.J. 1981. Fredericamycin A, a new antitumor antibiotic. II. Biological properties. J. Antibiot. **34**: 1402–1407.

Weber, T., Welzel, K., Pelzer, S., Vente, A. and Wohlleben, W. 2003. Exploiting the genetic potential of polyketide producing *Streptomycetes*. J. Biotechnol. **106**: 221–232.

Williams, P.G., Miller, E.D., Asolkar, R.N., Jensen, P.R. and Fenical, W. 2007. Arenicolides A–C, 26 membered ring macrolides from the marine actinomycete *Salinispora arenicola*. J. Org. Chem. **72**: 5025–5034.

Williams, P.G., Asolkar, R.N., Kondratyuk, T., Pezzuto, J.M., Jensen, P.R. and Fenical, W. 2007. Saliniketals A and B, bicyclic polyketides from the marine actinomycete *Salinispora arenicola*. J. Nat. Prod. **70**: 83–88.

Wu, S.J., Fotso, S., Li, F., Qin, S. and Laatsch, H. 2007. Amorphane sesquiterpenes from a marine *Streptomyces* sp. J Nat Prod. **70**: 304–306.

Zhanq, Q., Gould, SJ. and Zabriskie, T.M. 1998. A new cytosine glycoside from *Streptomyces griseochromogenes* produced by the use of *in vivo* of enzyme inhibitors. J. Nat. Prod. **61**: 648–651.

2013, Abiotic Stress and Biotechnology   Pages 169–189
Editor: T. Pullaiah
Published by: REGENCY PUBLICATIONS, NEW DELHI

Chapter 8

Production of Secondary Plant Products from Callus and Suspension Cultures

T. Sudhakar Johnson[1], D. Madhavi[2], M.S.L. Sunita[2]
P. Sita Kumari[3] and P.B. Kavi Kishor[2]

[1]Plant Metabolic Engineering, Reliance Life Sciences Pvt. Ltd.,
Thane-Belapur Road, Rabale, Navi Mumbai – 400 701, Mh., India
[2]Department of Genetics, Osmania University,
Hyderabad – 500 007, Andhra Pradesh, India
[3]Department of Chemistry, K.T.R. Women's College,
Gudivada – 521 301, Andhra Pradesh, India

Introduction

Over the ages, humans have relied on nature for their basic needs like food, shelter, clothing, means of transportation, fertilizers, flavours, fragrances and medicine. Plants have formed the basis of sophisticated traditional medicine system that has been in existence thousands of years in countries such as China, Japan and India. Due to their large biological activities, plant secondary metabolites have been used for centuries in traditional local medicine. They include valuable compounds such as pharmaceutics, cosmetics, fine chemicals, or more recently neutraceuticals. The ability to synthesize secondary metabolites has been selected through the course of evolution in different plant lineage depending upon the specific needs. For example, floral scent volatiles and pigments have evolved to attract insect pollinators and thus enhance fertilization; accumulation of toxic chemicals has evolved to ward off pathogens and herbivores or to suppress the growth of neighbouring plants. Secondary products often determine the quality of food (color, taste and aroma), flower color, smell and appearance. Secondary products are regarded for their crucial role in the survival of the plant in its ecosystem, often protecting plants against pathogen attack, insect bite, mechanical injury, biotic and abiotic stresses. Plant secondary compounds are usually classified according to their biosynthetic pathway (Harborne, 1999), three large molecule families are generally considered: phenolics, terpenes and steroids and alkaloids. Till date, about 100,000 plant secondary metabolites are already known. But, only a small percentage of all plant species have been studied to some extent for the presence of secondary metabolites. Based on the NAPRALERT database, it is estimated that about 15 per cent of the ca. 250,000 known plant species have been subjected for some sort of phytochemical analysis, but less than 5 per cent of them have been studied to the presence of biological activities (Verpoorte, 2000). About 25 per cent of all prescriptions sold in the USA and probably in Western Europe are from natural products, while another 25 per cent are structural modifications of the natural products. About one third of the ~980 new pharmaceuticals in the past two and half decades originated from or were inspired by natural products. Nearly, 119 drugs are still obtained commercially from higher plants and 74 per cent of them are found from ethanobotanical information. In the near future, plants will continue to provide novel products as well as chemical models for new drugs, because the chemistry of the majority of the plant products is complex or yet to be characterized. Despite advancement in synthetic chemistry, we shall depend largely on plant sources for a number of secondary metabolites including pharmaceuticals.

What are Secondary Plant Products

Enzyme mediated chemical reactions occur in plants and is known as metabolism. If these reactions are arranged together, it will form into a metabolic pathway that synthesizes many products like carbohydrates, amino acids, fatty acids, nucleotides and the polymers derived from them. This is usually termed as primary metabolism, and the compounds that are derived from such a pathway are described as primary metabolites. Over and above, plants also produce compounds that do not have a specific function. Such natural chemical products of plants, not normally involved in primary metabolic processes such as photosynthesis and cell respiration (metabolisms not directly related to maintaining life) are called as secondary plant products. The products formed by secondary metabolism are called as secondary metabolites. Elaborate biosynthetic pathways exist for the production of secondary plant products, but they are generally synthesized from basic primary metabolites such as sugars and amino acids. Secondary metabolites are not necessarily produced under all conditions. The biosynthetic pathways are often restricted to an individual species or genus and might only be activated during particular stages of growth and development or during periods of stress caused by attack by microorganisms (bacteria, yeast, fungi etc.) or limitation of nutrients. Many of the secondary plant products are unique to the plant kingdom and are not produced by microbes or animals. With the advancement in basic research, it is possible to produce compounds which are not originally found in plants (Verpoorte 2007). The broad bioactivity and increased usage by humans of late, resulted in an increase in the research activities of secondary

Table 8.1: Secondary metabolites isolated from plant cell and tissue cultures.

Name of Plant Species	Secondary Metabolite	Culture Type
Agave amaniensis	Saponins	Callus
Allium sativum L.	Alliin	Callus
Aloe saponaria	Tetrahydroanthracene glucosides	Suspension
Ambrosia tenuifolia	Altamisine	Callus
Brucea javanica (L.) Merr.	Canthinone alkaloids	Suspension
Bupleurum falcatum	Saikosaponins	Callus
Camellia sinensis	Theamine, γ-glutamyl derivatives	Suspension
Canavalia ensiformis	L-Canavanine	Callus
Capsicum annuum L.	Capsaicin	Suspension
Cassia acutifolia	Anthraquinones	Suspension
Catharanthus roseus	Indole alkaloids	Suspension
Cephaelis ipecacuanha A. Richard	Emetic alkaloids	Root
Chrysanthemum cinerariaefolium	Pyrethrins	Callus
Cinchona	Alkaloids	Suspension
Cinchona robusta	Robustaquinones	Suspension
Cinchona sp.	Anthraquinones	Suspension
Cinchona succirubra	Anthraquinones	Suspension
Citrus sp.	Naringin, Limonin	Callus
Coffea arabica L.	Caffeine	Callus
Cruciata glabra	Anthraquinones	Suspension
Digitalis purpurea L.	Cardenolides	Suspension
Dioscorea deltoidea	Diosgenin	Suspension
Dioscorea doryophora Hance	Diosgenin	Suspension
Ephedra spp.	L- Ephedrine	Suspension
Eriobotrya japonica	Triterpenes	Callus
Eucalyptus tereticornis	Sterols and Phenolic compounds	Callus
Fumaria capreolata	Isoquinoline alkaloids	Suspension
Gentiana sp.	Secoiridoid glucosides	Callus
Ginkgo biloba	Ginkgolide A	Suspension
Glehnia littoralis	Furanocoumarin	Suspension
Glycyrrhiza echinata	Flavanoids	Callus
Glycyrrhiza glabra var. glandulifera	Triterpenes	Callus
Isoplexis isabellina	Anthraquinones	Suspension
Linum flavum L.	5-Methoxypodophyllotoxin	Suspension
Lithospermum erythrorhizon	Shikonin derivatives	Suspension
Mentha arvensis	Terpenoid	Shoot

Contd...

Table 8.1–*Contd...*

Name of Plant Species	Secondary Metabolite	Culture Type
Morinda citrifolia	Anthraquinones	Suspension
Mucuna pruriens	L-DOPA	Suspension
Nandina domestica	Alkaloids	Callus
Nicotiana rustica	Alkaloids	Callus
N. tabaccum L.	Nicotine	Suspensions
Ophiorrhiza pumila	Camptothecin related alkaloids	Callus
Panax notoginseng	Ginsenosides	Suspensions
Rauwolfia sellowii	Alkaloids	Suspensions
Rauwolfia serpentina Benth.	Reserpine	Suspension
Rhus javanica	Gallotannins	Root
Ruta sp.	Acridone and Furoquinoline Alkaloids and coumarins	Callus
Salvia miltiorrhiza	Lithospermic acid B and Rosmarinic acid	Callus
Salvia miltiorrhiza	Cryptotanshinone	Suspension
Scopolia parviflora	Alkaloids	Callus
Scutellaria columnae	Phenolics	Callus
Solanum chrysotrichum Schldl.	Spirostanol saponin	Suspension
Solanum laciniatum Ait.	Solasodine	Suspension
Silybum marianum	Flavonolignan	Root
Solanum paludosum	Solamargine	Suspension
Tabernaemontana divaricata	Alkaloids	Suspension
Taxus spp.	Taxol	Suspension
Taxus baccata	Taxol baccatin III	Suspension
Thalictrum minus	Berberin	Suspension
Torreya nucifera var. *radicans*	Diterpenoids	Suspension
Trigonella foenum-graecum	Saponins	Suspension
Withaina somnifera	Withaferin A	Shoot

metabolites. However, so far, the progress in this field is tardy and limited. In most cases, very little is known about the biosynthesis of these compounds. The best studied systems so far include flavonoids and anthocyanins, where complete pathway has been mapped at the level of products, enzymes, and genes (Verpoorte, 2007). Similarly, only few plants have been studied in detail for several different secondary metabolite pathways. Examples are tobacco (anthocyanins/flavonoids, terpenoids, alkaloids), *Catharanthus roseus* (alkaloids, steroids, brassinolides, flavonoids) and cinchona (anthraquinones and alkaloids). A broad spectrum of secondary metabolites that are produced through plant cell, tissue and organ culture are shown in Table 8.1.

Secondary Products that are Useful to the Mankind

As mentioned earlier, many secondary products produced by plants are being utilized by man since early times. These include several pharmaceutically important compounds (alkaloids, terpenoids, steroidal glycosides, flavonoids, flavones, lignans, stilbenes, gums, food flavours, pigments and fragrances). Even today, majority of rural population rely on plants for traditional medicine. But getting ample supply of these compounds from natural resources has become a problem due to drastic decrease in plant resources (due to decline in forest area), ruthless exploitation, increasing labour costs, technical and economic difficulties in the cultivation of medicinal plants and disturbances of the natural habitats.

Plant Cell Cultures as a Source of Valuable Secondary Metabolites, Advantages and Disadvantages Expected from Callus and Suspension Cultures

Although micropropagation is well-established technique, evidence that plant cell cultures can produce secondary metabolites came quite late. Later, plant cell cultures became an attractive tool to produce secondary metabolites *in vitro* (Komariah *et al.*, 2004). Plant cells are chemically totipotent, which means that each cell in culture retains complete genetic information and hence is able to produce the range of chemicals found in the parent plant. Plant cell cultures are now known to produce many important and major alkaloids (berberine, palmatine, hyoscyamine, camptothecin, vinblastin etc.), terpenoids (carotenes, mono, di, tri and sesquiterpenes), saponins and sapogenins (ginsenosides etc.), quinones (anthraquinones, benzoquinones and naphthoquinones), steroids (cardiac glycosides), peptides, (protenase inhibitors, plant virus inhibitors), food additives (pigments, sweetening steviosides), phenylpropanoids (flavonoids, isoflavonoids, stilbenes, tannins, lignans, coumarins, anthocyanins), and also essential oils. The advantages of this technology over the conventional production are:

☆ Cultured cells would be free of microbes and insects.

☆ It is independent of geographical and seasonal variations and various environmental factors. Cells of any plants, tropical or alpine could be manipulated *in vitro* to yield specific metabolites.

☆ It offers a defined production system, which ensures the continuous supply of products, uniform quality and yield.

☆ It is possible to produce novel compounds that are not normally found in parent plant.

☆ Automated control of cell growth and rational regulation of metabolic processes is possible *in vitro*. It offers efficient downstream process.

☆ Rapidity of production.

☆ In addition, plant cells can perform stereo- and region-specific biotransformations for the production of novel compounds from cheap precursors.

Several plant cell cultures have been reported to produce higher amounts of secondary metabolites than in the intact plants (Table 8.2). However, there are several concerns that should be addressed to utilize plant cell culture technology for commercial production. Improper or lack of proper technologies for the production of secondary products from cultured cells; technologies that are not feasible in several cases; instability of cell lines, slow growth rate and scale-up problems and the economics of large scale production of metabolites by plant cell cultures are the major concerns. The following section takes a closer look at various attempts made by researchers to improve the productivity of plant cell cultures and also metabolite accumulation.

Table 8.2: High yields of secondary products *in vitro*.

Secondary Metabolite	Plant species	Yield per cent (DW)
Capsaicin	*Capsicum frutescens*	514
Rosmarinic acid	*Salvia officinalis*	36
Rosmarinic acid	*Coleus blumeii*	21.4
Anthroquinones	*Morinda citrifolia*	18
Shikonin	*Lithospermum erythrorhizon*	12.4
Berberine	*Thalictrum minus*	10.6
Jatrorhizine	*Berberis wilsonae*	10
Anthocyanains	*Perilla frutescens*	8.9
Berberine	*Coptis japonica*	7.5
Sanguinarine	*Papaver somniferum*	2.5
Plumbagin	*Plumbago rosea*	16

Factors that Affect Accumulation of Secondary Products in Cultured Cells

Environmental Factors

1. Media optimization and effect of nutrients (nitrogen and carbon sources)
2. The choice of culture system
3. Effect of plant growth regulators on the accumulation of secondary metabolites
4. Effect of light, pH, temperature and the gaseous environment on the accumulation of secondary products
5. Effect of precursor feeding

Biological Factors

1. Growth of callus and suspension cultures
2. Morphological differentiation and accumulation of secondary products
3. Variation in biosynthetic activity and cell line selection

Other Factors

1. Effect of elicitation on secondary plant product accumulation
2. Effect of immobilization
3. Effect of permeabilization

Environmental Factors

Media Optimization and Effect of Nutrients (Nitrogen and Carbon Sources)

Plant cells are generally grown in simple synthetic nutrient medium. The chemical composition of commonly employed media have been primarily devised for improving cell growth, and it is not necessarily best suited for the production of secondary metabolites. Unfortunately, studies on the effect of different nutrient media and their interactions on the accumulation of secondary products have been limited. However, it has been observed that both nitrogen and carbohydrate sources play a vital role in the accumulation of secondary products. Most of the secondary metabolites are accumulated late in the culture cycle after cell division has ceased; this is associated with the process of cell differentiation. This property can be exploited to maximize the yield of a designated substance. This can be achieved by manipulating phosphorous, nitrogen and carbon in the medium. Nitrogen is essential for the synthesis of proteins and nucleic acids and hence removing the source of nitrogen from the medium will reduce or stop growth. The plant tissue culture medium such as Murashige and Skoog, Linsmaier and Skoog or B5 has both nitrate and ammonium as sources of nitrogen. However, the ratio of the ammonium/nitrate–nitrogen and overall levels of total nitrogen have been shown to markedly affect the production of secondary plant products. For example, reduced levels of NH_4 and increased levels of NO_3 promoted the production of shikonin and betacyanins, whereas higher ratios of NH_4/NO_3 increased the production of berberine and ubiquinone. Formation of catechol tannins in sycamore suspension cultures is promoted by increasing the ratio of the carbon to nitrogen sources. This finding suggests that the production of polyphenolics is under an antagonistic regulation between sugar and nitrogen metabolism. Reduced levels of total nitrogen improved the production of capsaicin in *Capsicum frutescens*, anthraquinones in *Morinda citrifolia* and anthocyanins in *Vitis* species. It has also been shown that phosphate-free medium is especially conducive to the production of alkaloids and other secondary metabolites by callus cultures of *Peganum harmala*. Accumulation of cinnaoyl putrescines in cultures of tobacco was enhanced by phosphate limitation (Yeoman and Yeoman, 1996). Reduction in nitrogen level is associated with a concomitant increase in the level of secondary metabolite. However, complete elimination of nitrate in cultures of *Chrysanthemum cinerariaefolium* induced two-fold increases in pyrethrin accumulation in the second phase of culture (Rajasekaran *et al.*, 1991).

Generally, sucrose supports growth and also the secondary products *in vitro*, as shown in *Ipomoea batata* for the accumulation of lignans and plumbagin in suspensions of *Plumbago rosea*. The beneficial effects of increasing the concentration of sucrose have been observed for nicotine accumulation in tobacco; for polyphenol production by cells of *Rosa*; solasodine alkaloid in *Solanum elaegnifolium* and alkaloid production in *Catharanthus roseus* and anthocyanin accumulation in *Vitis* suspensions. Interactions between sucrose and nitrogen have also been reported. Anthocyanin levels increase in *Vitis* cell cultures when sucrose levels are increased. But, at lower sucrose concentrations, reduced nitrogen levels also resulted in higher amounts of anthocyanins. This suggests that there is an optimal C : N ratio for pigment production. This also supports the concept of an inverse relationship between major aspects of primary metabolism, *e.g.*, protein synthesis and the synthesis and accumulation of secondary metabolites. When different carbon sources were tested for their influence on berberine, glucose, fructose and sucrose supported good growth and berberine accumulation but not galactose and lactose in *Tinospora cordifolia* (Rao *et al.*, 1996; Rao and Ravishankar, 2002). Increase in the concentration of sucrose increased the fresh and dry weights. While only withaferin A was detected at 2 per cent sucrose (256 mg/g DW), almost equal amounts of withaferin A (890 mg) and withanolide A (886 mg) were noticed at 3 per cent sucrose in hairy root cultures of *Withania somnifera*. Still higher concentrations of sucrose (4 per cent) enhanced withaferin A, but decreased withanolide A. At 6 per cent sucrose, hairy roots turned brown and died. Upon increasing the glucose from 1 per cent, 696 and 2100 mg of withaferin A were observed respectively at 2 per cent and 4 per cent levels on dry weight basis. However, no withanolide A was detected in these cultures. Fructose at 3 per cent level stimulated the accumulation of withaferin A (370 mg/g DW), but did not support biomass much and also withanolide A. It appears from the above studies that both the quality and quantity of carbohydrate in the medium play an important role on the accumulation of different secondary plant products.

The Choice of Culture System

For the production of secondary products by cultured cells, several approaches are followed. (1). Traditionally, plant cells are cultured in bioreactors that are similar to fermentors. These are used in batch mode. In two-stage batch, a precursor to the product can be added at the second stage. Bioreactors that are used for culturing plant cells may be of vessels stirred with a paddle or propeller, *i.e.*, mechanically stirred or vessels in which the contents are mixed with a stream of air, *i.e.*, air-lift reactor. (2). Some people use culture systems in which the cells are immobilized in, or attached to, an inert matrix and the modules form, either a fluidized bed, or a fixed bed. In general, the cells either move slowly in relation to the culture medium (fluidized bed) or remain stationary (fixed bed). The product can be removed here continuously. Immobilized system can operate semi-continuously, or continuously. (3). It is also possible and many people grow whole organs, either shoots or roots in specially designed bioreactors; especially roots transformed with *Agrobacterium rhizogenesis* (hairy roots). However, air-lift reactors have their own intrinsic problems when used with plant cells. These include high rates of air flow through the cultures which affect biomass yield and lower substantially the biomass per volume of culture attainable with stirred tanks. Currently, the choice of a bioreactor design which is a 'hybrid' between air-lift and stirred tank appears to provide the best for culturing plant cells. These bioreactors give adequate mass transfer of nutrients (especially oxygen), reduction of cell breakage and lysis due to shear.

Effect of Plant Growth Regulators on the Accumulation of Secondary Metabolites

Plant growth regulators affect not only growth and differentiation of cultured cells but also secondary metabolism. Both auxin and cytokinin in the medium play an important role on the proliferation of callus as well as differentiation *in vitro*. Their effects on secondary metabolism vary greatly depending upon the concentration, and balance between growth regulators and also on kinds of metabolites. It was reported that nicotine synthesis is strongly inhibited by 2,4-D, whereas it is promoted by kinetin. High levels of auxin inhibit the activity of putrescine-N-methyl transferase catalyzing the N-methylation of putrescine, the key intermediate in the biosynthesis of nicotine. In *Lithospermum* cultures, formation of shikonin is completely inhibited by the synthetic auxin 2,4-D or NAA, whereas it is not affected by natural auxin IAA. Production of anthraquinones in *Morinda citrifolia* cultures occurs in the presence of NAA, but not in the presence of 2,4-D. Also, 2,4-D did not affect the production of emodin-type anthraquinones in *Cassia tora* cultures, but it stimulated the production of certain compounds such as diosgenin and L-Dopa. Increasing concentrations of 2,4-D and NAA along with BAP in Linsmaier and Skoog's medium not only increased growth of suspensions but also berberine over a period of 4-weeks. While 2,4-D and BAP combination was better for growth, NAA was found suitable for berberine production. But, a combination of auxins and cytokinins enhanced the production of plumbagin in plants such as *Drosophyllum* and *Plumbago rosea*. Generally, treatments which encourage structural differentiation, *i.e.*, shoot or root from callus also change the biochemical profile.

Effect of Light, pH, Temperature and the Gaseous Environment on the Accumulation of Secondary Products

Light generally stimulates the formation of compounds including carotenoids, flavonoids, polyphenols and plastoquinones. Large increases in the activities of all enzymes involved in the accumulation of flavones and flavonol glycosides occur upon illumination of the cultures, especially with UV-light. White or blue light completely inhibits the formation of shikonin derivatives in *Lithospermum*. Shikonin biosynthesis in light pre-treated cultures is promoted by the addition of flavin mononucleotide (FMN) to the medium, whereas it is inhibited by the addition of a blue-light treated solution of FMN. These results suggest that FMN, which is necessary as a coenzyme of the oxidation –reduction reaction involved in the biosynthetic pathway is decomposed by blue light, to yield a compound which is no longer active as a coenzyme (Tabata, 1977). The examples mentioned above indicate the important role of light in the regulation of secondary metabolism, but there are other metabolites whose biosynthesis is not significantly influenced by light. Although light is essential for the accumulation of anthocyanins, it is not essential for the production of betalains by most cell cultures. But the participation of light as an important factor in determining product yield is likely to present problems to those who are designing bioreactors because of the difficulty in providing adequate illumination without affecting the temperature of the culture.

Optimum growth of cultured plant cells is generally obtained at a pH between 5.5 and 6.0 and at a temperature of 20 - 25°C. It is known that pH of the culture medium can influence the uptake of nutrients and precursors, the permeability of the membranes and release of products from the vacuole to the culture medium. Cultures of *Ipomoea* accumulating tryptophol have been found to be sensitive to the medium pH. Similarly, the effect of temperature on product accumulation has not received much attention. In *Catharanthus* cell cultures, a dramatic effect on growth and product accumulation was observed in cultures grown under different temperature regimes. Over and above, aeration and mixing of plant cells is also important, especially in large scale production systems.

Effect of Precursor Feeding

Precursor feeding has been an obvious and popular approach to increase secondary metabolite production in plant cell cultures. The concept is based upon the idea that any compound, which is an intermediate, in or at the beginning of a secondary metabolite biosynthetic route, stands a good chance of increasing the yield of the final product. Attempts to induce or increase the production of plant secondary metabolites, by supplying precursor or intermediate compounds, have been effective in many cases. The addition of phenylalanine as a precursor led to improvement in rosmarinic acid yield in *Coleus blumei* cell cultures. Addition of phenylalanine to *Salvia officinalis* suspension cultures stimulated the production of rosmarinic acid and decreased the production time as well. Phenylalanine is also the precursor of the N-benzoylphenylisoserine side chain of taxol, and supplementation of *Taxus cuspidata* cultures with phenylalanine resulted in increased yields of taxol (Fett-Neto *et al.,* 1994). Use of the distant precursor phenylalanine and a near precursor such as isocapric acid resulted in enhanced capsaicin content in cell cultures of *Capsicum frutescens* (Lindsey and Yeoman, 1985). Feeding of ferulic acid to cultures of *Vanilla planifolia* resulted in increase in vanillin accumulation. Similarly, anthocyanin synthesis in carrot cultures was restored by the addition of a dihydroquarcetin (naringen). Furthermore, addition of leucine led to enhancement of volatile monoterpenes α- and β-pinine in cultures of *Perilla frutescens*, whereas addition of geraniol to rose cell cultures led to accumulation of nerol and citronellol. Using immobilized placental tissues of *C. frutescens*, Johnson *et al.* (1996; 1998) increased the levels of capsaicin and dihydrocapsaicin. Immobilized cell cultures of *C. frutescens* when treated with ferulic acid and vanillylamine were biotransformed to capsaicin and vanillin. The above results provide an example for alternate route to the formation of vanillin by *Capsicum* cell cultures. Greater accumulation of capsaicin and dihydrocapsaicin was noticed when immobilized placental tissues of *C. frutescens* were fed with intermediates of capsaicinoid pathway, t-cinnamic, p-coumaric, caffeic and ferulic acids. The percent conversions of externally supplied precursors of capsaicinoid pathway to capsaicin and dihydrocapsaicin in immobilized placental tissues of *C. frutescens* are shown in Table 8.3.

While using precursors, one has to keep in mind about the cost of precursors. The use of precursors even of modest cost can be counterproductive, and this problem is perhaps best appreciated when processes involving one-or-two-step biotransformation are contemplated. In general, precursors that are close to the product are likely to be expensive, whereas general precursors such as amino acids are likely to be much cheaper. But, if the chemical synthesis is difficult, then one has to fall back

Table 8.3: Per cent conversion of externally supplied precursors of capsaicinoid pathway to capsaicin and dihydrocapsaicin in immobilized placental tissues of *C. frutescens.*

Treatment	Capsaicin (per cent)	Dihydrocapsaicin (per cent)
t-Cinnamic acid	82.0	27.5
p-Coumaric acid	69.2	19.35
Ferulic acid	18.91	4.6
Caffeic acid	13.16	4.6
Vanillylamine	4.24	1.6
l-Valine	2.87	1.25
Phenylalanine	2.78	1.13

Source: Johnson and Ravishankar (1997). In: Ravishankar, G.A. and Venkataraman, L.V. eds. Advances in biotechnological applications of plant tissue and cell culture. Oxford and IBH, New Delhi, pp. 259-265.

on biological conversion even if the precursor is expensive. There are also other difficulties that are associated with precursor feeding experiments. In many cases, the precursor might not be taken up by the cells or it does not arrive at the appropriate location within the cells. This hampers the discovery of biosynthetic pathway in several plant systems. For example, highly ionized molecules tend not to be taken up at the near neutral pH used in the culture medium. Precursor molecules can be chemically modified, *e.g.*, acids esterified to facilitate entry, or solubilization by the addition of cyclodextrins, but this might be expensive and is not always effective. One more problem is that the added precursors might be toxic to the cells even at low concentrations. The timing of precursor feeding might be also important for optimum accumulation of secondary metabolites. For example tryptamine added during the second or third week of culture to *Catharanthus roseus* stimulated both cell growth and alkaloid metabolism, but not in others.

Biological Factors

Growth of Callus and Suspension Cultures

In a batch culture system, the biosynthetic activity of cultured cells usually varies with cell growth or substrate utilization. Little is known about the interrelation between the rate of product formation and the age of individual cells in plant cell culture. In the first type of growth pattern, product formation proceeds almost in parallel with cell growth. The production of nicotine, tropane alkaloids and morindone anthraquinones belong to this type. In the case of volatile oil production in *Ruta* callus cultures, maximum oil content was observed at the logarithmic phase of tissue growth. In the second type, product formation is delayed until cell growth declines or ceases. Polyphenols and shikonin production belong to this type. In the third type, production curve is diphasic and lags behind the growth curve as in the case of diosgenin production. In order to increase the production efficiency, it is necessary to shorten the lag period. For example, in anthocyanin as well as in shikonin production, the lag phase can be shortened effectively by decreasing the concentration of auxin in the culture medium.

Morphological Differentiation and Accumulation of Secondary Products

Production in Differentiated Tissues

In higher plants, there are certain compounds which are synthesized or accumulated only in particular organs or tissues. Examples for such localized substances are essential oils found in morphologically specialized structures such as glandular scales and secretory sacs or ducts, and latex components found in laticifers. Also, compounds like tropane alkaloids are primarily synthesized in the roots of tobacco. Callus and suspension cultures usually fail to produce compounds such as morphine, menthol and carvone. This reveals that morphological differentiation of specific organs or tissues are required for the accumulation of certain compounds in culture. In some cases, *in vitro* regenerated plants or shoots or naturally growing plants are more useful for the production of considerable amounts of secondary products, especially if the undifferentiated cell culture is either unable to or barely accumulating the required secondary metabolite. This has been reported for several classes of metabolites, especially for alkaloids. Ellagic acids present in *Rubus chamaemorus* plants was 3-times lower than in shoot cultures generated *in vitro* and over 10-times lower in callus. Similarly, in *Ocimum basilicum* cell cultures, the accumulation of rosmarinic acid was markedly lower than in regenerated plantlets. In *Salvinia officinalis* and rosemary *in vitro* cultures, the abietane diterpene antioxidants (carnosol and carnosic acid) were present only in shoot cultures and not in callus, suspension or hairy roots. On the other hand, undifferentiated cell suspensions are able to accumulate great amounts of phenolic acids. Moreover, the cells cultured in suspension tend to form larger aggregates upon differentiation, therefore, this increases their shear stress resistance. The potential of fast growing somatic embryo cultures can be also utilized for the production of medicinally important compounds, but reports on that topic are scarce and deal chiefly with alkaloid production such as paclitaxel. Recently, phenol glycosides, lignans and flavonoid accumulation have been reported in Siberian ginseng somatic embryos (Shoahel *et al.*, 2006). Even in dedifferentiated cells, some biosynthetic potential typical for the developed organs from which they were initiated can be conserved. In *Pueraria lobata* callus cultures, the bioactive isoflavonoid content depended on the source organ, reflecting relations in the mature plant. The selection of proper donor plants and organs should also be considered for proper accumulation while starting the culture. Shoots obtained from callus cultures of tobacco produce more nicotine. Roots initiated from *Scopalia* produce tropane alkaloids but not callus cultures. Callus cultures failed to produce saikosaponins (used as an anti-inflammatory, antipyretic and sedative) in *Buplerum*. Induction of roots in suspension cultures, however, increased the saponin content.

Organ Cultures

Plant organs represent an interesting alternative to cell cultures for the production of plant secondary metabolites. Two types of organ cultures have been well studied in order to obtain higher yields of secondary metabolites; viz., hairy root cultures and shoot cultures. The organ cultures are relatively more stable and hence they are preferred. The ability of *Agrobacterium rhizogenes* and *Agrobacterium tumifaciens* to induce hairy roots and shoots respectively in a range of host plants has led to studies on them as sources of root- and shoot- derived pharmaceuticals. There are several plants belonging to several families wherein hairy roots and shooty teratomas have been induced so far.

Hairy Root Cultures

Since production of secondary metabolites is generally higher in differentiated tissues, there are attempts to cultivate shoot and root cultures for the production of medicinally important compounds. Hairy roots are generally obtained after the successful transformation of a target plant tissue with *Agrobacterium rhizogenes*. They have several advantages over other cultures. They can be subcultured for indefinite period on a nutrient medium without phytohormones and display unusual growth characteristics due to profuse lateral root formation. Biomass doubling time has been very impressive ranging from less than one day to one week. Also several studies reported long-term genetic stability of hairy roots and also metabolite accumulation. Morphological differences in the growth of hairy roots exist even though they are derived from the same species. This could be because of the gene copy number and structure of the TL-DNA transferred to the host plant. Generally, hairy roots grow against the geotropism and this has been reported in several cases. This characteristic feature may be helpful since it increases the aeration in liquid medium. This may result in enhanced biomass and secondary product too. Hairy roots are usually 100 to 1000-fold more sensitive to exogenous auxin than normal roots. Therefore, in commercial cultures, hairy roots are preferred since one can save money. Maldonado-Mendoza *et al.* (1993) studied 500 hairy root lines from *Datura stramonium*. It was demonstrated that growth and tropane alkaloid production was stable over a period of 5 years. In addition to their growth capacities, hairy roots display interesting properties regarding the production of secondary metabolites. The metabolite profile was found in hairy roots to be similar to normal roots found in plants. A major characteristic of hairy roots is that they are able to produce secondary metabolites concomitantly with growth. Hence, it is possible to get a continuous source of secondary metabolites from actively growing hairy roots unlike low levels of production from cell suspension cultures. For further increasing the metabolite production, same strategies as developed for cell cultures, have been applied to hairy roots.

Transformed Shoot Cultures

Transgenic shoot cultures, called shooty teratomas, can be obtained after infection with *Agrobacterium tumefaciens*. Like hairy root cultures, shooty teratomas exhibit some comparable properties; such as genetic stability, accumulation of good amounts of secondary metabolites and faster biomass growth. However, there are some differences in the metabolic patterns, as some syntheses are specifically located in either roots or shoots. Many attempts have been made to improve secondary metabolites *in vitro* using shooty teratomas. However, the major changes in the area of plant secondary metabolites have probably been achieved thanks to the introduction of new field of molecular genetics, through the so-called metabolic engineering approach.

Variation in Biosynthetic Activity and Cell Line Selection

Once compound of interest has been identified from the target plant species, the first part of the work consists of collecting the largest genetic pool of various clonal lines. This screening allows identification of high-yielding clones for future experiments. Cellular variation is another factor which can regulate secondary metabolism and has potential use in improving biosynthetic capabilities of culture strains. Callus cultures sporadically give rise to variant subcultures showing different concentration levels of particular secondary metabolite. One of the procedures that is adapted for increasing productivity in plant cell cultures has been selection of individual cells which exhibit enhanced capacity for secondary product accumulation. In a heterogeneous population of plant cells, only about 10 per cent of cells are capable of producing the secondary compound of interest. Continuous selection and concentration of these elite cell lines can result in significant improvements in production rates and yields. It has been shown that productivity can be increased by an order of magnitude if the selection is performed on a continual basis. By eliminating unstable cells in repeated selections, cell lines with high and stable product contents can be obtained. Cells may get reverted back at some frequency always. Hence, selection must be carried out regularly as part of the productive strain improvement. Selection can be easily achieved if the product of interest is coloured, for *e.g.*, a pigment or a dye. In cultures of *Lithospermum erythrorhizon*, extensive screening of number of clones resulted in a 13-20-fold increase in shikonin production (Fujita *et al.*, 1984). In *Nicotiana rustica*, suspensions showed large variations in growth and nicotine production. Enhanced anthocyanin production by clonal selection and visual screening has been reported in *Euphorbia milli* and *Daucus carota*. Other techniques such as high performance liquid chromatography (HPLC) and radio-immuno assay (RIA) were also used to screen for high-yielding cell lines. Mutation studies have been undertaken in order to obtain overproducing cell lines. The use of selective agents has been employed as an alternative approach to select high-producing cell lines. In this method, large population of cells is exposed to a toxic (or cytotoxic) inhibitor or environmental stress and only cells that are able to resist the selection procedures will grow. P-Fluorophenylalanine (PFP), an analogue of phenylalanine, was extensively used to select high-yielding cell lines with respect to phenolics. Increase in capsaicin and rosmarinic acid in PFP resistant cell lines of *Capsicum* and *Anchusa* were reported (Johnson *et al.*, 1998; Quesnell and Ellis 1989). Other selective agents such as 5-methyltryptophan, glyphosate and biotin have also been used to select high-yielding cell lines.

Other Factors

Effect of Elicitation on Secondary Plant Product Accumulation

It is believed that secondary plant products have a role in the defense of the plant system. Hence, stress due to infection by microorganisms enhances the *de novo* synthesis of secondary metabolites. This phenomenon is known as elicitation. Accumulation of some secondary metabolites can be enhanced in plant cell cultures by certain compounds called elicitors. Elicitors can be either biotic (prepared from fungal extracts or toxins, fungal mycelial extracts, bacterial extracts or toxins, culture filtrates, fractions or compounds obtained from microbial cell walls, and yeast extracts) or abiotic in nature. Physical and chemical stresses such as ultraviolet light radiation, exposure to cold or heat, ethylene, fungicides, antibiotics, salts of heavy metals or high salt concentrations have been defined as abiotic elicitors. Endogenous elicitor molecules are often oligosaccharides, and can be produced in plant tissue cultures. Elicitors are signals triggering the formation of secondary metabolites. However, sensitivity of cell culture system, specificity of elicitors, concentration of elicitors and timing and duration of the elicitor exposure to the culture medium play a vital role on the secondary metabolism. It appears that the elicitors might bind to the receptors present on the plasma membrane and help in the signal transduction and thus elicit a response in the cells. Elicitors of fungal, bacterial and yeast origin, viz., polysaccharides, glycoproteins, inactivated enzymes, purified curdlan, xanthan and chitosan (extracted from the shell), and salts of heavy metals were reported for the production of various secondary metabolites (Rao *et al.,* 1996). Yeast extract has increased the synthesis of berberine in *Thalictrum minus* cultures. Brodelius (1988) conducted several experiments on the effects of yeast elicitors and demonstrated that benzophenanthridine alkaloids in cell cultures of *Thalictrum rugosum* and *Escholtzia californica* can be increased by many folds. Elicitors prepared from *Rhizoctonia solani* enhanced the solavetivone production in *Hyoscyamus muticus* and *Aspergillus flavus* mycelia extract elicited anthocyanin content in *Daucus carota* cell cultures. Cell cultures of *Plumbago rosea* when treated with the elicitors prepared from the fungi, bacteria, yeast extract and chitosan enhanced the synthesis of plumbagin, a naphthoquinone in suspension cultures. Dixon and Paiva (1995) reported that a number of phenylpropanoid compounds or their derivatives act as defense against biotic or abiotic stresses. Treatment of *Papaver somniferum* cell suspensions with a homogenate of *Botrytis mycelium* resulted in a remarkable accumulation of sanguinarine of up to 3 per cent of the cell dry weight. The treatment of root cultures of *Datura stramonium* with copper and cadmium salts has been found to induce the rapid accumulation of high levels of sesquiterpenoid defensive compounds. Elicitation can be applied on large scale to induce production of certain compounds. It is now clear that the flux of metabolites in a number of secondary pathways can be influenced by elicitor treatment, by induction or amplification of various enzymes associated in the biosynthetic pathway. If a product is extra cellular, it offers more advantages in carrying out continuous or semi-continuous production process. Kurtz *et al.* (1987) showed that for the production of sanguinarine in poppy cell cultures, that a continuous process is feasible in which a sequence of elicitation and medium change can be used to produce the alkaloid. Also, in case of the production of paclitaxel, elicitation has been shown to cause a clear increase in productivity. Thus, elicitation has been used in several tissue culture systems for increasing the product yield and has commercial potential.

Effect of Immobilization

Plant cells are generally grown in a liquid nutrient medium. But, while growing cells on an industrial or large scale in a bioreactor, cells are usually washed out during the operation. In such cases, the rates of reactions become slow and consequently the product rates. In a suspension culture, cells grow faster and hence, the secondary product accumulation is slow. Further, cell aggregates have the better ability to accumulate metabolites than the free floating suspensions. Hence, it is essential to anchor the plant cells on some support or entrap the cells in a matrix such as gel, agar, alginate, polyacrylamide or polyurethane foam etc. Such a process of anchoring plant cells is known as immobilization. In other words, immobilization means, keeping the cell in a place. Normally, in cell suspensions or bioreactors, cells are floating around in nutrient liquid medium. In an immobilized cell bioreactor, the cells are trapped - perhaps stuck to a sticky surface - while nutrient flows over them. Plant cells may be immobilized in sodium alginate beads by complexing with calcium. The sodium alginate bead formation is carried out in the growth chamber by dripping an alginate-cell gelatin suspension into a calcium solution contained in the growth chamber. Secondary metabolites of viable plant cells are produced with the cells immobilized in a porous inorganic support. Immobilization includes the steps of: (a) preparing a support comprising a substantially uniform and porous matrix of inorganic material having a tensile strength of at least 500 MPa; (b) introducing a culture of viable plant cells into the pores of said matrix; (c) entrapping the plant cells by coating the matrix with a solution or colloidal suspension not interfering with the cell viability; and (d) immobilizing the entrapped cells within the matrix with a reactive gas including a carrier gas saturated with volatile SiO_2 or organic modified SiO_2 precursors. Secondary metabolites are rarely released from the suspended or immobilized cell cultures, and they have very low solubility in water due to their hydrophobicity.

In situ Removal of the Product

In order to use immobilized cells more economically, it is necessary that the water insoluble products should be removed from the culture medium without disturbing the cell metabolic activity. The concentration of the product may be toxic to the cells (except in the cultures of *Lithospermum*, where the cultures can tolerate substantial amounts of the product in the medium) and hence, it must be kept always small in cultures. To avoid the toxicity, secondary metabolites are separated from the site of synthesis (usually by storage in the vacuole; sometimes in specialized cells) in many plants. It appears therefore, that end-product inhibition might be common in plants. In this connection, *in situ* adsorption or extraction of metabolites by using hydrophobic materials received great importance. Berlin *et al.* (1984) first reported the use of adsorbents to retain volatile compounds from cell cultures of *Thuja*. Enhanced shikonin was noticed by *in situ* extraction from calcium alginate immobilized cells of *Lithospermum*. In many cases, *in situ* product removal by charcoal, XAD-7 and other inert materials have enhanced secondary metabolite production, and the products were selectively released from the cells and dissolved in the solvents or adsorbents.

Effect of Permeabilization

Cell permeabilization means cell cracking. In most cases, the products formed by plant cell cultures are stored in the vacuoles. This problem of product storage within plant cells has led to the development of techniques for plant cell permeabilization. In order to release the products from vacuoles of plant cells, two membrane barriers - plasma membrane and tonoplast - have to be penetrated. Cell permeabilization depends on the formation of pores in one or more of the membrane systems of the plant cell, enabling the passage of various molecules into and out of the cell. The popular agents for accomplishing this is dimethyl sulfoxide (DMSO), and toluene since cells appear to survive its application and treated plant cells maintain a relatively high level of enzyme activity. In addition to DMSO, enzymes such as cellulase and pectinase can also be used for permeabilization of plant cells. The permeability of the cells by the polycation chitosan (diacetylated chitin molecule) could be because of its interaction with cell membrane. Chitosan binds to polygalacturonate, a plant cell wall component, and induces leakage of low molecular weight compounds as well as some proteins (>5000 D). The increase in cell permeability by chitosan may be due to disruption of the intermolecular bonding responsible for maintaining an intact membrane, changes in membrane fluidity or effects on the components associated with membrane transport. The permeability of the cells can be monitored by measuring the activity of enzymes of the primary metabolism, viz. hexokinase, glucose 6-phosphate dehydrogenase, isocitrate dehydrogenase, malic and citrate synthetase (Brodelius, 1988). Attempts have been made to permeabilize the plant cells transiently, to maintain the cell viability and to have short time periods of increased mass transfer of substrate and metabolites to and from the cell. A wide variety of permeabilizing agents are used to enhance the accessibility of enzymes or to provoke the release of intracellularly stored product. Organic solvents such as isopropanol, dimethylsulfoxide (DMSO) and polycations like chitosan have been used as permeabilizing agents in many plants. Other permeabilization methods include ultrasonication, electroporation and ionophoretic release, in which the cells are subjected to a low current in a specially designed device (Brodelius, 1988). In addition, using high electric field pulses and ultrahigh pressure has been reported for the recovery of secondary metabolites. Excretion of secondary products can also be enhanced by changing the composition of the medium and increased ionic strength of the medium. The pH of the medium can influence the excretion of secondary metabolites. Majerus and Parilleus (1986) observed a sharp increase of the excretion of alkaloids by *Catharanthus roseus* when the pH of the culture medium was changed from 9 to 4.3.

Application of high electric current field pulses also led to high levels of cell permeabilization but at field strengths beyond 0.75 kV/cm and constant amount of ten pulses, cell viability approached zero values. A new approach of surfactant-induced release of compounds was reported by Bassetti *et al.* (1995). They used surfactant pluronic F-68 to obtain non-lethal release of plant cell intracellular products. Pluronic F-68 is a copolymer surfactant having a central, hydrophobic polypropylene oxide group sandwiched between two lateral, hydrophilic polyethylene oxide groups. By using pluronic F-68 at 2 per cent, Bassetti *et al.* (1995) reported long term non-lethal release of anthraquinones from suspension cultures of *Morinda citrifolia*. An average of 55 per cent of the intracellularly stored anthraquinones were released by permeabilization. Some of the non-lethal treatments to achieve secondary metabolite release from plant cells have been provided in Table 8.4.

Plant Metabolic Engineering

With the advent of large quantity of data regarding the biosynthetic pathways leading to secondary metabolites, a new area called metabolic engineering has emerged.

Table 8.4: Non-lethal treatments for the release of secondary metabolites.

Treatment	System	Product	Release (per cent)
Lypozime	*Morinda citrifolia*	Anthraquinones	7.0
Hexadecane	*M.citrifolia*	Anthraquinones	37
Ultrasound (1.02 MHz)	*Beta vulgaris*	Pigments	5-10
Mild heat	*B. vulgaris*	Pigments	15.0
DMSO	*Catharanthus roseus*	Ajmalicine isomers	85-90
Triton X-100	*C. roseus*	Alkaloids	50.0
DMSO	*Coleus blumei*	Rosmarinic acid	66.0
O_2 starvation	*Rubia tinctorum*	Anthraquinones	5.0
Hexadecane	*Tagetus patula*	Thiophenes	30-70
Chitosan	*Capsicum frutescens*	Capsaicin	77.6
Laminarin	*C. frutescens*	Capsaicin	71.68
Pluronic F-68	*Morinda citrifolia*	Anthraquinones	55.0

Metabolic engineering is the improvement of cellular activities by manipulation of enzymatic, transport, and regulatory functions of the cell with the use of recombinant DNA technology. Bioactive compounds with privileged structures are highly sought paradigms in drug development. This structure is a molecular scaffold that can accommodate various pharmacophores arranged to promote interaction with biological targets (Leonard *et al.,* 2009). Plants produce very low amounts of pharmaceutically important compounds and some of them cannot be synthesized due to their structural complexities. Therefore, metabolic engineering has facilitated the development of plant cell and tissue systems as alternative production platforms that can be scaled. Changes in the secondary metabolism have been made without seriously impairing primary metabolism. Metabolic engineering has been quite successful for the production of pharmaceuticals in microorganisms, for example, for the increased production of known compounds or for the production of new compounds. Owing to the smaller genome size, the degree of complexity in microorganisms is lower than that of plants. Microorganisms have fewer intracellular organelles compared to plant cells; hence metabolite transport between enzymatic steps can be negligible. The bottom-up assembly of artificial biosynthetic pathways in *E.coli* and yeast enabled the biosynthesis of plant alkaloids in a short period of time. High level synthesis of plant flavonoid at ~400 mg l^{-1} from engineered *E. coli* could be facilitated by redirecting various metabolic fluxes from glucose toward malonyl-CoA (a flavonoid building block). This titer was further improved up to ~700 mg l^{-1} by partially repressing fatty acid metabolism in the *E.coli* hosts. In the case of the production of plant natural products in *Saccharomyces cerevisiae*, the synthesis of ~100 mg l^{-1} artemisinic acid from glucose could be achieved by the upregulation of the mevalonate pathway and the down-regulation of a competing pathway, *i.e.*, sterol biosynthesis (Ro, 2006; Leonard, 2008). Presently, the introduction of new genes into plants has become more or less routine. The particle gun and the *Agrobacterium* mediated genetic transformation are the most successful methods. So far, in medicinal plants, developmentally regulated genes have been manipulated successfully. A close correlation has been shown between differential expression of tyrosine/dopa decarboxylase gene and the organ dependent accumulation of alkaloids in opium poppy. Over expression of biosynthetic pathway genes in *Petunia* and also down regulation of specific genes using antisense RNA technology has been reported by van der Krol *et al.* (1995) in *Petunia*. However, the biosynthetic pathways are complicated in many cases and the genes or enzymes have not been identified.

Future Prospects

So far, attempts to increase the yields of particular secondary products have largely used simple empirical methods in which the culture conditions have been varied or deliberately manipulated and these effects on the accumulation of final product is noted. The utility of these systems was proved by the industrial scale production of scopolamine and berberine from cell cultures by Sumitomo Chemical Industries and Mitsui Petrochemical Industries (McCoy and O'Connor, 2008; Roberts, 2007). Although metabolic engineering could certain extent help to improve the yields of secondary metabolites, pathway compartments, existence of multiple biosynthetic pathways and regulatory control mechanisms such as feedback regulation, product degradation are among the few factors that have discouraged the metabolic engineering efforts in cell cultures. Although plant cell/tissue culture system offers tremendous advantages as scalable alkaloid production platforms, many opportunities still lie in the cellular and metabolic engineering sectors to create the multifaceted phenotypic traits (for example, high productivity, product tolerance and stability) required for use in industrial bioprocesses. The much awaited breakthrough to reach an industrial accomplishment is probably still to come, and this is going to happen with a blend of both metabolic engineering and plant cell, tissue and organ culture technologies.

References

Bassetti, L., Hagendoorn, M. and Johannes, T. 1995. Surfactant–induced non–lethal release of anthraquinones from suspension cultures of *Morinda citrifolia*. J. Biotechnol. **39**: 149–155.

Berlin, J., Martin, B., Nowak, J., Witte, L., Wray, V. and Strack, D. 1984. Effects of permeabilization on the biotransformation of phenylalanine by immobilized tobacco cell cultures. Z. Naturforsch. **44c**: 249–254.

Brodelius, P. 1988. Permeabilization of plant cells for release of intracellularly stored products viability studies. Appl. Microb. Biotechno. **27**: 561–566.

Dixon, R.A. and Paiva, N.L. 1995. Stress induced phenylpropanoid metabolism. Plant Cell. **7**: 1085–1097.

Fujita, Y., Takahashi, S. and Yamada, Y. 1984. Selection of cell lines with high productivity of shikonin derivatives through protoplasts of *Lithospermum erythrorhizon*. Proc. Euro. Cong. Biotechnol., 3 ', 1. 101–100.

Fett–Neto, A.G., Melanson, S.J., Nicholson, S.A., Pennington, J.J. and DiCosmo, F. 1994. Improved taxol yield by aromatic carboxylic and amino acid feeding to cell cultures of *Taxus cuspidata*. Biotechnol. Bioeng. **44**: 967–971.

Harborne, J.B. 1999. Classes and functions of secondary products, In : N.J. Walton and D.E. Brown (Eds.). Chemicals from plants, perspectives on secondary plant products, Imperial College Press, pp. 1–25.

Johnson, T.S. and Ravishankar, G.A. 1996. Precursor biotransformation in immobilized placental tissues of *Capsicum frutescens* Mill: I. Influence of feeding intermediate metabolites of the capsaicinoid pathway on capsaicin and dihydrocapsaicin accumulation. J. Plant Physiol. **147**: 481–485.

Johnson, T.S., Sarada, R. and Ravishankar, G.A. 1998. Capsaicin formation in p–fluorophenylalanine resistant cell cultures of *Capsicum frutescens* and activity of phenylalanine ammonia lyase. J. Biosci. **23**: 209–212.

Komaraiah, P., Jogeswar, G., Ramakrishna, S.V. and Kavi Kishor, P.B. 2004. Acetylsalicylic Acid and ammonium–induced somatic embryogenesis and enhanced plumbagin production in suspension cultures of *Plumbago rosea* L. In Vitro Cell. Dev. Biol.–Plant **40**: 230–234.

Kurtz, W.G.W., Constable, F., Eilert, U. and Tyler, R.T. 1987. Elicitor treatment: a method for metabolite production by plant cell culture in vitro. In: Breimer, D.D. and Speiser, P. (Eds.). Topics in Pharmaceutical Science, Elsevier Science Publisher, Amsterdam, Netherlands. pp. 283–290.

Leonard, E., Yan, Y., Fowler, Z.L., Li, Z., Lim, C.G., Lim, K.H. and Koffas, M.A. 2008. Strain improvement of recombinant *Escherichia coli* for efficient production of plant flavonoids. Mol. Pharm. **5**: 257–265.

Leonard, E., Runguphan, W., O'Connor, S. and Prather, K.J. 2009. Opportunities in metabolic engineering to facilitate scalable alkaloid production. Nature Chemical Biol. **5**: 292–300.

Lindsey, K. and Yeoman, M.M. 1985. Immobilized plant cells. In: Yeoman M.M. (Editor). Plant cell culture technology. Springer–Verlag, Berlin, pp. 229–267.

Maldonado–Mendoza, T., Ayora–Talavera, V.M. and Loyola, V. 1993. Establishment of hairy root cultures of *Datura stramonium*, Plant Cell Tissue and Organ Culture **33**: 321–329.

Majerus, F. and Pareilleus, A. 1986. Production of indole alkaloids by gel–entrapped cells of *Catharanthus roseus* cells in a continuous flow reactor. Biotech. Lett. **8**: 863–866.

McCoy, E. and O'Connor, S.E. 2008. Natural products from plant cell cultures. Prog. Drug Res. **65**: 331–370.

Quesnell, A.A. and Ellis, B.E. 1989. Comparison of UV–irradiation and p–fluorophenylalanine as selective agents for the production of aromatic compounds in plant cell culture. J Biotechnol. **10**: 27–38.

Rajasekaran, T., Rajendran, L., Ravishankar, G.A. and Venkataraman, L.V. 1991. Influence of nutrient stress on pyrethrin production by cultured cells of pyrethrum (*Chrysanthemum cinerariaefolium*). Curr. Sci. **60**: 705–707.

Rao, S.R., Sarada, R. and Ravishankar, G.A. 1996. Phycocyanin, a new elicitor of capsaicin and anthocyanin accumulation in plant cell cultures. Appl Microbiol Biotechnol. **46**: 619–621.

Rao, S.R. and Ravishankar, G.A. 2002. Plant cell cultures: Chemical factories of secondary metabolites. Biotechnology Advances **20**: 101–153.

Ro, D.K., Paradise, E.M., Ouellet, M., Fisher, K.J., Newman, K.L., Ndungu, J.M., Ho, K.A., Eachus, R.A., Ham, T.S., Kirby, J., Chang, M.C.Y., Withers S.T., Yoichiro Shiba, Richmond Sarpong and Keasling, J.D. Production of the antimalarial drug precursor artemesinic acid in engineered yeast. Nature **440**: 940–943.

Roberts, S.C. 2007. Production and engineering of terpenoids in plant cell culture. Nat. Chem. Biol. **3**: 387–395.

Shoahel, A.M., Ali, M.B., Yu, K.W., Hahn, E.J., Islam, R. and Paek, K.Y. 2006. Effect of light on oxidative stress, secondary metabolites and induction of antioxidant enzymes in *Eleutherococcus senticosus* somatic embryos in bioreactor. Process Biochem. **41**: 1179–1185.

Tabata, M. 1977. Recent advances in the production of medicinal substances by plant cell cultures. In: Proceedings in Life Sciences, Plant Tissue Culture and Its Biotechnological Applications, W. Barz, E. Reinhard and M.H. Zenk (Eds.). Springer–Verlag, Berlin, Heidelberg, pp.3–16.

van der Krol, A.R., Mur, L.A., Beld, M., Mol, J.N.M. and Stuitje, A.R. 1990. Flavonoid genes in *Petunia*: addition of a limited number of gene copies may lead to a suppression of gene expression. The Plant Cell **2**: 291–299.

Verpoorte, R. 2000. Pharmacognosy in the new millennium: lead finding and biotechnology. J. Pharm. Pharmacology **52**: 253–262.

Verpoorte, R. 2007. Introduction, In: Applications of plant metabolic engineering. Verpoorte R., Alfermann A.W. and Johnson T.S. (Eds.), Springer Verlag, Dordrecht, 2007, pp. XI–XXI.

Yeoman, M.M. and Yeoman, C.L. 1996. Manipulating secondary metabolism in cultured plant cells. New Phytologist **134**: 553–569.

2013, Abiotic Stress and Biotechnology
Editor: T. Pullaiah
Published by: REGENCY PUBLICATIONS, NEW DELHI

Pages 191–208

Chapter 9

Conservation of Biodiversity Through Plant Tissue Culture

Y.K. Bansal and M. Chacko

*Department of Biological Sciences,
R.D. University, Jabalpur – 482 001, M.P., India*

ABSTRACT

Biodiversity brings enormous benefits to mankind by direct harvesting of plants for several purposes. However, overexploitation of plants from their natural habitats has exerted immense pressure on the existence of many plant species. Consequently many of them have become endangered or critically endangered. Exploitation of clonal propagation and plant tissue culture technology enables conservation of valuable biodiversity. Micropropagation is used routinely to generate a large number of high-quality clonal plants, including medicinal, agricultural, ornamental and vegetable species and in some cases also plantation crops, fruits and vegetable species.

The most striking feature of the earth is the existence of life and the most striking feature of life is its diversity. Topography, soil, climate and geographical location of a region influence the vegetation diversity of the ecosystem. Biodiversity is the variation of life forms within a given ecosystem, biome or on the entire earth. Biodiversity is often used as a measure of the health of biological systems (Wilson, 1985).

In the modern era, due to human activities, species and ecosystems are threatened with destruction to an extent rarely seen in earth history. Habitat loss, degradation, fragmentation is an important cause of known extinctions. All species have specific food and habitat needs. The more specific these needs and localized the habitat, the greater the vulnerability of species to loss of habitat to agricultural land, livestock, roads and cities. In future, the only species that will survive are likely to be those whose habitats are highly protected or whose habitat corresponds to the degraded state associated with human activity. Habitat damage, especially the conversion of forestland to agriculture, changing global climate, increasing human population and their interruption in natural ecosystem and pollution threatens species and ecosystems. The conditions have become irreversible. Thus, in addition to conservation of such plant resources systematic and concerted efforts should be undertaken for the yield and quality improvement through modern biological tools (www.biodiversity hot spot.org).

Genetic resources once lost are not only irretrievable but it has multifarious effects on the environment, since plants are the prime life support system of the biosphere. To restore endangered plants to their natural habitat and to explore these plants for ornamental value as well as for identification of desirable gene pool/traits and extraction of valuable products especially of medicinal value from these plants, conventional methods might not be successful (Dunstan and Thorpe, 1986). To meet the commercial requirement as well as to rescue the endangered plant species biotechnology offers several advantages to achieve the targets in a much shorter time than that taken by conventional techniques. For this purpose, it is essential to have a highly efficient production technology including high yielding certified planting material. In this context, plant tissue culture and other biotechnological tools can play an important role in boosting production (Bapat *et al.,* 2008).

Plant cell, tissue and organ culture is an important component in the current status of biotechnology since its potential to revolutionize agriculture, horticulture, forestry and other related areas have been fully exploited. The main idea of plant tissue culture is based on the fact that cells and tissues in culture are free from certain complicated, correlative influences present in the intact plant and show maximum flexibility with regard to the differentiation and regeneration. Uniform propagules, improved progeny, enhanced vigour and superior quality are some of the characteristics of tissue culture raised plant (Jasrai and Wala, 1997).

Plant cell and tissue culture is fundamental to most aspects of plant biotechnology whose applications include plant propagation, germplasm maintenance and storage, which is crucial for retaining the gene pools of plants that are not under active cultivation (Cocking, 1989; Schmauder and Boebel, 1990; Bansal, 2005). Within the last few decades plant tissues and cell culture has grown immensely and has carved out a niche of its own as a separate branch of biological sciences. Recent advances in the techniques and applications of plant cell culture and plant molecular biology have created unprecedented opportunities for genetic manipulation of agriculture, horticulture and forest plant species (Bonga, 1982; Nirmala and Kaul, 2005; Pareek and Mathur, 2005). It has become major tool in the study of an increasing number of fundamental and applied programmes in the plant sciences. Its increasing use to investigate cell and developmental Biology, Biochemistry, Physiology, Genetics and Molecular Biology is providing new knowledge about fundamental characteristics of plants. Furthermore, it has become an integral part of Plant Biotechnology Research (Tarun *et al.,* 2005).

One of the most exciting and important aspects of *in vitro* cell and tissue culture approach is the capability to regenerate and propagate plants from cultured cells and tissues. Micropropagation involves the production of plants from very small plant parts, tissues or cells grown aseptically in test tubes or other containers where the environment and nutrition can be rigidly controlled (Murashige and Skoog, 1962; Hartmann and Kester, 1989). The inherent capacity of a plant cell to give rise to a whole plant, a capacity which is often retained even after a cell has undergone final differentiation in the plant body is described as "cellular totipotency" (Steward, 1963; Bhojwani and Razdan, 1992). For a differentiated cell to express its totipotency it first undergoes dedifferentiation followed by redifferentiation. Tissue culture techniques offer not only an excellent opportunity to study the factors that elicit the totipotentiality of cells but also allow investigation of factors controlling cytological, histological and organogenetic differentiation.

Micropropagation has significant advantages over traditional clonal propagation techniques (Karnosky, 1981). These include the potential of combining rapid large-scale propagation of new genotypes, the use of small amounts of original germplasm and the generation of pathogen-free propagules.This impressive application of the principles of plant cell division and regeneration to practical plant propagation is the result of continuous tedious studies in hundreds of laboratories worldwide, many of them in developing countries, on the standardization of explant sources, media composition and physical state, environmental conditions and acclimatization of *in vitro* plants (Shrivastava *et al.,* 1996). Particularly noteworthy are the many recent studies on the molecular aspects of organogenesis and somatic embryogenesis. However, further practical applications of micropropagation, which is also commercially viable, depends on reducing the production costs such that it can compete with seed production or traditional vegetative propagation methods (*e.g.*, cuttings, tubers and bulbs, grafting). Indian tissue culture laboratories are among the pioneering ones in establishing technologies for *in vitro* multiplication of several species (Shreedhar *et al.,* 1998).

The application of micropropagation techniques provides several benefits viz.:

☆ It enhances the rate of rapid multiplication of plants, producing thousands of plantlets in a matter of months.

☆ It provides the availability of plants throughout the year.

☆ It helps the conservation of genetic resources of threatened plants.

☆ Plant improvement can be brought about by regeneration techniques.

☆ Healthy plant material is ensured since soil and disease-causing organisms are excluded during the propagation cycle.

☆ The method is programmable to meet specific targets of time and quantity because it is independent of seasonal changes and the weather.

☆ Micropropagation saves an enormous amount of care usually required by cuttings and seedlings (watering, weeding, spraying etc.).

☆ Excess material produced can often be stored over long periods.

☆ Species and cultivars can be stored in small spaces.

The techniques that have been developed for micropropagation are described in greater detail.

Clonal Propagation

Propagation by seeds is the major and most efficient method by which plant reproduce in nature (Hartmann and Kester, 1989, Naidu *et al.*, 2006). However, forest species do not always have a successful sexual reproduction and for many species vegetative propagation is difficult to achieve. Test tube plant breeding (Martin, 1985) as one of the methods for forest species propagation has been used to carry out clonal breeding and afforestation.

Genetically uniform (identical) plants derived from a single individual or explant are called clones (Reiger *et al.*, 1996). Processes that produce clones can be put under the term cloning'. This includes all the methods of vegetative propagation such as cutting, layering, and grafting. Propagation by tissue culture also helps in producing clones. Using the shoot tip, it is possible to obtain a large number of plantlets. This technique is used extensively in the commercial field for micropropagation of ornamental plants like *Chrysanthemum, Gladiolus*, etc. and also crops such as sugar cane, tapioca, and potato. Thus an unlimited number of plants that is genetically similar or is clones can be produced in a short span of time by tissue culture.

Single-cell Culture

Establishment of a single-cell culture provides an excellent opportunity to investigate the properties and potentialities of plant cells. Such studies contribute to our understanding of the interrelationships and complementary influences of cells in multicellular organisms. Several workers have successfully isolated single-cell division and even raised complete plants from single-cell cultures. Using cell cultures in studies designed to describe the attention of plant biologists. It was soon realized that single-cell systems have great potential for crop improvement. Free cells in cultures permit quick administration and withdrawal of diverse chemicals and substances, thereby making them easy targets for mutant selection.

Apart from this, the individual cells with a population of cultured cells invariably show cytogenetic and metabolic variations depending on the stage of the growth cycle and culture conditions. Such variability, termed spatial heterogeneity, has been the subject of much interest since differences between cells in their karyotype and the ability to accumulate secondary metabolites are manifested during morphogenesis in the clones regenerated from single cells. In this way the cell line selection technique can be usefully applied to produce high-yielding cultures as well as plants with superior agronomic traits.

Organogensis

Organogenesis is an outcome of the process of dedifferentiation followed by redifferentiation of cells. Dedifferentiation favors unorganized cell growth and the resultant developed callus has meristems randomly divided. Most of these meristems, if provided appropriate *in vitro* conditions, would redifferentiate shoot buds and roots.

As early as 1939 it was observed by White that by submerging callus from a tobacco hybrid in a stationary liquids medium, leaf buds and shoot like structures could be induced. He interpreted this to be the result of limited air supply. Shoot-bud differentiation in cultured tissues is dependant on the auxin/cytokinin ratio in the medium. Skoog and his coworkers have done detailed and comprehensive work on this subject. Skoog and Miller (1957) rejected the concept of organ-forming substances (rhizocaulines and caulocalines) proposed by Went (1938) and instead suggested that organ formation is controlled by quantitative interaction (ratio rather than absolute concentration) of substances in growth and development.

Embryogenesis

Somatic embryogenesis is the phenomenon of *in vitro* production of embryo like structures, from the somatic cells of the plant without the intervention of sexual fusion as occurs in germ cells (Raghavan, 1986). Somatic embryogenesis is the process of a single cell or group of cells initiating the developmental pathway that leads to reproducible regeneration of non-zygotic embryos capable of germinating to form complete plants. Under natural conditions this pathway is not normally followed but from tissue cultures somatic embryogenesis occurs more often and as an alternative to organogenesis for regeneration of whole plants. According to Sharp *et al.* (1980), somatic embryogenesis is initiated either by "pre-embryogenically determined cells (PEDCs)" of by "induced embryogenically determined cells (IEDCs)".

In pre-embryogenically determined cells, the embryogenic pathway is predetermined and the cells appear to only wait for the synthesis of an inducer (or removal of an inhibitor) to resume independent mitotic divisions in order to express their potential. Induced embryogenic determined cells, on the other hand, require re determination to the embryogenic state by exposure to specific growth regulators (Etienne *et al.,* 1993 a,b). These cells are dedifferentiated generally in microspore (anther) cultures and callus cultures. Once the embryogenic state has been reached both cell types proliferate in a similar manner as embryogenically determined cells. Following the full embryogenic pathway as a coordinated group of embryogenic determined cells then directly produces plantlets. Sometimes individual cell or cells from the group may escape and give rise to either embryoids or nodular embryogenic callus consisting of proembryoids. These are embryo like structures which are bipolar units and germinate into full plantlets under suitable culture conditions.

Haploid Production

Haploids are sporophytes of higher plants with gametophytic chromosome number. Ever since Guha and Maheshwari discovered haploid plants in 1964 (Guha and Mukharjee, 1999), plant breeders have worked intensively to obtain haploids either *in vitro* or *in vivo*. The *in vivo* techniques employed are gynogenesis, androgenesis, genome elimination by distant hybridization, semigamy, chemical treatment, temperature shocks and irradiation effects. *In vivo* methods using anther culture or pollen culture yield spectacular results.

As a result of haploid induction followed by chromosome doubling, homozygosity can be achieved in the quickest possible way making genetic and breeding research much easier. The genetic segregation is simplified in homozygotes, recessive genes not being masked by dominant ones.

Somatic Hybridization and Cybridization

Another process, other than the sexual cycle that has become available for higher plants, which can lead to genetic recombination. This non-conventional genetic procedure involving fusion between isolated somatic protoplasts under *in vitro* conditions and subsequent development of their product (heterokaryon) to a hybrid plant is known as somatic hybridization. Ever since the first report on protoplast fusion-derived somatic hybrid plants of *Nicotiana glauca + N. langsdorffii* by Carlson *et al.* (1972), somatic hybridization has opened up several possibilities for the parasexual manipulation of plants.

Plastids and mitochondrial genomes (cytoplasmically encoded traits) are inherited maternally in sexual crossings. Through the fusion process the nucleus and cytoplasm of both parents are mixed in the hybrid cell (heterokaryon). This results in various nucleocytoplasmic combinations. Sometimes interactions in the plastome and genome contribute to the formation of cybrids (cytoplasmic hybrids). Cybrids, in contrast to conventional hybrids, possess a nuclear genome from only one parent but cytoplasmic genes from both parents. The process of protoplast fusion resulting in the development of cybrids is known as cybridization. In cybridization, heterozygosity of extrachromosomal material can be obtained, which has direct application in plant breeding. Production of synthetic artificial seeds from somatic embryos/shoot buds have recently resulted in low cost, high volume propagation, rapid multiplication, incorporation of nutrients etc.

Somaclonal Variations

Somaclonal variation is a term coined by Larkin and Scowcroft (1981) to cover all types of variations which occur in plants regenerated from cultured cells or tissues. It is well known that genetic variations occur in undifferentiated cells, isolated protoplasts, calli, tissues and morphological traits of regenerated plants. The cause of variation is attributed to changes in the chromosome number and structure. Genetic heterogeneity in cultures arises mainly due to such factors as: a) the expression of genetic disorders in cells of the initial explants, and b) new irregularities brought about by culture conditions through spontaneous mutants.

In vitro Regeneration of *Chlorophytum borivilianum* Sant. et Fernand.

Chlorophytum borivilianum Sant. et Fernand is an important endangered medicinal plant. The species is endemic to India distributed throughout India especially in the dry hilly regions of Rajasthan, Madhya Pradesh, Jharkand, Chattisgarh, Maharashtra and Gujarat (Oudhia, 2000; Yadava and Pandey, 2004). Among the medicinal *Chlorophytum* species reported in India, each species has a specific area of occurence. The species has recently gained prominence as safed musli because on processing it yields milky white tubers which have been used as nutritive tonic and aphrodisiac (Chadha *et al.,* 1980, Oudhia and Tripathi, 1999; Kothari, 2004).

The species was on the verge of near extinction a few years ago due to large scale collection from the forest and destructive nature of harvesting (tuber being economic organ the whole plant is uprooted), shy flowering behaviour and low seed formation, low seed germination (10-32 per cent) (Verma *et al.,* 2000; Shrivastava *et al.,* 2000) and low tuber multiplication ratio (4-5 times per annum) (Kothari and Singh, 2001). Therefore, it was declared as an endangered medicinal plant (Nayar and Sastry, 1988).

Material and Methods

Organogenic and embryogenic studies were developed for evolving an effective and efficient somatic embryogenic schedule. Explant *viz.,* stem disc sections was used for *in vitro* regeneration. Explants inoculated on Murashige and Skoog media supplemented with different auxins and cytokinins. The cultures were maintained in culture tubes and conical flasks and were kept in the culture room at a temperature of 25±2°C, relative humidity (RH) of 60-70 per cent and a light intensity of approx. 1500 lux provided by cool, white, fluorescent tubes under a photoperiod of 16/8 hr (light/dark). The liquid cultures were kept on an orbital shaking incubator with built in light, temperature and speed control and were constantly agitated at 150 rpm at a temperature of 25±2°C. Experiments with static liquid cultures with filter paper support were also carried out.

The frequency of shoot initiation, mean number of shoots and mean shoot length was calculated from the buds showing signs of differing morphogenesis. The frequency of root initiation, mean number of roots and mean root length was calculated from the elongated shoots. The frequency of embryogenesis and mean number of embryos was calculated on the basis of visual observation of explants exhibiting embryogenic characteristics- Frequency of somatic embryogenesis (FSE), Mean number of somatic embryos (MNSE), Frequency of conversion into plantlets (FC), frequency of greening (FGr), Frequency of normal germination (FG).

Hardening and Acclimatization

When the shoots of the *in vitro* regenerated plantlets attained a height of 8-10 cm., bearing healthy leaves and a good root system they were transferred into small plastic cups containing presterilized sand:soil (1:1) holed at bottom with a thin layer of soil and upper layer of sand and were covered with bell jars and transparent polythene to protect excessive water loss from leaves in natural conditions. After a week the mouth of the bell jar was opened for an hour each day and after 15-20 days the bell jar was gradually removed. As the plantlets started growing in height they were transferred into bigger pots by carefully removing the plastic cup.

Results and Discussion

Somatic Embryogenesis

Initiation

Somatic embyos have been obtained in the present study from stem disc explants of *C. borivilianum.* Somatic embryogenesis from seedlings has earlier been reported from seedling explants of *C. borivilianum* (Ramawat *et al.,* 1997).

The application of various combinations and concentrations of plant growth regulators simultaneously or sequentially has been known to trigger embryogenesis *in vitro* (Gray, 1995). The mechanism involved is the alteration of cell polarity and promotion of subsequent asymmetrical, periclinical and oblique divisions (DeJong *et al.*, 1992). Induction of the process of somatic embryogenesis was characterized by the formation of a soft, transparent, shiny and highly fragile proliferating tissue giving rise to somatic embryos from its superficial layers. Initiation of somatic embryos involved the formation of small, globular, shining structures on the stem disc explants in the present study. The expression of somatic embryogenesis in tissue has been regarded as an outcome of genetic, epigenetic and physiological changes at least to

Table 9.1: Effect of cytokinin with auxin on somatic embryogenesis in *C. borivilianum*.

PGR conc. (μM)	2,4-D	Medium			
		MS		B5	
		FE	MNE	FE	MNE
BA					
0.44		25±1.2	14±0.7	29±1.21	18±0.86
	0.45	21±1.1	18±0.78	25±1.08	20±0.83
	2.26	33±1.4	16±0.8	38±1.58	15±0.65
	4.52	19±0.76	11±0.55	21±0.84	14±0.61
2.2		23±1.15	23±1.09	27±1.35	20±0.83
	0.45	21±0.88	20±0.04	29±1.45	19±0.79
	2.26	25±1.11	24±0.96	41±1.64	20±1.0
	4.52	17±0.68	21±0.91	19±0.95	24±0.96
4.4		21±0.8	12±0.57	24±1.0	16±0.8
	0.45	31±1.29	20±0.87	28±1.17	17±0.68
	2.26	34±1.7	25±1.25	37±1.61	21±1.05
	4.52	23±0.9	19±0.79	20±0.83	25±1.08
KN					
0.46		29±1.4	18±0.9	30±1.43	24±1.0
	0.45	32±1.33	20±0.8	37±1.61	29±1.21
	2.26	40±2.0	19±0.76	46±2.3	31±1.55
	4.52	21±0.9	15±0.65	28±1.12	20±0.87
2.32		51±2.1	28±1.33	69±3.45	34±1.62
	0.45	59±2.5	38±1.9	74±3.1	42±1.68
	2.26	68±3.4	40±1.67	86±3.74	51±2.13
	4.52	57±2.5	25±1.25	61±2.44	34±1.48
4.65		23±1.1	11±0.46	29±1.21	20±1.0
	0.45	27±1.1	17±0.68	34±1.7	26±1.04
	2.26	32±1.33	21±0.91	39±1.95	28±1.17
	4.52	20±0.8	20±0.83	23±0.92	16±0.8

some degree (Narayanswamy 1994). The embryoids in the present study appeared towards the periphery of the explant. Embryogenic tissue was characterized by cell aggregate (Emons *et al.,* 1992) consisting of small thin walled cytoplasmic-rich dividing cells located at the periphery was evident from histological sections of *C. borivilanum* too. B5M was found to be better than MSM in initiating and proliferation of somatic embryos (Muraldharan and Mascarenhas, 1995).

Table 9.2: Effect of abscisic acid on maturation of somatic embryos in *C. borivilianum* on selected medium.

SM +Abscisic Acid (mg/l)	I		
	MNE	FGr	FPG
0.01	48±2	42±1.68	29±1.2
0.05	30±1.5	41±1.64	32±1.6
0.1	23±1	39±1.95	38±1.52
	II		
	MNE	FGr	FPG
0.01	51±2.1	69±2.87	17±0.68
0.05	32±1.6	38±1.65	25±1.25
0.1	20±0.94	32±1.33	34±1.47
	III		
	MNE	FGr	FPG
0.01	50±2.0	88±3.83	11±0.46
0.05	24±1.0	34±1.7	24±1.2
0.1	14±0.56	29±1.16	32±1.39

Note: Selected Medium - B5 medium supplemented with 2.26 µM 2,4-D +2.32 µMKn.

Of all the auxins employed in the present work, 2, 4-D (2.26 µM) has proved to be the most effective in initiating embryogenic cultures as also observed earlier (Evans *et al.,* 1981; Litz. *et al.,* 1995). On repeated subculture to the media supplemented with auxin (2,4-D) alone, the embryos turned yellowish brown and lost their initial embryogenetic potential. Reversion of globular stage proembryos to calli has earlier been observed in *Hevea brasiliensis* (Carrons *et al.,* 1995).

Repetitive Embryogenesis

PGRs and stress play a central role in mediating the signal transduction cascade leading to the reprogramming of gene expression (Dudit *et al.,* 1995; Arnold *et al.,* 2003). Some primary somatic embryos fail to mature and instead give rise to successive cycles of new embryo production. The cyclic production of new generations of somatic embryos is known as recurrent, secondary or repetitive somatic embryogenesis (Merkle *et al.,* 1990). In the present study, repetititve embryogenesis was observed on B5M +KN (2.32µM) +2,4-D (2.26 µM) exhibited the best frequency of somatic embryogenesis (86) and number of embryos (51). This medium was, therefore, designated as the Maintenance Medium (MM) for maintenance of somatic embryos (Table 9.3).

Table 9.3: Effect of cytokinings on caulogenesis from stem disc explant of *C. borivilianum.*

S.No.	PGR Conc (µM)	FSI	MNS	MLS
1.	Control	6.6 ± 0.44	1.5±0.5	0.5±0.12
2.	BA			
	0.44	14±0.1	2.5±0.5	1±0.21
	2.2	30±1.63	3.2±0.42	1.5±0.15
	4.44	25.6±0.1	4.1±0.49	1.7±0.37
	22.2	16±0.32	23.6±1.3	1.4±90.06
	44.4	12±0.35	14.7±0.85	1.5±0.46
3.	Kn			
	0.46	34±0.78	2.3±0.35	1±0.1
	2.32	23±1.45	2.9±0.59	2.5±0.23
	4.65	21±3.06	3.8±0.76	5±0.42
	23.23	11±1.11	5±1.16	2.5±0.53
	46.46	8±0.5	1.6±0.4	1.5±0.13

A similar report of proliferation of somatic embryos was also reported by Arora *et al.* (1999). The loss of integrated control of a group of cells and the tendency of single embryonic cells to break away and express their potential, allows continuous proliferation of globular pro-embryoid like nodules as reported in past (Litz and Conover, 1983; Kononwicz *et al.,* 1984). A single culture undergoing repetitive embryogenesis is theoretically capable of generating an unlimited number of somatic embryos and thus has a great potential for mass propagation and gene transfer, in a short period of time (Durham and Parrot, 1994; Baker and Wetzstein, 1995) (Figures 9.1–9.11).

Maturation and Germination

Maturation of somatic embryos was brought about when MM was supplemented with ABA (0.01 mg/l). The proportion of embryos to obtain maturity on ABA-containing media was greater than on ABA free medium. Similar observations were reported by Muralidharan and Mascarenhas (1995). ABA in the medium is known to favor the standard development and singulation of the embryos (Eitenne *et al.,* 1993 b,c). ABA reduced undesired features such as abnormal development, callus growth, secondary embryo production and root hair development in *Salix* (Gronroos, 1995). It also brought a drastic decline in the frequency of precocious germination. ABA treated somatic embryo showed an enhanced frequency of conversion compared to non-treated ABA cultures (Table 9.4).

Conversion and Plantlet Formation

The conversion of embryos into plantlets was increased when ABA treated somatic embryos transfer to B5 supplemented with KN (2.32 µM). The frequency of conversion of somatic embryos into plantlets was 81 per cent with high frequency of germination and reduced precocious germination.

Figures 9.1–9.3: Initiation of embryogenesis in *C. borivilianum*.

Figures 9.4–9.6: Proliferation of embryos on MM (Maintenance Medium).

Figures 9.7–9.9: Germinating somatic embryos.

Organogenesis

Shoot Initiation

None of the explants except the stem disc responded to the *in vitro* conditions. Stem disc (0.5–1 cm in diameter) and 0.5 cm in width/height with greenish meristematic zone was the most responding explant in the present study. The cytokinin (BAP) at its moderate concentrations. (2.2 µM and 4.44 µM) proved effective in the emergence of shoots from the apical and lateral portions of the explant. However, number of shoots was found to be highest at higher concentrations of BAP (22.2 µM). Multiple shoots were induced from all parts of the explant accompanied by a

Figures 9.10–9.16: Multiple shoot formation at various stages.
Figures 9.17–9.23: Elongated shoots separated and subjected for rooting

precursory proliferation of explant tissue which served as initials for further multiple shoot proliferation (Figures 9.12–9.23). The treatment involving combination of cytokinins proved ineffective in terms of all the parameters taken into consideration except the length of shoots. The presence of BAP in the medium was most effective in inducing shoot initiation from cultured stem disc explant of *Chlorophytum* as compared to KN. This is in confirmation to the earlier reports of many workers who found BAP to be more suitable than KN (Kothari and Chandra, 1984; Sen and Sharma, 1991; Harada and Murai, 1996; Komalavalli and Rao, 1997).

Table 9.4: Effect of auxins on rooting of elongated shoots of *C. borivilianum.*

S.No.	PGR Conc (µM)	FRI	MNR	MLR
1.	Control	40±2.67	4.75±0.34	2.5±0.17
2.	IBA			
	0.44	92±4.8	5.67±0.81	2.1±0.23
	2.2	80±4.7	5.9±0.66	1.98±0.33
	4.4	72±4.5	6.2±1.03	1.8±0.26
3.	NAA			
	0.54	90±5.3	7.5±0.83	1.94±0.28
	2.69	78±4.8	6.9±0.98	1.82±0.20
	5.37	62±3.3	6.8±1.13	1.74±0.29
4.	IAA			
	0.57	74±4.6	5.2±0.74	1.71±0.19
	2.85	68±3.6	4.8±0.8	1.5±0.21
	5.71	58±3.4	5.2±0.58	1.23±0.2

Rooting of Microshoots

The rooting of the microshoots was achieved in the medium having reduced strengths (full, half and quarter) with all the PGRs (alone and in combination). Highest FRI and MLR was achieved on Half MS+IBA (0.44 µM). Best MNR (8-9) was produced on MSM supplemented with NAA (0.54 µM).

The roots developed varied in their morphology from thick and fleshy to thin, white and hairy. An unusual occurrence of thick brown tuberous roots was observed from the base of the elongated shoots in the present study.

Hardening and Acclimatization

About 10 cm old plantlets bearing 3-4 cm long roots were used for hardening and acclimatization. The plantlets on transfer from controlled culture conditions to soil exhibited an initial shock in the form of temporary wilting. To prevent this plantlets were kept under polythene covers and inverted beakers for 2 weeks. Maintenance of soil – transferred plantlets under polythene cover enabled them to overcome the shock caused due to low humidity conditions. The plantlets were gradually exposed to the natural environment and irrigated regularly with sterile distilled water for a month. The plantlets were successfully hardened and acclimatized within a span of

15-21 days. The successfully acclimatized plantlets exhibited increase in size of leaf and developed tubers similar to the naturally grown plants These were capable of producing new offshoots at the onset of favorable environmental conditions (monsoon). A total of 70 per cent plantlets were found surviving upon transfer to soil.

Prolific and vigorous growth of multiple shoots and somatic embryos, formation of well developed healthy plantlets, their high survival rate (60-80 per cent) in soil and morphological similarity with the natural plants and increased presence of saponins in tubers formed on hardened and acclimatized *in vitro* plantlets are the major achievements.

References

Arnold, Sara von, Sabala, I., Bozhkov, P., Dyachok, J. and Filonova, L. 2003. Develpmental pathways of somatic embryogenesis. Plant Cell Tiss Org Cult **69**: 233–249.

Arora D.K., Suri S.S. and Ramawat, K.G. 1999. Factos affecting somatic embryogenesis in long term callus cultures of safed musli (*Chlorophytum borivilianum*) an endangered wonder herb. Indian Journ. Exp. Biol. **37**: 75–82.

Baker, C.M. and Wetzstein, H.Y. 1995. Repetitive somatic embryogenesis in peanut cotyledon cultures by continual exposure to 2,4–D. Plant Cell.Tiss.Org.Cult. **40**(3): 249–254.

Bansal, Y.K. 2005. Micropropagation of Minor Forest Produce (MFP) species: A review In: Advances in Biotechnology (Ed. P.C. Treivedi). Agrobios (India) Jodhpur, pp 327 – 339.

Bapat V.A., Yadav S.R. and Dixit G.B. 2008 Rescue of endangered plants through biotechnological applications. Science Letters **31**(7–8): 201–210.

Bhojwani, S.S. and Razdan, M.K. 1992. Plant Tissue Culture: Theory and Pracrice, Elsiever, Amsterdam, The Netherlands.

Bonga, J.M. 1982. Vegetative propagation in relation to juvenility, maturity and rejuvenation. In: Tissue Culture in Forestry (Eds. J. M. Bonga and D.J. Durzan). Martinus Nijhoff/Junk, The Hague, pp 4–35.

Carlson, P.S., Smith, H.H. and Dearing, R.D. 1972. Parasexual interspecific plant hybridization. Proc. Natl. Acad. Sci. **69**: 2292–2294.

Carron, M.P., Etienne, H. and Michaux–Ferriere, N. 1995. Somatic embryogenesis in rubber tree (*Hevea brasiliensis* Mull.). In: Biotechnology in Agriculture and Forestry. Vol.30, Springer Verlag, Heidelberg, pp.353–369.

Chadha, Y.R., Gupta, R. and Nagarjun, S. 1980. Scientific appraisal of some commercially important plants of India. Ind. Drugs Pharm Ind **15** : 7.

Cocking, E.C. 1989. Plant cell and tissue culture In: A Revolution in Biotechnology (Ed. J.L. Mark). Cambridge Univ Press, Cambridge.

De Jong, A.K., Cordewener, J., Lo Schiavo, F., Terzi, M., Vandekerkhove, J., Van Kammen, A. and De Vries, S.C. 1992. A carrot somatic embryo mutant is rescued by chitinase. Plant Cell **4**: 425–433.

Dudit, D., Gyorgyey, J., Borge, L. and Bako, L. 1995. Molecular biology of somatic embryogenesis. In: Thorpe, T.A. (ed) *In vitro* embryogenesis in plants (pp 267–308). Kulwer Academic Publishers Dordrecht, Boston, London.

Dunstan, D.I. and Thorpe, T.A. 1986. Regeneration in Forest Trees. In: Cell Culture and Somatic Cell Genetics of Plants Vol. 3. (Ed.) I.K. Vasil. Academic Press, New York, pp. 223–241.

Durham, R.E. and Parrot, W.A. 1994. Repetitive somatic embryogenesis from peanut embryos in liquid medium. Plant Cell Rep. **13**: 122–125.

Emons, A.M.C., Vos, J.W. and Kieft, H. 1992. A freeze fracture analysis of the surface of embryogenic and non–embryogenic suspension cells of *Daucus carota*. Plant Sci. **87**: 85–97.

Etienne, H., Montoro, P., Michaux – Ferriere, N. and Carron, M.P. 1993a. Effects of dessication medium osmolarity and abscisic acid on the maturation of *Hevea brasiliensis* somatic embryos. J. Exp. Bot. **44**(267): 1613–19.

Etienne, H., B. Sotta, P., Montoro, Miginiac, E. and Carron, M.P. 1993b. Comparison of endogenous ABA and IAA contents in somatic and zygotic embryos of *Hevea brasiliensis* Mull. Arg. during ontogenesis. Plant Sci. **92**: 111–119.

Evans, D.A., Sharp, W.R. and Flick, C.E. 1981. Growth and behaviour of cell culture embryogenesis and organogenesis. In: Plant Tissue Culture Methods and Applications in Agriculture (Ed.) T. A. Thorpe. Academic Press, London, New York, pp.75–113.

Gray, D.J. 1995. Somatic embryogenesis in grapes. In: Somatic embryogenesis in woody plants. Vol. 2 (Eds.) S. Jain, P. Gupta, R. Newton. Kluwer Acad. Publ., The Netherlands, pp. 191–218.

Gronroos, L. 1995 Somatic embryogenessis in *Salix*. In: Somatic embryogenessis in woody plants (Eds.) S. Jain, P. Gupta and R. Newton. Kluwer Academic Publishers, The Netherlands pp. 219–234.

Guha-Mukharjee, S.1999. The discovery of haploid production by anther culture. *In vitro* cell Dev. Biol. Plant **35(5)**: 357–360.

Harada, H. and Murai, Y. 1996. Micropropagation of *Prunus mume*. Plant Cell Tiss. Org. Cult. **46** (3): 265–267.

Hartman, H.S. and Kester, D.E. 1989. Plant propagation, Principle and Practice. Prentice Hall of India Pvt. Ltd. New Delhi.

Jasrai, Y.T. and Wala, B.B. 1998. Creation is God's territory, should tresspassers be prosecuted? Case of plant tissue culture. Journal of M.S. University of Baroda **44–45 (3)**: 1–6.

Karnosky, D.F. 1981. Potentials for forest tree improvement via tissue culture. Biosci. **31(2)**: 114–120.

Komalavalli, N. and Rao, M.V. 2000. *In vitro* micropropagation of *Gymnema sylvestre* – A multipurpose medicinal plant. Plant Cell, Tiss. Org. Cult. **61**: 97–105.

Kononwicz, H. and Janick, J. 1984. Response of embryogenic callus of *Theobroma cacao* L. to gibberellic acid and inhibitors gibberellic acid synthesis. Z. Pflanzenphysiol. **133**: 359–366.

Kothari S.L. and Chandra, N. 1984. *In vitro* propagation of African marigold. Hort. Sci **19**: 703–705.

Kothari, S. K. and Singh, K. 2001. Evaluation of safed musli (*Chlorophytum borivilianum* Santapau and Fernandes) germplasm. J. of Spices and aromatic crops **10**: 147–149.

Kothari, S.K. 2004. Safed musli (*Chlorophytum borivilianum*) revisited. J Med. Arom. Pla. Sci. **26**: 60–63.

Larkin, P.J. and Scowcroft, W.R. 1981. Somaclonal variation – a novel source of variability from cell cultures for plant improvement. Theor. Appl. Genet. **60**: 197–214.

Litz, R.E. and Conover, R. 1983. High frequency somatic embryogenesis from *Carica* suspension cultures. Ann. Bot. **51**: 683–686.

Litz, R.E., Moon, P.A. and Chavez, V.M. 1995. Somatic embryogenesis of leaf callus derived from mature trees of the cycad *Ceratozamia hildae*. Plant Cell Tiss. Org. Cult. **40**(1): 25–31.

Martin, C. 1985. Plant breeding *in vitro*. Endeavour **9(2)**: 81 – 86.

Merkle, S.A., Parrot, W.A. and Williams, E.G. 1990. Application of somatic embryogenesis and embryo cloning. In: Plant Tissue Cultures, Application and Limitation. (Ed.) S.S. Bhojwani. Elsevier, Amsterdam, pp. 67–101.

Muralidharan, E.M. and Mascarenhas, A.F. 1995. Somatic embryogenesis in *Eucalyptus*. In: Somatic embryogenessis in woody plants (Eds.) S. Jain, P. Gupta and R. Newton. Kluwer Academic Publishers, The Netherlands, pp.23– 40.

Murashige, T. and Skoog, F. 1962. A revised medium for rapid growth and bioassays with tobacco tissue cultures. Physiol. Plant **15**: 473 – 497.

Naidu, C.V., Josthna, P., Swamy, P.M. and Naidu, M.B. 2006. Studies on some pretreated methods to improve seed germination in four selected *Terminalia* species. J. Indian Bot. Soc. **85**: 35 – 36.

Narayanswamy, S. 1994. Plant Cell Tissue Culture. Tata McGraw Hill Publishing Co., New Dehli.

Nayar M.P. and Sastry, A.R.K. 1988. *Chlorophytum borivilianum*. In: Red Data Book of Indian Plants; Vol 2, (Eds.) M.P. Nayar and A.R.K. Sastry (Botanical Survey of India, Calcutta) pp142.

Nirmala, C. and Kaul, M.L.H. 2005. Biotechnology miracle of mirage II Remarkable genetic system of Agrobacterium In: Trends in Plant Tissue Culture and Biotechnology (Ed. L.K. Pareek). Agrobios, India, Jodhpur.

Oudhia, P. 2000. Problems perceived by safed moosli (*Chlorophytum borivilianum*) growers of Chattisgarh (India) region: A study. J. Med. Arom. Plant. Sci 22–23(4A–1A): 396–399.

Oudhia, P. and Tripathi, R.S. 1999. Scope of cultivation of important medicinal plants in Chattisgarh plains. In: Proceedings Of National Conference on Health care and Development of Herbal Medicines. Indira Gandhi Agricultural University, Raipur (India). pp. 71–78.

Pareek, L.K. and Mathur, S. 2005. Regeneraqtion of plantlets through tissue culture methods in dicotyledonous fruit trees – A brief review. In: Trends in Plant Tissue Culture and Biotechnolgy (Ed. L.K. Pareek), Agrobios, Jodhpur, India.

Raghavan, V. 1986 Somatic embryogenesis In: Embryogenesis in Angiosperms: A Developmental and Experimental Study. Development Cell Biology Series (Eds.) P. W. Barlow, P. B. Green and C. K. Wylie, New York Cambridge University Press. pp 115–151.

Ramawat, K.G., Jain, S. and Sonie, K.C. 1997. In: Biotechnological Applications of Plant Tissue and cell culture, (Eds.) Ravishankar, G.A. and Venkataraman, L.V. (Oxford and IBH Publishing Co. Pvt. Ltd.) pp. 199–203.

Reiger, R., Michaelis, A. and Green, M.M. 1996. Glossary of genetics (5th Edition) Narosa Publishing House, New Delhi.

Schmauder, H.P. and Boebel, P. 1990. Plant cell cultivation as a biotechnological method. Acta Biotechnol. **10**(6): 501 – 506.

Sen, J. and Sharma, A.K. 1991. Micropropagation of *Withania somnifera* from germinating seeds and shoot tips. Plant Cell Tiss Org. Cult. **26** : 71 – 73.

Sharp, W.R., Sondahl, M.R., Caldas, L.S. and Maraffa, S.B. 1980. The physiology of *in vitro* asexual embryogenesis. Hortic Rev. **2**: 268–310.

Shreedhar, D., Rao, M.M., Ramesh, C. and Lawrence, L. 1998. Commercial scale micropropagation of *Eucalyptus camaldulensis*. Indian For. **124**: 217 – 224.

Shrivastava, A., Kumbhare, V., Pandey, D.K. and Shrivastav, S. 1996. Preliminary phytochemical and insecticidal studies of leaf extract of *Adhatoda vasica* on *Albizia lebbeck* Defoliator, *Rhesala imparatawak*. (Lepidoptera: Noctuidae). Indian J. For. **19**(1): 83–86.

Shrivastava, D.K., Mishra, P.K., Verma, S. and Gangrade, S.K. 2000. Studies on propagation methods and dormancy in safed musli (*Chlorophytum* spp.). J Med. Arom. Plant Sci **22–23**(4A–1A): 275–276.

Skoog, F. and Miller, C.O. 1957. Chemical regulation of growth and organ formation in plant tissue cultured *in vitro*. Symp. Soc. Exp. Biol. **11**: 118–131.

Steward, F.C. 1963. Totipotency and variation in cultured cells: some metabolic and morphogenetic manifestation. In: Plant Tissue and Organ Culture – A Symposium. (Eds. P. Maheshwari and N.S. Rangaswamy). Catholic Press, Ranchi, India, pp. 1 – 25.

Tarun K., Tomar U.K. and Emmanuel C.J.S.K. 2005 Recent advances in tree biotechnology and the associated environmental concerns. In: Advances in Biotechnology (Ed. P.C. Trivedi), Agrobios, India pp 307–316.

Verma, S., Sharma, R.K. and Shrivastava, D.K. 2000. Seed germination, viability and invigoration studies in medicinal plants of commercial value. J. Med. Arom. Pla. Sci **22–23**(4A–1A): 426–428.

Went, F.W. 1938. Specific factors other than auxin affecting growth and root formation. Plant Physiol. **13**: 55–80.

White, P.R. 1939. Potentially unlimited growth of excised plant callus in an artificial nutrient. Am. J. Bot. **26**: 59 – 64.

Wilson, E.O. 1992. The Diversity of Life. W.W. Norton and Co., New York.

Yadava, A.K. and Pandey, R. 2004. *Chlorophytum borivilianum* Baker: A Potent medicinal plant. Indian For. **130** (3): 340–342.

2013, Abiotic Stress and Biotechnology
Editor: T. Pullaiah
Published by: REGENCY PUBLICATIONS, NEW DELHI

Pages 209–221

Chapter 10

Secondary Metabolites Enhacement through Mediation of *Agrobacterium rhizogenes*

H.P. Sharma

Laboratory of Biotechnology, University Department of Botany, Ranchi University, Ranchi – 834 008, Jharkhand, India

Introduction

Medicinal plants are unique source of pharmaceutically significant secondary metabolites which includes alkaloids glycosides, and flavonoids etc. (Raskin *et al*, 2002; Newman *et al.*, 2003; Samuelson, 2004; Balunas and Kinghorn, 2005). Demands of these secondary metabolites are not readily accomplished because of their low yield in intact plants, environmental, geographical and/or governmental restrictions. Chemical synthesis or semi-synthesis of these metabolites is either extremely difficult or economically infeasible because of their highly complex structure and stereospecific chemical nature. On account of over exploitation of these plant resources for extraction purpose will put the plants under various degree of threats; specially plants where roots are the potential source of medicines. Such plants will be at high risk of extinction as their extraction requires uprooting resulting death of the whole plant. At this point biotechnological approaches hold much promise as an alternative tool facilitating in conservation of the donor plants and to give scale-up production of secondary metabolites (Verpoorte *et al.*, 2002). One of the important tools is micropropagation, where large number of plants could be produced in small time and the second method is production of hairy roots through the mediation of *Agrobacterium rhizogenes*. The hairy roots produced by *A. rhizogenes* infection are characterized by high growth rate, genetic stability and growth in hormone free media. Hairy root cultures offer promise for high production and productivity of valuable secondary metabolites (used as pharmaceuticals, pigments and flavours; Table 10.1) in many plants (Flores and Filer, 1985; Hamill *et al.*, 1987; Giri and Narasu, 2000; Oksman-Caldentey and Arroo, 2000). Besides, these transformed roots have the potential to generate whole viable plant and maintain their genetic stability during further sub-culturing and regeneration. In addition to this recent development, bioreactor systems provide ways to scale up hairy root cultivation from small amount. In addition to hairy root induction there are other methods which are equally effective for the scaling up the production of useful products, such a cell culture of selected elite lines, biotransformation, elicitation and entrapping methods etc.

Table 10.1: *A. rhizogenes* mediated production of secondary metabolites.

Species	Transformation Method	Metabolite	References
Psoralea corylifolia	*A. rhizogenes*	Phytoestrogenic isoflavones daidzein and angenistein	Shinde *et al.*, 2009
Centaurium erythraea	*A. rhizogenes*	Secoiridoids	Piatczak *et al.*, 2006
Rubia akane	*A. rhizogenes*	Anthraquinone	Park and Lee, 2009
Glycyrrhiza glabra	*A. rhizogenes*	triterpenoid saponin Glycyrrhizin	Mehrotra *et al.*, 2008
Plumbago rosea	*A. rhizogenes*	Plumbagin	Satheeshkumar *et al.*, 2009
Camellia sinensis	*A. rhizogenes*	Polyphenols, Catechins, Caffeine	John *et al.*, 2009
Plumbago rosea	*A. rhizogenes*	Plumbagin	Yogananth and Basu, 2009
Psoralea corylifolia	*A. rhizogenes*	Psoralen	Baskaran and Jayabalan, 2009
Nicotiana benthamiana	*A.rhizogenes*	Human acetylchol-inesterase	Woods *et al.*, 2008
Podophyllum hexandrum	*A. rhizogenes*	Podophyllotoxin	Li *et al.*, 2009
Thalictrum flavum	*A. rhizogenes*	Antimicrobial properties	Samanani *et al.*, 2002
Saussurea involucrata	*A. rhizogenes*	Rutin, hispidulin and syringin	Fu *et al.*, 2006
Taxus brevifolia	*A. rhizogenes*	Taxol	Han *et al.*, 1994
Rauvolfia micrantha	*A. rhizogenes*	Ajmalicine, ajmaline	Sudha *et al.*, 2003
Pueraria phaseoloides	*A. rhizogenes*	Puerarin	Shi and Kintzios, 2003
Cichorium intybus	*A. rhizogenes*	Esculin and esculetin	Bais *et al.*, 2000
Panax ginseng	*A. rhizogenes*	Ginsenosides	Bapat and Ganapathi, 2005
Glycyrrhiza uralensis	*A. rhizogenes*	Glycyrrhizin	Saito *et al.*, 1991
Azadirachta indica	*A. rhizogenes*	Insecticide	Jung and Tepfer 1987
Picrorhiza kurroa	*A. rhizogenes*	Picroside-I (hepatoprotective)	Sood and Chauhan, 2010

Molecular Structure of Ri Plasmid

A. rhizogenes, the causative agent of hairy root syndrome, is a common soil bacterium (Gram negative) capable of entering a plant through a wound and causing a proliferation of secondary roots. The underlying mechanism of hairy root formation is the transfer of several bacterial genes to the plant genome (Zupam and Zambryski, 1997). The observed morphogenic effects in the plants after infection have been attributed to the transfer of part of a large plasmid known as the Ri (root-inducing) plasmid. The symptoms observed with *A.rhizogenes* are suggestive of auxin effects resulting from an increase in cellular auxin sensitivity rather than auxin production (Bonhomme *et al.,* 2000a; Hong *et al.,* 2006).

Ri plasmids are large (200 to greater than 800 kb) and contain one or two regions of T-DNA and a vir (virulence) region, all of which are necessary for tumor genesis. The Ri-plasmids are grouped into two main classes according to the opines synthesized by hairy roots. **First**, agropine-type strains induce roots to synthesize agropine and the related acids. **Second**, mannopine-type strains induce roots to produce mannopine and the corresponding acids. The agropine-type Ri-plasmids are very similar as a group and a quite distinct group from the mannopine-type plasmids. Perhaps the most studied Ri-plasmids are agropine-type strains, which are considered to be the most virulent and therefore, more often used in the establishment of hairy root cultures.

The T-DNA of the agropine-type Ri-plasmid consists of two separate T-DNA regions designed the TL-DNA and TR-DNA. Each of the T-DNA fragments spans a 15 - 20 kb region, and they are separated from each other by at least 15 kb of non-integrated plasmid DNA. These two fragments can be transferred independently during the infection process. The genes encoding auxin synthesis (tms1 and tms2) and agropine synthesis (ags) have been localized on the TR-DNA of the agropine type Ri-plasmid. The mannopine type Ri-plasmids contain only one T-DNA that shares considerable DNA sequence homology with TL of the agropine-type plasmids.

Mutation analysis of the TL-DNA has led to identification of four genetic loci, designed locus *rolA, rolB, rolC,* and *rolD*, which affect hairy root induction. The complete nucleotide sequence of the TL-region revealed the presence of 18 open-reading frames (ORFs), 4 of which, ORFs 10, 11, 12 and 15, respectively, correspond to the *rolA, rolB, rolC,* and *rolD* loci.

It was also shown that *rolA, rolB,* and *rolC* play the most important role in hairy root induction. In particular, *rolB* seems to be the most crucial in the differentiation process of transformed cells, while *rolA* and *rolC* provide with accessory functions. *rolA* is associated with internode shortening and leaf wrinkling; *rolB* is responsible for protruding stigmas and reduced length of stamens; *rolC* causes internode shortening and reduced apical dominance.

Although the TR-DNA is not essential for hairy root formation it has been shown that the aux1 gene harbored in this segment provides to the transformed cells with an additional source of auxin. It has been shown that the *Agrobacterium* plasmid carries three genetic components that are required for plant cell transformation. The first component, the T-DNA that is integrated into the plant cells, is a mobile DNA element. The second one is the virulence area (vir), which contains several vir genes. These genes do not enter the plant cell but together with the chromosomal DNA (two loci), cause the transfer of T-DNA. The third component, the so-called border sequences (25 bp), resides in the *Agrobacterium* chromosome. The mobility of T-DNA is largely determined by these sequences, and they are the only *cis* elements necessary for direct T-DNA processing.

Hairy roots are fast growing and laterally highly branched, and are able to grow in hormone-free medium. Moreover, these organs are not susceptible to geotropism anymore. They are genetically stable and produce high contents of secondary metabolites characteristic to the host plant. The secondary metabolite production of hairy roots is stable compared to other types of plant cell culture. The alkaloid production of hairy roots cultures has been reported to remain stable for years. The secondary metabolite production of hairy roots is highly linked to cell differentiation. Alkaloid production decreased clearly when roots were induced to form callus, and reappeared when the roots were allowed to redifferentiate. An interesting characteristic of some hairy roots is their ability to occasionally excrete the secondary metabolites into the growth medium. However, the extent of secondary product release in hairy root cultures varies among plant species.

The average growth rate of hairy roots varies from 0.1 to 2.0 g dry weight/litre/day. This growth rate exceeds that of virtually all-conventional roots and is comparable with that of suspension cultures. However, the greatest advantage of hairy roots compared to conventional roots is their ability to form several new growing points and, consequently, lateral branches. The growth rate of hairy roots may vary greatly between species, but differences are also observed between different root clones of the same species. The pattern of growth and secondary metabolite production of hairy root cultures can also vary. Secondary production of the hairy roots of *Nicotiana rustica* L. was strictly related to the growth, whereas hairy roots of *Beta vulgaris* L. exhibited non-growth-related product accumulation. The case of the hairy roots of *Scopolia japonica* and *Hyoscyamus muticus,* the secondary products only started to accumulate after growth had ceased. Secondary metabolite synthesis dissociated from growth would be desirable for commercial production, as it would allow the use of continuous systems.

Hairy Root Induction

Surface-sterilized cotyledons were wounded and infected with *A. rhizogenes* strains. The inoculated cotyledons were co-cultivated with *A. rhizogenes* strains for 2 days at 25°C with a 16 hours photoperiod. The experiment was designed to be completely randomized with four replicates. Forty explants were used for each population. After co-cultivation, explants were transferred to hormone-free growth mediums (High salt media such as MS favors hairy root formation in some plants). Low salt media such as B5 favor excessive bacterial multiplication in the medium and therefore the explant needs to be transferred several times to fresh antibiotic containing medium before incubation. Therefore, semi-solid MS (Murashige and Skoog) medium solidified with 0.8 per cent agar, and contained 3 per cent sucrose, plus 0.4g/l augmentin (antibiotic) to kill the bacteria (pH: 5.7) at a density of 10 explants per plate (9 cm petri dish), and cultured at 25°C, with a 16 hr photoperiod. Frequency of hairy root formation for each treatment was scored 30 days after co-cultivation.

Individual roots emerged from the wound sites were excised and subcultured onto the same medium. Forty days after co-cultivation, hairy roots were weighed out and transferred to 50 ml of MS liquid medium (pH= 5.7) containing 3 per cent sucrose, and shaken in an orbital shaker at 120 rev/min at 25°C in the dark. The roots were then subcultured onto the same medium every 4 weeks. After 4 months in liquid culture, hairy roots from each explant were weighed out and the mean weight for each treatment was calculated.

The susceptibility of plant species to *Agrobacterium* strains varies greatly. Significant differences were observed between the transformation ability of different strains of *Agrobacterium* (Park and Facchini, 2000). The age and differentiation status of plant tissue can affect the chances of successful transformation.

The level of tissue differentiation determines the ability to give rise to transformed roots after *A. rhizogenes* inoculation. In this case, successful infection of some species can be achieved by the addition of acetosyringone. The ability to exploit plant root cultures as a commercial source of natural products depends on the development of suitable bioreactors. A large number of reactor configurations have been tried for 'hairy root' culture (Curtis, 1993; Cetin *et al.*, 2005; Mehrotra *et al.*, 2008). The standard tank reactor is not suitable for mass cultivation of roots because the contact of the impeller not only damages the root tips but also induce callus formation leading to decrease in productivity. Therefore, a modified stirred tank developed, where a stainless steel mesh around the impeller was fitted to prevent root damage. Hilton and Rhodes (1990) have shown such type to enhance three-fold hyoscyamine production from hairy root culture of *Datura stramonium*. Innovative strategies have been developed for operation of reactor for enhanced hairy root proliferation and secondary metabolic productivity (Huang *et al.*, 2004; Su and Lee, 2007; Mehrotra *et al.*, 2008).

Transformation Detections

In order to detect the success of genetic transformation in plant cells, there are 3 ways.

Determination of Opine Contents

The transformation of a plant cell with *A. rhizogenes* can be confirmed by transformed root morphology of hairy roots obtained after infection and their transformed regenerants. Transformed roots have an altered phenotype, profusion of laterals, and show lack of geotropism (Tepfer, 1984). Also, the transformed regenerants of hairy roots inherit an aberrant phenotype in having wrinkled leaves and shortened internode compared to their normal counterparts (Guerche *et al.*, 1987).

Since the opine synthesis in *A. rhizogenes* infected plant cells is encoded by T-DNA of Ri plasmid (White *et al.*, 1982, 1985), its detection serves as an effective biochemical marker in elucidating the transformed nature of the cultured root tissue (Petit *et al.*, 1983; Tepfer, 1984). Transformation of hairy root culture is also confirmed by the detection of opines through paper electrophoresis (Bais *et al.*, 2001). Although synthesis of opines is a firm indication that hairy roots are indeed transformed, the expression of opine genes in hairy root tissue may become unstable with time (Kamada *et al.*, 1986).

T-DNA Detection by Southern Blot Hybridization

Genomic DNA from transformed and non-transformed soil-grown plants was extracted using the CTAB extraction method. Approximately 10 mg of DNA from each sample were digested with *HindIII, BamHI, EcoRI*, respectively. Non-transformed plant was digested with EcoRI, then separated by electrophoresis 0.8 per cent (w/v) agarose gel, transferred from the agarose gel to Hybond+ nylon membrane and cross-linked to the membrane by UV light for 3 min. Probe were labeled with a ^{32}P labeled probe specific to the coding sequence of the introduced rolB, or rolA, or rolC gene for Southern hybridization. Filters were pre-hybridized in 5 ¡ SSC, 5 ¡ Denhardt's solution, 0.5 per cent SDS, 20 mg/ml denatured salmon sperm DNA at 65°C and subsequently hybridized overnight with labeled probe. After stringent washing (0.1×SSC, 0.1 per cent SDS, 65°C) filters were autoradiographed at -70°C for 3 days with an intensifying screen.

Bacterial Gene by PCR

The polymerase chain reaction was used to confirm the presence of rolB, or rolA, or rolC gene in roots by their primers. The PCR reactions were carried out in a total volume of 30 ml: 1 ml samples of the transformed plant genomic DNA, 20 pmol of each primer, 200 m M each dNTPs, 0.5 units Taq DNA polymerase and 3 ml 10 ×– PCR buffer. Cycling conditions were: denaturation at 94°C for 1 min, annealing at 55°C for 1 min and extension at 72°C for 3 min.

Samples were subjected to 30 cycles of PCR programming. Amplified products were analyzed by electrophoresis on 0.8 per cent agarose gels and detected straining with ethidium bromide.

Advantages

Secondary Metabolite Production

Hairy root cultures are characterized by a high growth rate and are able to synthesize root derived secondary metabolites. Normally, root cultures need an exogenous phytohormone supply and grow very slowly, resulting in poor or negligible secondary metabolite synthesis. However, the use of hairy root cultures has revolutionized the role of plant tissue culture for secondary metabolite synthesis. These hairy roots are unique in their genetic and biosynthetic stability. Their fast growth, low doubling time, ease of maintenance, and ability to synthesize a range of chemical compounds offers an additional advantage as a continuous source for the production of valuable secondary metabolites. To obtain a high-density culture of roots, the culture conditions should be maintained at the optimum level. Hairy root cultures follow a definite growth pattern, however, the metabolite production may not be growth related.

Hairy roots also offer a valuable source of root derived phytochemicals that are useful as pharmaceuticals, cosmetics, and food additives. These roots can also synthesize more than a single metabolite and therefore, prove economical for commercial production purposes. Transformed roots of many plant species have been widely studied for the *in vitro* production of secondary metabolites. Transformed root lines can be a promising source for the constant and standardized production of secondary metabolites.

Hairy root cultures produce secondary metabolites over successive generations without losing genetic or biosynthetic stability. This property can be utilized by genetic manipulations to increase biosynthetic capacity. Secondary metabolite biosynthesis in transformed roots is genetically controlled but it is influenced by nutritional and environmental factors. The composition of the culture medium affects growth and secondary metabolite production. The sucrose level, exogenous growth hormone, the nature of the nitrogen source and their relative amounts, light, temperature and the presence of chemicals can all affect growth, total biomass yield, and secondary metabolite production (Ford *et al.,* 1998; Bensaddek, 2001; Sivakumar *et al.,* 2005; Weathers *et al.,* 2005; Kim *et al.,* 2007).

Sucrose is the best source of carbon and is hydrolyzed into glucose and fructose by plant cells during assimilation; its rate of uptake varies in different plant cells. In hairy roots the source of new cells are in the tips so proliferation occurs only at the apical meristem and laterals form behind the elongation zone. Such a defined growth pattern leads to steady accumulation of biomass in root cultures. To obtain a high density root culture, the culture conditions should be maintained at the optimum level. Hairy root cultures are able to synthesize stable amounts of phytochemicals but the desired compounds are poorly released into the medium and their accumulation in the roots can be limited by feedback inhibition. Media manipulations have been reported to aid in the release of metabolites.

Plant Regeneration

Transformed roots are able to regenerate whole viable plants; hairy roots as well as the plants regenerated from hairy roots are genetically stable. However, in some instances transgenic plants have shown an altered phenotype compared to controls.

Plants can be regenerated from hairy root cultures either spontaneously (directly from roots) or by transferring roots to hormone-containing medium. The advantage of Ri plasmid-based gene transfer is that spontaneous shoot regeneration is obtained avoiding the callus phase and somaclonal variations. Ri plasmid-based gene transfer also has a higher rate of transformation and regeneration of transgenic plants; transgenic plants can be obtained without a selection agent thereby avoiding the use of chemicals that inhibit shoot regeneration; high rate of co-transfer of genes on binary vector can occur without selection. Further, *Agrobacterium tumefaciens* mediated transformation results in high a frequency of escapes; whereas *Agrobacterium rhizogenes* mediated transformation consistently yields only transformed cells that can be obtained after several cycles of root tip cultures.

These hairy roots can be maintained as organ cultures for a long time and subsequent shoot regeneration can be obtained without any cytological abnormality. Rapid growth of hairy roots on hormone-free medium and high plantlet regeneration frequency allows clonal propagation of elite plants. In *in vitro* cultures, the hairy root regenerants show rapid growth, increased lateral bud formation, and rapid leaf development; these regenerants are useful for micropropagation of plants that are difficult to multiply. Altered phenotypes are produced from hairy root regenerants and some of these have proven to be useful in plant breeding programs. Morphological traits with ornamental value are abundant adventitious root formation, reduced apical dominance, and altered leaf and flower morphology. Dwarfing, altered flowering, wrinkled leaves, or increased branching may also be useful for ornamentals. Dwarf phenotype is an important characteristic for flower crops such as *Eustoma grandiflorum* and *Dianthus.*

Tree Improvement

A major limitation of tree improvement programs is their long generation cycle. Classical breeding programs in trees are slow and tedious and it is difficult to introduce specific genes for genetic manipulation by crossing parental lines. *Agrobacterium rhizogenes* mediated transformation can be a useful alternative, as a rapid and direct route for introduction and expression of specific traits. The ability to manipulate tree species at cellular and molecular level shows great potential and *in vitro* transformation and regeneration from hairy roots facilitates application of biotechnology to tree species. This significantly reduces the time necessary for tree improvement and gives rise to new gene combinations that cannot be obtained using traditional breeding methods. In some tree species root initiation limits vegetative propagation. However by using *A. rhizogenes* rooting of cuttings from recalcitrant woody species have been improved.

Genetic Manipulation

Transformed roots provide a promising alternative for the biotechnological exploitation of plant cells. *A. rhizogenes* mediated transformation of plants may be used in a manner analogous to the well-known procedures employing *A. tumefaciens.* *A. rhizogenes* mediated transformation has also been used to produce transgenic hairy root cultures and plantlets have been regenerated. With the exception of the border sequences, none of the other T-DNA sequences are required for the transfer. The rest of the T-DNA can be replaced with the foreign DNA and introduced into cells from which whole plants can be regenerated. These foreign DNA sequences are stably inherited in a Mendelian manner. The *A. rhizogenes* mediated transformation has the advantage that any foreign gene of interest placed in binary vector can be transferred to the transformed hairy root clone.

It is also possible to selectively alter some plant secondary metabolites or to cause them to be secreted by introducing genes encoding enzymes that catalyze certain hydroxylation, methylation and glycosylation reactions.

Future Prospects and Conclusion

In the last several years, many efforts have been made in the area of genetic manipulation of secondary metabolism for useful products in general and medicinal plants in particular. Transgenic techniques have definitely been offering promising possibilities and indications for future research. However, we need more detailed knowledge, in particular, on basic plant biology. The genetic manipulation of the flavanoid pathway, as one of the examples, aimed at change of floral color has been successful in past because of the long term accumulation of basic knowledge of chemistry, biochemistry and molecular biology of flavanoid biosynthesis. The following guidelines could assist this area of future research:

1. Isolation and characterization of enzymes and genes for regulatory steps of each secondary pathway such as those which exist as in the case of anthocyanin biosynthetic pathway.

2. Clarification of cell type specific secondary metabolite expression, with reference to the developmental stage, where these genes are expressed. Identification of *cis* and *trans* acting factors that regulate the temporal and spatial gene expression of each secondary pathway.

3. For highly precise genetic manipulation, an extremely specific but powerful promoter is preferable. Otherwise, some gene products under non- specific promoter may show adverse effect on the normal physiological homeostasis in plant cells. It may be possible to use enhancers from strong promoters such as CaMV35S in conjunction with DNA motifs from the tissue specific promoter to achieve these goals.

4. Reproducible method for regeneration of the whole plant species is prerequisite. *Agrobacterium*-based transformation is suitable for some but not all plants. This application might be overcome by application of gene delivery technique, *i.e.*, microprojectile bombardment and other innovative methods which allow genes to be introduced in to plant tissues without the mediation of *Agrobacterium rhizogenes.*

5. Exploitation of the novel biosynthetic potential of transformed plants may be carried out using regenerated whole plants grown in the soil or possibility using novel bioreactor procedures. This will depend, in the final analysis, upon the commercial viability of either production method.

References

Bais, H. P., George, J., Ravishankar, G. A., and Sudha, G., 2001. Influence of exogenous hormones on growth and secondary metabolite production in hairy root cultures of *Cichorium intybus* L. cv. Lucknow local. *In vitro* Cell. Dev. Biol–Plant. **37**: 293–299.

Balunas M. J. and Kinghorn, A.D. 2005. Drug discovery from medicinal plants. Life Sci. **78**: 431–441.

Bapat, V.A. and Ganapathi, T.R. 2005. Hairy roots–a novel source for plant products and improvement. Natl. Acad. Sci. Lett., India **28 (3–4)**: 61–69.

Baskaran, P. and Jayabalan, N. 2009. Psoralen production in hairy roots and adventitious roots ultures of *Psoralea corylifolia* P. Biotechnol. Lett. **31**: 1073–1077.

Bensaddek L., Gillet, F., Nava–Saucedo, J.E. and Fliniaux, M. A. 2001. The effect of nitrate and ammonium concentrations on growth and alkaloid accumulation of *Atropa belladonna* hairy roots. J. Biotechnol. **85**: 35–40.

Bonhomme, V, Laurain–Mattar, D. and Fliniaux MA 2000. Effects of the *rol*C gene on hairy root : induction development and tropane alkaloid production by *Atropa belladonna*. J. Nat. Prod. **62**: 1249–1252.

Cetin, B., Gurel, A., Bedir, E., Akkaya, M., Ay, G., 2005. Formation of *Agrobacterium rhizogenes*–mediated transformed hairy root cultures of *Rubia tinctorum* L. and analysis of secondary metabolites. XIV. National Biotechnology Congress, August 31–September 2: 40–44, Eskisehir/Turkey.

Curtis, W.R., 1993. Cultivation of roots in bioreactors. Curr. Opinion Biotechnol., **4**: 205–210.

Flores, H.E. and Filner. 1985. Metabolic relationships of putrescine, GABA and alkaloids in cell and root cultures of Solanaceae. In: Neumann, K.H., Barz, W., Reinhard, E. (eds.) Primary and secondary metabolism of plant cell cultures. Springer, Berlin, pp 174–185.

Ford, R.K.P., Schlntes, P.J., Mitchell, E.A., Taykor, B.J., Scragg, R. and Stward, A.W. 1998. Hevay caffeine intake in pregnancy and sudden infant death syndrome. Newzealand cot death's study groups, **78**: 9–13.

Fu, C–X., Xu, Y–J., Zhao, D–X. and Ma, F.S. 2006. A comparison between hairy root cultures and wild plants of *Saussurea involucrata* in phenylpropanoids production. Plant Cell Rep. **24**: 750–754.

Giri, A. and Narasu, M.L. 2000. Transgenic hairy roots: recent trends and applications. Biotechnol. Adv. **18**: 1–22.

Guerche, P., Jouanin, L., Tepfer, D. and Pelletier, G.,1987. Genetic transformation of oilseed rape (*Brassica napus*) by the Ri T–DNA of *Agrobacterium rhizogenes* and analysis of inheritance of the transformed phenotype. Molecular and General Genetics. **206**: 382– 386.

Han, K.H., Fleming. P., Walker, K., Loper, M., Chilton, W.S., Mocek, U., Gordon, M.P. and Floss, H.G. 1994. Genetic–transformation of mature taxus–an approach to genetically control the *in vitro* production of the anticancer drug, taxol. Plant. Sci. **95 (2)**: 187–196.

Hamill, J.D., Parr, A.J., Rhodes, M.J.C., Robins, R.J. and Walton, N.J. 1987. New routes to plant secondary products. BioTechnol. **5**: 800–804.

Hilton, M. G. and Rhodes M. J. C. 1990. Growth and hyoscyamine production in hairy root culture of *Datura stramonium* in a modified stirred tank reactor. Appl. Microbiol. Biotechnol. 00: 100 100.

Hong, S.B., Peebles, C.A., Shanks, J.V., San, K.Y. and Gibson, S.I. 2006. Terpenoid indole alkaloid production by *Catharanthus roseus* hairy roots induced by *Agrobacterium tumefaciens* harboring *rol* ABC genes. Biotechnol. Bioeng. **93**: 386–390.

Huang, S–Y., Hung, C–H. and Chou, S–N., 2004. Innovative strategies for operation of mist trickling reactors for enhanced hairy root proliferation and secondary metabolite productivity. Enzyme and Microbial Technology, **35**: 22–32.

John, K.M., Joshi, S.D., Kumar, S.R. and Kumar, R.R. 2009. *A. rhizogenes* mediated Hairy root production in tea leaves [*Camellia sinensis* (L.) O. Kuntze]. Indian Journal of Biotechnology, **8**: 430–434.

Jung, G. and Tepfer, D. 1987. Use of genetic–transformation by the Ri–TDNA of *Agrobacterium rhizogenes* to stimulate biomass and tropane alkaloid production in *Atropa belladonna* and *Calystegia sepium* roots grown *in vitro*. Plant Sci. **50(2)**: 145–151

Kamada, H., Okamura, N., Satake, M., Harada, H., and Shimomura, K., 1986. Alkaloid production by hairy root cultures in *Atropa belladonna*. Plant Cell Rep. **5**: 239–242.

Kim, O.T., Bang, K.H., Shin, Y.S., Lee, M.J., Jung, S.J., Hyun, D.Y., Kim, Y.C., Seong, N.S., Cha, S.W. and Hwang, B. 2007. Enhanced production of asiaticoside from hairy root cultures of *Centella asiatica* (L.) Urban elicited by methyl jasmonate. Plant Cell Rep. **26**: 1941–1949.

Li, W., Li, M., F., Yang,D., L., Xu, R. and Zhang, R., Y. 2009. Production podophyllotoxin by root culture of *Podophyllum hexandrum* Royle. Electronic Journal of Biology, **5(2)**: 34–39.

Mehrotra, S., Kukreja, A. K., Khanuja, S. P. S. and Mishra, B. N. 2008. Genetic transformation studies and scale up of hairy root culture of *Glycyrrhiza glabra* in bioreactor, Electronic Journal of Biotechnology, **11**(2).

Newman, D.J., Cragg, G.M. and Snader, K.M. 2003. Natural products as sources of new drugs over the period 1981–2002. J. Nat. Prod. **66**: 1022–1037.

Oksman–Caldentey, Arroo R 2000. Regulation of tropane alkaloid metabolism in plants and plant cell cultures. In: Verpoorte, R. and Alfermann, A.W. (eds.) Metabolic engineering of plant secondary metabolism. Kluwer Academic Press, Dordrecht, pp. 253–281.

Park, U. S. and Facchini, P.J. 2000. *Agrobacterium rhizogenes*mediated transformation of opium poppy, *Papaver somniferum* L., and California poppy, *Eschscholzia californica* Cham., root cultures. J. Exp. Bot. **347**: 1005–1016.

Park U.S and Lee Y. S., 2009 Anthraquinone production by hairy root culture of *Rubia akane* Nakai: Influence of media and auxin treatment. Scientific Research and Essay **4(7)**: 690–693.

Petit, A., David, C., Dahl, G., Ellis, J., Guyen, P., Casse-Delbart, F., and Tempe, J., 1983. Further extension of the opine concepts: plasmids in *Agrobacterium rhizogenes* cooperate for opine degradation. Mol. Gen. Genet. **190**: 204–214.

Piatczak, E., Krolicka, A. and Wysokinska, H. 2006 Genetic transformation of *Centaurium erythraea* Rafn by *Agrobacterium rhizogenes* and the production of secoiridoids. Plant Cell Rep. **25**: 1308–1315.

Raskin I, Ribnicky, D.M., Komarnytsky, S., Ilic, N., Poulev, A., Borisjuk, N., Brinker, A., Moreno, D.A., Ripoll, C., Yakoby,N., O'Neal, J.M., Cornwell, T., Pastor, I. and Fridlender, B. 2002. Plants and human health in the twenty-first century. Trends Biotechnol. **20**: 522–531.

Saito, K., Yamazaki, T., Okuyama, E., Yoshihira, K., and Shimomura, K 1991. Anthraquinone production by transformed root cultures of *Rubia ticntorum*: Influence of phytohormone and sucrose concentration. Phytochem. **30**: 2977–2980.

Samanani, N., Park, S.U. and Facchini, P.J. 2002. In vitro regeneration and genetic transformation of the berberine–producing plant, *Thalictrum flavum* ssp. *glaucum*. Physiol. Plant. **116(1)**: 79–86.

Samuelsson, G. 2004. Drugs of Natural Origin: A Textbook of Pharmacognosy, 5th edition Swedish Pharmaceutical Press, Stockholm.

Satheeshkumar, K., Jose, B., Sonia, E.V. and Seeni, S. 2009. Isolation of morphovariants through plant regenation in *A. rhizogenes* induced hairy root cultures of *Plumbago rosea* L. Indian J. Biotechn., **8**: 435–441.

Shi, H.P. and Kintzios, S. 2003. Genetic transformation of *Pueraria phaseoloides* with *Agrobacterium rhizogenes* and puerarin production in hairy roots. Plant Cell Rep. **21**: 1103–1107.

Shinde, A.N., Malpathak, N. and Fulzele, P.D. 2009. Enhanced production of phytoestrogenic isoflavones from hairy root cultures of *Soralea corylifolia* L. using elicitation and precursor feeding. Biotechnology and Bioprocess Engineering, **14**: 288–294.

Sivakumar, G., Yu, K.W., Hahn, E.J. and Paek, K.Y. 2005. Optimization of organic nutrients for ginseng hairy roots production in large-scale bioreactors. Current Sci. **89**: 641–649.

Sood, H. and Chauhan, R.S. 2010. Biosynthesis and accumulation of a medicinal compound, Picroside–I, in cultures of *Picrorhiza kurroa* Royle ex Benth. Plant Cell Tissue Organ. Cult. **100**: 113–117.

Su, W.W. and Lee, K.T., 2007. Plant cell and hairy root cultures -process characteristics, products, and applications bioprocessing for value-added products from renewable resources. New Technologies and Applications, pp. 263–292.

Sudha, C.G., Obul Reddy, B., Ravishankar, G.A. and Seeni, S. 2003. Production of ajmalicine and ajmaline in hairy root cultures of *Rauvolfia micrantha* Hook. f., a rare and endemic medicinal plant. Biotechnol. Lett. **25**: 631–636.

Tepfer, D. 1984. Transformation of several species of higher plants by *Agrobacterium rhizogenes*: sexual transmission of the transformed genotype and phenotype. Cell. **37**: 959–967.

Verpoorte, R., Contin, A. and Memelink, J. 2002. Biotechnology for the production of plant secondary metabolites. Phytochem. Rev. **1**: 13–25.

Wheathers, P.J. and Bunk, G. and McCoy, M.C. 2005. The effect of phytohormones on growth and artemisinin production in *Artemisia annua* hairy roots. *In vitro* Cell Dev. Biol.–Plant. **41**: 47–53.

White, F.F., Ghidossi, G., Gordon, M.P. and Nester, E.W., 1982. Tumor induction by *Agrobacterium rhizogenes* involves the transfer of plasmid DNA to the plant genome. Proc. Nat. Acad. Sci. USA. **79**: 3193–3197.

White, F.F., Taylor, G.H., Huffmann, G.A., Gordon, M.P. and Nester, E.W. 1985. Molecular and genetic analysis of the transferred DNA of the root inducing plasmid of *Agrobacterium rhizogenes*. J. Bacter. **164**: 33–44.

Woods R.R., Geyer C.B. and Mor, S. 2008. Hairy-root organ cultures for the production of human acetylcholinesterase. BMC Biotechnology, **8**: 95.

Zupan, J. and Zambryski, P. 1997. The *Agrobacterium* DNA transfert complex. Crit. Rev. Plant Sci. **16**: 279–295.

2013, Abiotic Stress and Biotechnology
Editor: T. Pullaiah
Published by: REGENCY PUBLICATIONS, NEW DELHI

Pages 223–234

Chapter 11

In vitro Regeneration and Multiplication of *Dendrobium farmeri* Paxt.: A Floriculturally and Medicinally Important Species

*Saranjeet Kaur**

National Institute of Pharmaceutical Education and Research, Mohali – 160 062, Punjab

ABSTRACT

An efficient protocol was established for *in vitro* regeneration and multiplication of *Dendrobium farmeri* through protocorm segments using organic growth supplements such as banana homogenate, coconut water, and peptone. The effect of growth supplements was tested on multiplication (*neo*-formation of secondary protocorms from primary protocorms) of protocorms, development of maximum number of shoots and early plantlets in M medium (Mitra medium). The explants regenerated regardless of growth adjuncts in the medium but the regeneration frequency was found significantly higher in organic growth supplement-enriched medium than control. The protocorm segments (primary protocorm) regenerated into protocorm-like bodies (PLBs/secondary protocorms) which eventually differentiated into shoots. Among the treatments, the highest regeneration frequency was obtained in coconut water (10 per cent)

* E-mail: sarana_123@rediffmail.com

whereas shoots and roots were robust in appearence in banana homogenate (10 g/l). Higher concentration of BH proved detrimental to the survival of cultures. Peptone (1.5 g/l) supplemented medium also supported dense multiplication of protocorm-like bodies (PLBs) and their early growth into plantlets. The treatments with organic growth supplements surely gave better results as compared to control. Among these tested organic growth supplements, coconut water (10 per cent) supported maximum regeneration and number of shoots formation and, peptone (1.5 g/l) also proved beneficial for multiplication of regenerants and more development of shoots as compared to BH; whereas banana homogenate (10 g/l) favoured healthy growth of plantlets.

Keywords: *Banana homogenate, Coconut water, Orchids, Peptone, Protocorms, Regeneration.*

Abbreviations

BH: Banana homogenate

CW: Coconut water

M: Mitra *et al.,* 1976 medium

PL: Peptone

PLBs: Protocorm-Like bod(ies).

Introduction

The orchids represent one of the largest, highly evolved, diverse, and successful family of flowering plants – the orchidaceae. Being cosmopolitan in distribution, major speciation has occurred in tropical and subtropical regions except Antarctica. Their growth is particularly influenced by thick vegetation, high humidity and suitable microclimatic conditions. Orchid seeds are microscopic, non-endospermic and enclosed within a transparent testa cover. They are produced in large numbers but few of them germinate successfully, innumerable are lost since they require appropriate mycorrhizal associations and environmental conditions to germinate. The orchids are habitat specific which make them excellent indicators of environmental degradation, they need congenial conditions to survive in nature. Conservation of biodiversity (plants and animals) is a major issue round the globe. It is strongly emphasized to preserve the valuable species for healthy functioning of ecosystem. Extinction of a specific species has deep influence on the ecological equilibrium to such an extent that it can endanger species which are dependent on it. Integrative multidisciplinary approaches are required for conservation (*in situ* and *ex* situ) and sustainable use of critically endangered, floriculturally important orchid germplasm. Efforts are needed to conserve and eco-restore orchids through the development of effective biotechnological techniques. Use of *in vitro* techniques has emerged as one of the successful trends of biotechnology for *ex-situ* conservation and reintroduction of rare, endangered, and threatened orchids. Tissue culture techniques offer a viable system for true-to-type rapid mass multiplication and germplasm conservation of rare, endangered, threatened, floriculturally important, aromatic and medicinal plants and orchids.

The addition of organic additives to the culture medium to promote *in vitro* growth and proliferation of *neo*-formations is a common practice (cf. Kaur and Bhutani, 2012). Presence of organic additives may contribute towards the development of a simple and economical plant culture media and at the same time minimize the use of exogenous PGRs, thereby possibly reducing the occurrence of undesired somaclonal variation (Lee and Phillips, 1988). It is of utmost importance to maintain genetic uniformity of *in vitro* raised progenies of medicinal and floriculture importance. The use of PGRs in the culture medium should be minimized as possible and the culture medium should be boosted with organic and inorganic nutrients to accomplish regeneration and multiplication *in vitro*. Ever since Arditti (1967) suggested that complex growth supplements have the ability to influence *in vitro* regeneration, multiplication of protocorm-like body(s) and growth of orchid seedlings. Since then a variety of organic growth supplements such as apple juice, banana homogenate (BH), beaf extract, casein hydrolysate, coconut water (CW), corn extract, tomato juice, peptone, yeast extract etc. were tested for promoting multiplication, growth and development of *in vitro* cultures. Growth rate of the tissues can be increased by the addition of organic supplements and plant extracts (Fonnesbech, 1972). Effects of organic growth additives are tested in a large number of orchids such as *Paphiopedilum* species (Flamee, 1978), *Vanda* hybrids (Mathews and Rao, 1980), *Acampe praemorsa* (Krishnamohan and Jorapur, 1986), *Aranda* (Goh and Wang, 1990), *Cattleya, Encyclia, Oncidium* and *Stanhopea* (Villolobus and Munoz, 1994), *Dendrobium* species (Sudeep *et al.*, 1997), *Geodorum densiflorum* (Bhadra and Hossain, 2003), *Doritaenopsis* (Chowdhury *et al.*, 2003), *Cypripedium formosanum* (Lee and Lee, 2003), *Dendrobium* hybrid (Lekha Rani *et al.*, 2005), *Phalaenopsis gigantea* (Murdad *et al.*, 2006), *Dendrobium* species (Aktar *et al.*, 2008), *Zygopetalum mackayi* (Hong *et al.*, 2010), and in *Dendrobium chrysotoxum* (Kaur and Bhutani, 2011), *Cymbidium pendulum* (Kaur and Bhutani, 2012).

The technique of protocorm slices, to regenerate *in vitro* giving a large number of plantlets within a short span of time through repeated slicing and culture of daughter PLBs, has emerged as an important and high frequency multiplication methodology for orchids (cf. Kaur and Bhutani, 2011). The utility of PLB segment culture in achieving multifold increase in the rate of multiplication is reported in several orchids. The induction of protocorms like- bodies from protocorms and various other explants such as shoot-tip, stem-node, root-tip, root tubers is a reliable method for regeneration and multiplication of orchids (Park *et al.*, 2003, Kosir *et al.*, 2004, Anjum *et al.*, 2006, Kalimuthu *et al.*, 2007, Hong *et al.*, 2008, Medina *et al.* 2009, Ng and Saleh, 2011, Kaur and Bhutani, 2011, 2012). Due to having the ability of proliferation of protocorms into primary PLBs further into secondary PLBs, these are the most sought-after tissues used for genetic transformation studies (Sreeramanan *et al.*, 2008) and cryopreservation studies (Yin and Hong, 2009). PLBs which develop directly from meristem tissues are more genetically stable than those generated from callus (Lee and Philips, 1988).

In orchids, the explants either undergo caulogenesis *i.e.*, callus formation as an intermediary phase just prior to somatic embryogenesis or they regenerate directly in to somatic embryos *i.e.*, either secondary protocorms or shoot buds. To maintain genetic integrity in the neo-formations, callus induction should be avoided as it leads to somaclonal variations. Direct regeneration without undesirable callus formation also shortens the time period needed for regeneration and reduces the possibility of the occurrence of somaclonal variability. In the present study, medium fortified with different organic growth supplements was tested for induction of neo-formations in protocorm slices in order to develop an efficient and rapid propagation method for *Dendrobium* species. The orchids have become rare, endangered and threatened species all over the world. Several factors such as deforestation, fragmentation of habitat especially in tropical regions, increased use of fertilizers, excessive exploitation of soil, and over collection are responsible for the present day conservation status of orchids and *Dendrobium farmeri* does not seem to be an exception.

Dendrobium farmeri Paxt. is an epiphytic, pseudobulbous, evergreen orchid species, which bear delightful fragrant flowers. The species is native to summer moist forests including India, Burma, Thailand, Malaysia, Laos and Vietnam. It is a typical showy household plant and is also used to create beautiful landscapes along with bromeliads and other orchid species in many parts of the world. Due to its bewitchingly beautiful blossoms the species has emerged as one of the horticulturally important species, used in floriculture trade. Besides being floriculturally important it is also known for its anti-bacterial activity against *Klebsiella pneumoniae*, *E. coli*, and *Solmonella typhi* (Hossain, 2011). The species is also known to harbour a variety of chemical compounds as flavonoids (6,7-methylenedioxy and 6,7-dimethoxycoumarins) (cf. Hossain, 2011). Due to its horticultural importance and therapeutic properties, the species has been subjected to commercial collections which far exceed its natural regeneration. It is, therefore, strongly emphasized to save the species through tissue culture techniques. Therefore, studies were conducted with the aim to multiply the species *in vitro* using protocorm segments and testing the efficacy of growth supplements on *in vitro* multiplication of protocorms and early development of *D. farmeri* plantlets without the use of any growth regulators.

Materials and Methods

Explant Material, Culture Medium and Procurement of Growth Supplements

Dendrobium farmeri protocorms (1.5 mm long) were procured from 25-week old *in vitro* cultures which were raised through asymbiotic seed germination technique. The protocorms were maintained for 2 weeks on basal M medium (Mitra *et al.*, 1976 medium) supplemented with 2.0 per cent (w/v) sucrose (Daurala sugar works, Daurala, India) as source of nutrition and gelled with 0.9 per cent (w/v) agar (Hi-Media, Mumbai, India). The complex growth supplements such as banana homogenate (10, 20 and 30 g/l w/v), coconut water (10, 20 and 30 per cent v/v), peptone (1.0, 1.5 and 2.0 g/l) were used individually in the medium.

Banana homogenate was obtained from ripened fruits purchased from market. They were peeled, cut into small slices, homogenised in a blender and the resultant pulp was added to the medium according to the quantity required. Coconut water, from fresh green coconuts was drawn off, collected, filtered through a sieve and used as such in the medium. Peptone (Hi-Media, Mumbai, India) was used in the medium.

The pH of medium was adjusted to 5.8 after adding the organic growth supplements. The medium was dispensed in test tubes of size (25 mm × 150 mm) and autoclaved at 121°C at pressure of 1.06 kg/cm² for 15 min. Autoclaved medium was kept at 37°C to check any further contamination.

Inoculations and Incubation Conditions

The inoculations were done under aseptic conditions in a laminar air flow cabinet. All the cultures were incubated at 25 ± 2°C under 12 h photoperiod of 3,500 lux light intensity (Fluorescent tubes 40 W; Philips India, Ltd., Mumbai, India). Subculturings were done as and when required. The experiment was repeated twice.

Observations and Statistical Analysis

The cultures were observed at regular intervals of time under binocular microscope (Olympus SZX10, Tokyo, Japan) and data recorded accordingly. The results were tested using one-way ANOVA test and were analysed using the Tukey Multiple Comparison at $p \leq 0.05$ using SPSS (version 16) software package (SPSS Inc., Chicago, USA).

Results

The clonal propagation of orchids is predominantly based on the phenomenon of regeneration of new protocorms by protocorm sections. Presently, the system was effectively utilized to regenerate and multiply the cultures of *Dendrobium farmeri* in M medium and its combinations with growth supplements (Tables 11.1–11.3).

Table 11.1: The effect of different concentrations of coconut water (per cent) on *in vitro* multiplication of *D. farmeri* proto-corms in M medium.

Additive	Regeneration Response (per cent)	Number of PLBs/Explant	Number of Shoots/Explant	Complete Plantlets (wks)
M	10.25 ± 1.20^{bcd}	1.00 ± 0.20^{b}	1.00 ± 0.20^{b}	18.75 ± 1.50^{b}
M + CW$_{10\ per\ cent}$	83.05 ± 0.10^{acd}	5.00 ± 1.10^{acd}	10.00 ± 0.50^{acd}	10.00 ± 0.40^{acd}
M + CW$_{20\ per\ cent}$	58.75 ± 1.25^{abd}	1.21 ± 0.37^{b}	2.50 ± 0.28^{ab}	21.25 ± 0.70^{abd}
M + CW$_{30\ per\ cent}$	31.25 ± 1.18^{abc}	0.75 ± 0.25^{b}	1.50 ± 0.20^{b}	17.00 ± 0.40^{bc}

M: Mitra medium; PLBs: Protocorm-like bodies; CW: Coconut Water [concentration = per cent (v/v)]; Subscript are the conc.; Values in a column with similar superscripts are not significantly different at $p \leq 0.05$ according to Tukey's test.

In control (M medium) only 10.25 ± 1.20 per cent explants regenerated and developed into a single plantlet per explant after 18.75 ± 1.50 weeks. Addition of organic growth supplements supported early initiation of response in the cultures and multiplication of the *neo* - formations. In coconut water (10 per cent) enriched medium the explants regenerated with 83.05 ± 0.10 per cent frequency. The rate of PLB multiplication was significantly higher in CW 10 per cent as compared to control and other concentrations of CW used (Table 11.1). The regenerants multiplied profusely in this combination (Figure 11.1A). Almost 10.00 ± 0.50 shoots per explant

Table 11.2: The effect of different quantity of peptone (g/l) on *in vitro* multiplication of *D. farmeri* protocorms in M medium.

Additive	Regeneration Response (per cent)	Number of PLBs/Explant	Number of Shoots/Explant	Complete Plantlets (wks)
M	10.25 ± 1.20^{bcd}	1.00 ± 0.20^{c}	1.00 ± 0.20^{c}	18.75 ± 1.50^{cd}
$M + P_1$	22.50 ± 1.40^{acd}	1.50 ± 0.20^{c}	2.00 ± 0.40^{c}	19.25 ± 0.47^{cd}
$M + P_{1.5}$	51.25 ± 2.30^{abd}	2.75 ± 0.25^{abd}	8.25 ± 0.60^{abd}	11.25 ± 0.75^{abd}
$M + P_{2.0}$	31.25 ± 1.20^{abc}	1.75 ± 0.28^{c}	2.70 ± 0.50^{ac}	16.75 ± 0.85^{abc}

M: Mitra medium; PLBs: Protocorm-like bodies; P: Peptone [concentration g/l (w/v)]; Subscript are the conc.; Values in a column with similar superscripts are not significantly different at $p \leq 0.05$ according to Tukey's test.

Table 11.3: The effect of different quantity of banana homogenate (g/l) on *in vitro* multiplication of *D. farmeri* proto-corms in M medium.

Additive	Regeneration Response (per cent)	Number of PLBs/Explant	Number of Shoots/Explant	Complete Plantlets (wks)
M	10.25 ± 1.20^{bcd}	1.00 ± 0.20^{bc}	1.00 ± 0.20^{bcd}	18.75 ± 1.50^{bcd}
$M + BH_{10}$	25.75 ± 2.10^{ac}	1.75 ± 0.20^{a}	3.25 ± 0.20^{acd}	21.25 ± 0.70^{ad}
$M + BH_{20}$	21.50 ± 1.10^{abd}	1.50 ± 0.28^{a}	1.75 ± 0.20^{abd}	21.50 ± 1.00^{ad}
$M + BH_{30}$	25.00 ± 2.00^{ac}	1.25 ± 0.20^{b}	0.00^{abc}	0.00^{abc}

M: Mitra medium; PLBs: Protocorm-like bodies; BH: Banana homogenate [concentration g/l (w/v)]; Subscript are the conc.; Values in a column with similar superscripts are not significantly different at $p \leq 0.05$ according to Tukey's test.

were formed and the developmental processes leading to plantlet formation were also accelerated. Plantlets with 2–3 leaves and 1–2 roots were obtained within 10.00 ± 0.40 weeks (Figure 11.2A). Interestingly, the cultures showed simultaneous multiplication of PLBs and development of shoots in CW (10 per cent) enriched medium (Figure 11.1D). Peptone (1.5 g/l) in the medium also favoured multiplication of the regenerants (Figure 11.1B) and early plantlet development within 11.25 ± 0.75 weeks of culture (Figure 11.2C); a maximum of 8.25 ± 0.6 shoots per explant developed in the cultures (Table 11.2). Although addition of BH could slightly increase the percentage of responding explants than control but it showed decrease in the percentage of responding explants in comparison to the response of the explants in medium supplemented with other growth adjuncts. The resultant PLBs (Figure 11.1C) and developing shoots (Figure 11.2B) were robust in appearance Higher concentrations of BH not even dropped the regeneration frequency but the regenerated PLBs failed to differentiate into shoots and roots (Table 11.3); they turned brown and perished soon after.

Figure 11.1(A-D): *In vitro* multiplication of *Dendrobium farmeri* protocorm segments in M medium and its combinations with growth adjuncts. A: Initial stage of PLBs multiplication in CW (10 per cent); B: PLB budding and further growth in P (1.5 g/l) enriched medium; C: PLB development in BH (10 g/l) supplemented medium; D: Cultures showing simultaneous multiplication of PLBs and development of shoots in CW (10 per cent) fortified medium.

Discussion

Possibilities of inducing daughter PLBs in protocorm sections offer an exciting system for clonal propagation of desired orchid genotype. Presently, the system was utilized to multiply the cultures using the protocorms. In this experiment, the regeneration potential of proto-corm explants (in terms of percentage of regeneration, average number of PLBs and shoots/explant) was markedly influenced by quality and quantity of organic growth supplements added to the medium. The explants responded to regeneration in the basal medium with a very low regeneration frequency. The percentage of regeneration response was increased by addition of organic growth supplements in the cultures.

Figure 11.2 (A-C): Organic growth supplements and *in vitro* plantlet development of *D. farmeri* in M medium. A: Plantlet development in CW (10 per cent) supplemented medium; B: Healthy growth of plantlets in BH (10 g/l); C: Growth of plantlets in P (1.5 g/l).

In an earlier investigation in *Cymbidium pendulum* addition of growth supplements showed their beneficial effect in increasing the percentage of responding explants, enhancing the multiplication of regenerants and accelerated development of plantlets (Kaur and Bhutani, 2012). Arditti (1979) also made similar conclusions indicating the beneficial effects of organic growth supplements on growth and differentiation of proto-corms and further seedling development. It was also earlier reported by Chen and Chang (2002) that at an efficient concentration, organic and in-organic nitrogen sources can promote the growth of explants.

In the present experiment, 10 g/l of banana homogenate proved beneficial for development of healthy shoots growth of *D. farmeri*. The positive effects of BH (10 g/l), in the present cultures, in development of healthy shoot system (robust shoots in appearence) from PLBs, could be attributed to the higher sucrose concentrations in BH as earlier also suggested by Aktar *et al.,* (2008) in enhancing *in vitro* regeneration of *Dendrobium* orchid PLBs. In *Vanda* species BH (10 per cent) increased the shoot length. Beneficial effects of BH (10 per cent) on leaf size of *Spathoglottis kimbal-lianai* is reported by Minea *et al.,* (2004). Banana homogenate significantly increased the number of leaves in *Dendrobium nobile* cultures (Sudeep *et al.,* 1997). A higher dose of BH 30 g/l proved detrimental to the growth of our cultures. Coconut water (10 per cent) was seen to be most effective in eliciting maximum regeneration response in the explants. The secondary protocorms/PLBs multiplied vigorously. Literature studies reveal the beneficial effect of CW in enhancing foliar growth of orchids in *in vitro* grown seedlings and it is correlated to the fact that presence of a cytokinin (Kinetin) in coconut water is the probable cause of growth of cultures (Bhasker, 1996). Growth promotory nature of CW is related to its ability of inducing cell divisions in non-dividing cells hence promoting early protocorm differentiation (Intuwong and Sagawa, 1973). Earlier addition of CW (15 per cent) to the basal medium increased growth of the cultures and the shoots vigorously rooted in epiphytic orchids (McIntyre *et al.*, 1974). CW successfully initiated differentiation of shoots from PLBs of *Vanda teres* (Sinha and Roy, 2004) and has helped in achieving high-frequency multiplication of PLBs in *Phalaenopsis gigantea* (Murdad *et al.*, 2006), *Cymbidium pendulum* (Kaur and Bhutani 2012). Overbeek *et al.* first introduced coconut water as a new component of the nutrient medium for its ability of promoting growth of the cultures. Coconut water is a combination of sugars, sugar alcohols, amino acids, nitrogenous compounds, organic acids, a variety of enzymes, plant growth regulators mainly cytokinins together with IAA, GA_1; GA_3 and ABA. Among cytokinins trans – zeatin riboside has the greatest concentration followed by trans – zeatin O– glucoside and dihydrozeatin O– glucoside (Jean *et al.*, 2009). The growth promotory effects of CW cannot be solely credited only to the cytokinins, as the effect could be due to the synergistic effect of all components of CW.

In the present experiment, peptone also favoured multiplication of PLBs. A perusal of literature reveals that peptone being water soluble protein hydrolysate with very high amino acid content promotes growth of cultures. Similar effect of peptone was observed in our *Cymbidium pendulum* cultures (Kaur and Bhutani, 2012). Literature studies also report its beneficial effects in inducing protocorm multiplication in *Cymbidium macrorhizon* and *Cymbidium* species (Kusumoto and Furukuwa 1977). Peptone is also known to have stimulated callus growth in *Phalaenopsis, Doritaenopsis,* and *Neofinetia* (Ichihashi and Islam 1999). It also supported better seedling growth in *Paphiopedilum, Phaius,* and *Vanda* (Curtis, 1947). According to Amaki and Higuchi (1989) peptone promoted growth of PLBs. It also favoured early and healthy growth of seedlings in *Peristeria elata* (Bejoy *et al.*, 2004) and *Cymbidium pendulum* (Kaur and Bhutani, 2012).

Plant growth regulators (PGRs) are known to induce somaclonal variations. For the multiplication of desired species by using different explants (protocorms/ leaves/ stem- nodes/ roots/ inflorescence/ floral organs etc) instead of PGRs, organic growth supplements should be used in the nutrient regime. Inclusion of organic additives contributes towards the development of a simple and economical plant tissue culture media which reduces the risk of occurrence of somaclonal variations. It is of utmost importance to maintain genetic uniformity of *in vitro* raised progenies of conservation and floriculture importance.

Conclusion

Through the investigation of the effects of different organic growth supplements on the regeneration and multiplication of cultures, results indicated that CW (10 per cent) in M medium was the most effective in initiating maximum regeneration/ explants and profuse multiplication of PLBs, Peptone (1.5 g/l) also favoured multiplication of regenerants simultaneously supporting early plantlet development, Banana homogenate encouraged healthy growth of plantlets. Thus, organic growth supplements proved beneficial overall for the multiplication of the cultures besides enhancing growth of plantlets as compared to control. This is a simple, efficient low-cost medium for *D. farmeri* tissue culture without the use of plant growth regulators.

References

Aktar, S., Nasiruddin, K. M. and Hossain, K. 2008. Effects of different media and organic additives interaction on *in vitro* regeneration of *Dendrobium* orchid. J. Agri. Rur. Dev. **6**: 69 - 74.

Amaki,W. and Higuchi, H. 1989. Effects of dividing on the growth and organogenesis of protocorm-like bodies in *Doritaenopsis*. Sci. Hort. **39**: 63-72.

Anjum, S., Zia, M. and Chaudhary, M.F. 2006. Investigations of different strategies for high frequency regeneration of *Dendrobium malones* 'Victory'. Afr. J. Biotechnol. **5**:1738–1743.

Arditti, J. 1967. Factors affecting the germination of orchid seeds. Bot. Rev. **3**: 1 - 97.

Arditti, J. 1979. Aspects of physiology of orchids. Adv. Bot. Res. **7**: 421 - 655.

Bejoy, M., Kumar, S. C., Radhika, B. J. and Joemon, J. 2004. Asymbiotic seed germination and early seedling development of the dove orchid *Peristeria elata* Hook. J. Orchid Soc. India **17**: 75 - 79.

Bhadra, S. K. and Hossain, M. M. 2003. *In vitro* germination and micropropagation of *Geodorum densiflorum* (Lam.) Schltr. an endangered orchid species. Plant Tiss. Cult. Biotech. **13**: 165 - 171.

Bhasker, J. 1996. Micropropagation of *Phalaenopsis*. Ph.D. Thesis, Kerala Agricultural University, Thrissur, Kerala, India.

Chen J.T. and Chang W.C. 2002. Effects of tissue culture conditions and explants characteristics on direct somatic embryogenesis in *Oncidium* 'Gower Ramsey'. Plant Cell Tissue Organ Cult. **69**: 41–44.

Chowdhury, I., Rahman, A.R.M., Islam, M.O. and Matsui, S. 2003. Effect of plant growth regulators on callus proliferation, plantlet regeneration and growth of plantlets of *Doritaenopsis* orchid. Biotech. **3**: 214-221.

Curtis, J.T. 1947. Studies on nitrogen nutrition of orchid embryos. I. Complex nitrogen sources. Amer. Orchid Soc. Bull. **16**: 654 - 660.

Flamee, M. 1978. Influence of selected media on the germination and growth of *Paphiopedilum* seedlings. Amer. Orchid Soc. Bull. **47**: 419 - 423.

Fonnesbech, M. 1972. Growth hormones and propagation of *Cymbidium in vitro*. Physiol. Plant. **27**: 310 - 316.

Goh, C. and Wang, P.F. 1990. Micropropagation of the monopodial orchid hybrid *Aranda Deborah* using inflorescence explants. Sci. Hort. **44**: 315 - 321.

Hong, P.I., Chen, J.T. and Chang, W. C. 2008. Plant regeneration via protocorm-like body formation and shoot multiplication from seed-derived callus of maudiae type slipper orchid. Acta Physiol. Plant. **30**:755–759.

Hong, P.I., Chen, J. and Chang, W. 2010. Shoot development and plant regeneration from protocorm-like bodies of *Zygopetalum mackayi*. *In Vitro* Cell. Dev. Biol.– Plant **46**: 306-311.

Hossain, M.M. 2011. Therapeutic orchids: traditional uses and recent advances - An overview. Fititerapia **82**: 102-140.

Ichihashi, S. and Islam, M.O. 1999. Effect of complex organic additives on callus growth in three orchid genera, *Phalaenopsis, Doritaenopsis* and *Neofinetia*. J. Jap. Soc. Hort. Sci. **68**: 269 - 274.

Intuwong, O. and Sagawa, Y. 1973. Clonal propagation of Sarcanthine orchids by aseptic culture of inflorescences. Amer. Orchid Soc. Bull. **42**: 264 - 270.

Jean, W. H., Yong, L. Ge., Yan, F. Ng. and Swee N. T. 2009. The chemical composition and biological properties of coconut (*Cocos nucifera* L.) water. Molecules **14**: 5144-5164.

Kalimuthu, K., Senthilkumar, R. and Vijayakumar, S. 2007. *In vitro* micropropagation of orchid, *Oncidium* spp. (Dancing Dolls). Afr. J. Biotechnol. **6**:1171–1174.

Kaur, S. and Bhutani, K.K. 2011. *In vitro* propagation of *Dendrobium chrysotoxum* Lindl. Floricult Ornamental Biotech. **5**: 50-56.

Kaur, S. and Bhutani, K.K. 2012. Organic growth supplement stimulants for *in vitro* multiplication of *Cymbidium pendulum* (Roxb.) Sw. Hort. Sci. **39**: 47-52.

Kosir, P., Skof, S. and Luthar, Z. 2004. Direct shoot regenerated from nodes of *Phalaenopsis* orchids. Acta Agric. Slovenica **83**:233–242.

Krishnamohan, P.T. and Jorapur, S.M. 1986: *In vitro* seed culture of *Acampe praemorsa* (Roxb.) Blatt. & McC. In: S. P. Vij (ed.) Biology, Conservation and Culture of Orchids. East –West Press, New Delhi, pp. 437 - 439.

Kusumoto, M. and Furukuwa, J. 1977. Effect of organic matter on growth of *Cymbidium* protocorms cultured *in vitro*. J. Jap. Soc. Hort. Sci. **45**: 421 - 426.

Lee, Y.L. and Lee, N. 2003. Plant regeneration from protocorm derived callus of *Cypripedium formosanum*. *In vitro* Cell. Dev. Biol. **39**: 475 - 479.

Lee, M. and Phillips, R.L. 1988. The chromosomal basis of somaclonal variation – a review. J. Plant Physiol. Plant Mol. Biol. **39**: 413-437.

Lekha Rani, C., Vidya, C., Rajmohan, K. and Mercy, S. T. 2005. Protocorm differentiation and seedling growth in *Dendrobium* hybrid seed cultures as influenced by organic additives. J. Orchid Soc. India **19**: 67 - 70.

Mathews, V. H. and Rao, P. S. 1980. *In vitro* multiplication of *Vanda* hybrids through tissue culture technique. Plant Sci. Lettters **17**: 383 - 389.

McIntyre, D. K., Veitch, G. J. and Wrigley, J.W. 1974. Australian orchids from seeds II. Improvement in techniques and further successes. Amer. Orchid Soc. Bull. **43**: 52 -53.

Medina, R.D., Flachsland, E.A., Gonzalez, A.M., Terada, G., Faloci, M.M. and Mroginski, L.A. 2009. *In vitro* tuberization and plant regeneration from multimodal segment culture of *Habenaria bractescens* Lindl., an Argentinean wetland orchid. Plant Cell Tissue Organ Cult. **97**:91–101.

Minea, M., Piluek, C., Menakanit, A. and Tantiwiwat, S. 2004. A study on seed germination and seedling development of *Spathoglottis kimballiani*. Kasetsalt. J. Nat. Acad. Sci. **38**: 141 - 156.

Mitra, G. C., Prasad, R. N. and Chowdhary, A. R. 1976. Inorganic salts and differentiation of protocorms in seed callus of an orchid and correlated changes in its free amino acid content. Ind. J. Exp. Biol. **14**: 350 - 351.

Murdad, R., Hwa, K.K., Seng, C. K., Latip, A.M., Aziz, A.Z. and Ripin, R. 2006. High frequency multiplication of *Phalaenopsis gigantea* using trimmed bases protocorms technique Sci. Hort. **111**: 73 - 79.

Ng, Chyuam-Y., and Saleh, N.M. 2011. *In vitro* propagation of *Paphiopedilum* orchid through formation of protocorm-like bodies. Plant Cell Tissue Organ Cult. **105**: 193-202.

Park, S.Y., Murthy, H.N. and Paek, K.Y. 2003. Protocorm-like body induction and subsequent plant regeneration from root tip cultures of *Doritaenopsis*. Plant Sci. **164**: 919–923.

Sinha, P. and Roy, S. K. 2004.Regeneration of an indigenous orchid, *Vanda teres* Roxb. Plant Tiss. Cult. Biotech. **13**: 165 - 171.

Sreeramanan, S., Vinod, B., Sashi, S. and Xavier, R. 2008. Optimization of the transient Gusa gene transfer of *Phalaenopsis violacea* orchid via *Agrobacterium tumefaciens*: an assessment of factors influencing the efficiency of gene transfer Mechanisms. Adv. Nat. Applied Sci. **2**:77–88.

Sudeep, R., Rajeevan, P. K., Valasalakumari, P. K. and Geetha, C. K. 1997. Influence of organic supplements on shoot proliferation in *Dendrobium*. J. Hort. **3**: 38 - 44.

Villolobus, L. and Munoz, A. M.1994. Tissue culture of orchids *Cattleya*, *Encyclia*, *Oncidium* and *Stanhopea*. Orchid Rev. **102**: 58 - 62.

Yin, M. and Hong, S. 2009. Cryopreservation of *Dendrobium candidum* Wall. ex Lindl. protocorm-like bodies by encapsulation-vitrification. Plant Cell Tissue Organ Cult. **98**:179–185.

2013, Abiotic Stress and Biotechnology
Editor: T. Pullaiah
Published by: REGENCY PUBLICATIONS, NEW DELHI

Pages 235–283

Chapter 12

Real-Time PCR: A Reliable Method for Gene Expression Studies in Plants

Krishnaveni Thiruveedhi[1], Chandra Obul Reddy Puli[2]*,
Chandra Sekhar Akila[1], Chakradhar Thammineni[3],
Sravani Konduru[2], Jayanna Naik Banavath[2]
and Varakumar Pandit[2]

[1]Department of Biotechnology, [2]Department of Botany,
School of Life Sciences, Yogi Vemana University,
Vemanapuram Kadapa – 516 003, A.P., India
[3]Centromere Biosolutions, ICRISAT, Hyderabad, India

ABSTRACT

There are several ways to look for changes in cellular transcript levels (*e.g.,* Northern blotting, RNase protection assays and RT-PCR). However substantial problems are associated with the true sensitivity, reproducibility and specificity of the methods. The recent introduction of fluorescence-based kinetic RT-PCR procedures significantly simplifies the process of producing reproducible quantitation of mRNAs and promises to overcome these limitations. This chapter discusses the theory behind real-time PCR, its principle and the technical aspects including sample preparation, isolation of RNA, reverse transcription, primer

* Corresponding Author: E-mail: coreddy@yogivemanauniversity.ac.in

design, data analysis of absolute versus relative quantitation, types of normalization or data correction, and detection chemistries. A special emphasis has also been given about the relevance of this technique in plant science especially on abiotic stress studies.

Keywords: *Real-time PCR, Fluorescent chemistries, Syber Green-I, Relative quantification, Reference genes, Abiotic stress.*

Introduction

Gene expression analysis is an essential method for understanding many processes of plant biology, increasing our knowledge of the signaling and metabolic pathways that underlie environmental responses and development. The expression level of a gene can be studied with great precision through the detection and/or quantification of functional messenger RNAs (mRNAs) in the cytoplasm. Completion of genomic sequencing in the last few years from *Arabidopsis* and several other crops through the Arabidopsis Genome Initiative, 2000 and related projects has opened up a new era where almost every gene can be targeted and its expression will be quantified and analysed. Two types of genes can be found in the genome. They can either be transcripted into messenger ribonucleotidic acids (mRNAs) that will be translated into proteins or be transcripted into non-coding RNAs. Non-coding RNAs includes ribosomal RNAs (rRNAs), transfer RNAs (tRNAs) and pseudogene RNAs, microRNAs (miRNAs), small nuclear RNAs (snRNAs), small nucleolar RNAs (snoRNAs), piwi-interacting RNAs (piRNAs) and small interfering RNAs (siRNAs). Most of these non-coding RNAs exert regulatory functions. When considering the expression of genes coding for proteins their quantification has to be performed on mRNA or on protein levels. mRNA quantification presents several advantages. For example, this method avoids the need to use suitable antibodies and has the ability to detect very small amounts of target molecule, depending on the technique used. mRNA quantification can be a very powerful and reliable method for investigating gene expression but only if handled thoughtfully.

Initially, nucleic acid quantification meant the addition of radiolabeled UTP or deoxythymidine triphosphate (dTTP) to cell cultures or one of many possible *in vitro* experimental preparations and measuring their incorporation into nucleic acids by TCA (trichloroacetic acid) precipitation. Although radioactive incorporation is a quantitative technique and gave the investigator an idea of the global changes in the nucleic acid population of their experimental system, it was not satisfactory for identifying or quantifying specific genes or transcripts. The first breakthrough in the identification of specific genes came with the development of the Southern transfer method (Southern, 1975). In 1977, the Northern blot was introduced for RNA analysis and named after the DNA Southern blot invented by Sir Edwin Southern (Alwine *et al.,* 1977). Northern blotting uses denaturing gel electrophoresis and blotting with fluorescence-based hybridization probe dependent detection of target RNAs. RNA levels can be quantified and directly compared between multiple samples on a single membrane, although this method lacks the accuracy. The Northern blot however is still quite useful in the study of RNA degradation and transcript size, although nonspecific hybridization can confound data interpretation and the use of radioactivity is frequently necessary (Lee *et al.*, 2005; Maderazo *et al.*, 2003). The RNase protection assay is also used to detect, quantify and characterize RNA species due to its sensitivity and specificity (Melton *et al.*, 1984; Zinn *et al.*, 1983). Hybridization of antisense RNA corresponding to a known complementary target sequence prevents target digestion by single strand specific RNase activity. This process results in the degradation of all remaining single stranded RNAs (*i.e.*, those not hybridized to the probe sequence), enabling the accurate quantitation of specific target sequences. These techniques, while revolutionary at the time of their inception, are long and involved protocols that are not amenable to the examination of many different transcripts in a high throughput manner. Furthermore, the required amount of RNA can be quite large.

The polymerase chain reaction (PCR) is one of the most powerful technologies in molecular biology and was first introduced by Kerry Mullis in 1983 (Mullis, 1990) for which he won the Nobel Prize in 1993. PCR can amplify small amount nucleic acid sequence present in a complex sample into large quantities in a few hours, which can

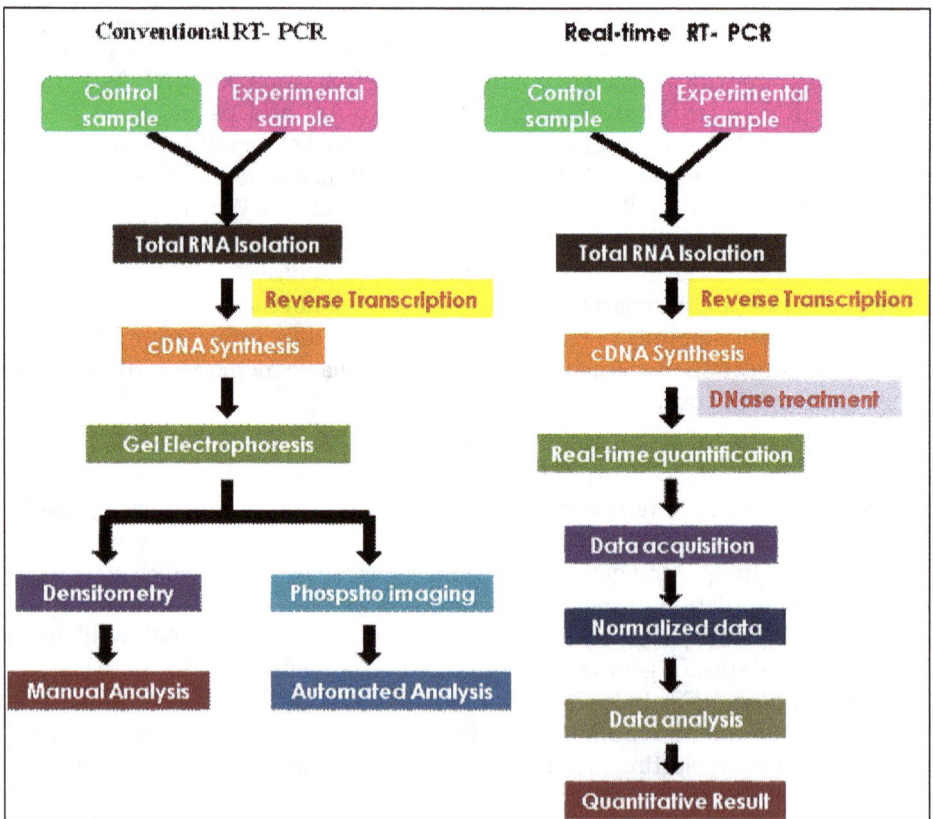

Figure 12.1: Schematic representation of differences between traditional RT-PCR and real-time PCR.

be used further to manipulate the DNA for cloning, genetic engineering and sequencing. PCR, combining with the reverse transcriptase (RT) reaction, vanishingly convert small amounts of RNA into cDNA and identification of a specific RNA transcript was now possible from very low copy numbers of starting material. This amplification, however, poses a challenge for accurately quantifying the initial amount of cDNA. In PCR, detection and quantification of the amplified sequences are performed at the end of the reaction after the last PCR cycle and involve post-PCR analysis such as gel electrophoresis and image analysis. PCR theoretically amplifies DNA exponentially, doubling the number of molecules present with each amplification cycle, but numerous factors complicate the reaction and makes end point analysis as semi quantitative. PCR may not be exponential for the several cycles and the reaction eventually plateaus so the amount of DNA should be measured while reaction is still in the exponential amplification phase, which can be difficult to determine in endpoint analysis. This limitation was resolved by the development of real-time PCR by Higuchi in 1993. Real-time PCR assays are 10,000- to 100,000-fold more sensitive than RNase protection assays (Wang T *et al.*, 1999), 1000-fold more sensitive than dot blot hybridization (Malinen *et al.*, 2003), and can even detect a single copy of a specific transcript (Palmer *et al.*, 2003). In addition, real-time PCR assays can reliably detect gene expression differences as small as 23 per cent between samples (Gentle *et al.*, 2001) and also discriminate between messenger RNAs (mRNAs) with almost identical sequences requires much less RNA template than other methods of gene expression analysis (Figure 12.1).

Principle of Real-Time PCR

Real-time PCR is the state-of-the-art technique to quantify nucleic acids for gene expression and genotyping purpose. The principle of real-time PCR is a standard PCR reaction carried out in the presence of a dye, 'SYBR', which fluorescence when

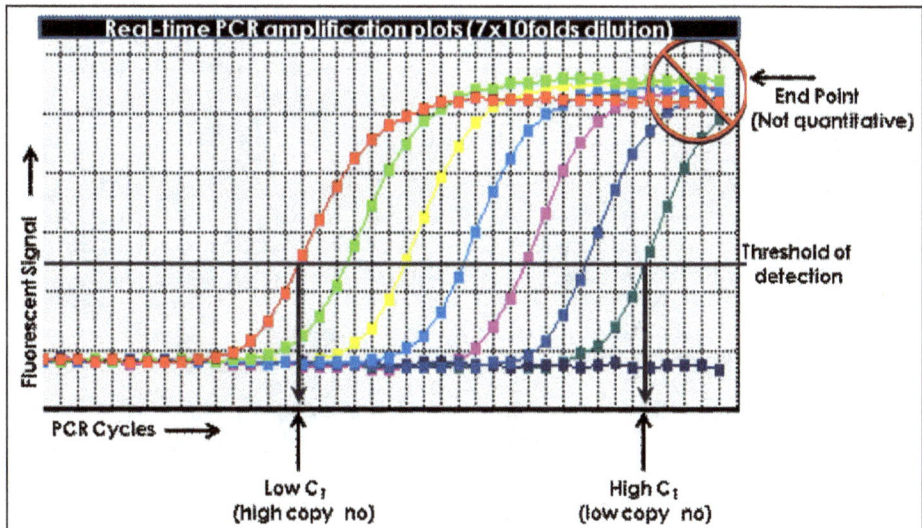

Figure 12.2: Principle of real-time PCR.

intercalated in the DNA helix. The fluorescence will increase as the amount of the PCR product increases and is quantified after each completed PCR cycle. Contrary to end-point PCR techniques, the result is independent from the plateau corresponding to the saturation of the reaction, the latter leading to inaccurate quantification (Gachon *et al.,* 2004) (Figure 12.2). If a particular sequence is abundant in the sample, amplification is observed in earlier cycles and if the sequence is scarce, amplification is observed in later cycles. The changes in fluorescence over the course of the reaction are measured by the instrument that combines thermal cycling and fluorescence detection camera. By plotting fluorescence against the cycle number, the real time PCR instrument generates an amplification plot that represents the accumulation of product over the duration of the entire PCR reaction.

Prior to embarking on a real-time experiment, it is useful to acquire an understanding of the basic principles underlying the polymerase chain reaction. PCR can be broken into four major phases (Figure 12.3): 1) the linear ground phase, 2) early exponential phase, 3) log-linear (also known as exponential) phase, and 4) plateau phase. During the linear ground phase (usually the first 10-15 cycles), PCR is just at beginning, and fluorescence emission at each cycle has not yet risen above background. Baseline fluorescence is calculated at this time. At the early exponential phase, the amount of fluorescence has reached a threshold where it is significantly higher (usually 10 times the standard deviation of the baseline) than background levels. The cycle at which this occurs is known as Ct in ABI Prism® literature (Applied Biosystems, Foster City, CA, USA) or crossing point (CP) in LightCycler® literature (Roche Applied Science, Indianapolis, IN, USA) (Heid *et al.,* 1996; von Ahsen *et al.,* 1999). This value is representative of the starting copy number in the original template

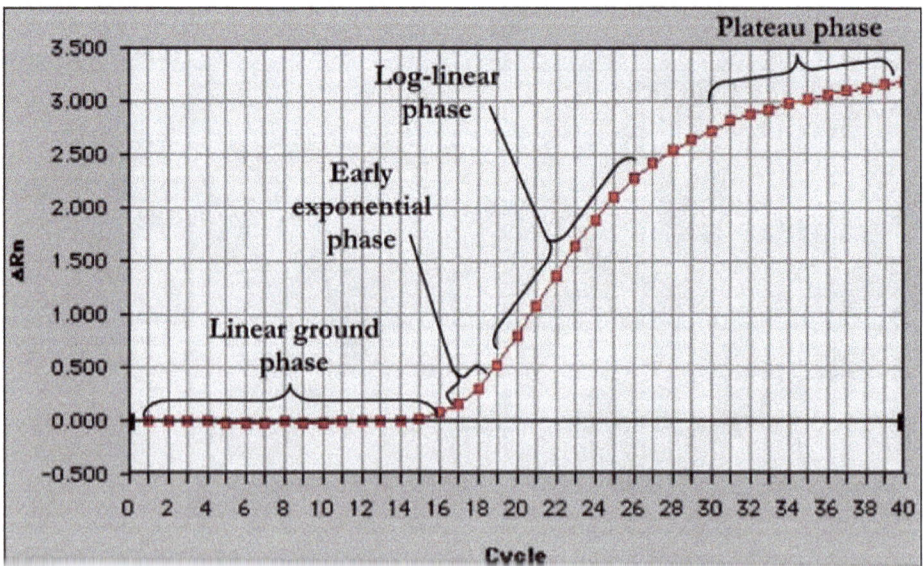

Figure 12.3: Phases of the PCR amplification curve (Adopted from Wong and Medrano, 2005).

and is used to calculate experimental results (Heid *et al.,* 1996). During the log-linear phase, PCR reaches its optimal amplification period with the PCR product doubling after every cycle in an ideal reaction conditions. Finally, the plateau stage is reached when reaction components become limited and the fluorescence intensity is no longer useful for data calculation (Bustin, 2000).

Real Time PCR Steps

RT-PCR is a combination of five steps: (i) Real- time PCR primer design, (ii) isolation of high quality RNA and DNase-I treatment, (iii) the reverse transcriptase (RT)-dependent conversion of RNA into cDNA, (iv) the amplification of the cDNA using the PCR, and (v) the detection and quantification of amplification products in real time.

(i) Primer Designing

Primer designing is one of the most important parameters in real-time PCR. When designing gene-specific real time PCR primers, keep in mind that the amplicon length should be approximately 80-250bp, since longer products do not amplify as efficiently. Primers should be designed according to standard PCR guidelines. For real-time PCR, design primers that anneal to exons on both sides of an intron or span an exon/exon boundary of the mRNA to allow differentiation between amplification of cDNA and potential contaminating genomic DNA. A BLAST search may be performed against public databases to make sure that the designed primers recognize the target of interest. Many primer design software packages are available on databased to design the real-time PCR primers (primer 3; Fast PCR etc.).

(ii) Isolation of High Quality RNA and DNASe-I Treatment

In general, any experiment involving the PCR that begins with converting an RNA template to cDNA will be more challenging than those that begin with DNA templates. Undoubtedly, the most important part of any RT-PCR experiment is the isolation and purification of the RNA template before the experiment is initiated. There are many methods for the isolation of nucleic acids. Most are variants on a few basic isolation themes tailored to meet the special needs of the particular organism and/or target nucleic acid of interest (Sambrook and Russell, 2001; Ausubel, 2001). There are a large number of kits on the market for RNA or DNA isolation from different organisms, tissue or cellular sources. Many plants contain carbohydrates that will co-purify with nucleic acids. In these instances, methods based on CTAB (Cetyltrimethylammonium Bromide) can be used to isolate RNA or DNA (Liao *et al.,* 2004). Another method for further purifying RNA is precipitation using LiCl (Lithium Chloride) instead of salt plus alcohol (Liao *et al.,* 2004). One important factor is to minimize the amount of DNA that is carried over with the RNA during isolation. This is important because not all real-time assays cross exon junctions. Even if the assay does cross an exon junction, pseudogenes may be amplified by the assay anyway as they are spliced but imperfect copies of the original transcript reinserted into the genome. One sure way around this problem in all cases is to DNase I treatment to total RNA preparation.

(iii) cDNA Conversion

As RNA cannot serve as a template for PCR, the first step in an RT-PCR assay is the reverse transcription of the RNA template into cDNA, followed by its exponential amplification in a PCR reaction. Usually, this involves the use of dedicated RNA- and DNA-dependent DNA polymerases, either in separate ('two-enzyme/two-tube') or in single ('two-enzyme/one-tube') reactions, as the use of dedicated enzymes with different properties allows for the customization of individual RT-PCR assays. Separation of the RT and PCR steps has the advantage of generating a stable cDNA pool that can be stored virtually indefinitely. Alternatively, a single polymerase able to function both as an RNA and DNA-dependent DNA polymerase can be used in a 'one-enzyme/one-tube' reaction and minimises hands-on time in addition to the risk of contamination.

RNA transcripts can exhibit significant secondary structure that affects the ability of the RNA dependent DNA polymerase (reverse transcriptase, RT) to generate transcripts (Buell *et al.*, 1978). This can affect RT-PCR quantification and should be minimised when comparing or quantifying diverse mRNA populations (Shimomaye and Salvato, 1989). The two commonly used RTs are avian myeloblastosis virus reverse transcriptase (AMV-RT) and Moloney murine leukaemia virus reverse transcriptase (MMLV-RT). AMV-RT is more robust than MMLV-RT (Brooks *et al.*, 1995), retains significant polymerisation activity up to 55°C (Freeman *et al.*, 1996) and can help eliminate problems associated with RNA secondary structure. In contrast, MMLV-RT and engineered derivatives (Kotewicz *et al.*, 1988) have significantly less RNase H activity than AMV-RT (Gerard *et al.*, 1997). As even reduced RNase H activity can interfere with the synthesis of long amplicons (DeStefano *et al.*, 1991), MMLV-RT may be a better choice if the aim of the experiment is the amplification of full-length cDNA molecules. The RT step can be primed using specific primers, random hexamers or oligo-dT primers, and the choice of primers requires careful consideration. The use of mRNA-specific primers decreases background priming, whereas the use of random and oligo-dT primers maximises the number of mRNA molecules that can be analysed from a small sample of RNA. Primers can cause marked variation in calculated mRNA copynumbers and, compared with specific downstream primers, random hexamers can overestimate mRNA copy numbers up to 19-fold (Zhang and Byrne, 1999).

iv) Amplification of the cDNA

The final step in the real time PCR is amplification, the polymerase chain reaction. The heart of the modern PCR is the addition of a thermostable DNA polymerase. The most commonly used enzyme is from *Thermus aquaticus* or *Taq*. Wild type *Taq* is a 5'–3' synthetic DNA-dependent DNA polymerase with 3'–5' proof reading activity and 5'-nuclease activity. Although wild-type *Taq* can be purchased commercially, the enzyme most commonly used for real-time PCR is a cloned version of the enzyme that has been mutated to remove the 3'–5' proof reading activity. *Taq* is a DNA-dependent DNA polymerase and is a very processive enzyme compared to other commercially available thermostable DNA dependent DNA polymerases, which accounts for its popularity for use in real-time PCR. There are two versions of *Taq* commercially available, the engineered polymerase and a hot-start enzyme. Hot-start enzymes may have added components that keep the polymerase from working until they are inactivated at a high temperature over a period of 1–15 minutes. Common additives are either one or more antibodies directed against the active site of the polymerase or one or more heat labile small molecules that block polymerase activity. Also, there are enzymes with an engineered mutation(s) that requires heating for enzymatic activation. Hot-start enzyme activation is accomplished by the high temperature (94–95°C) of the first denaturing step in the PCR cycle. The difference among them is in how long the enzyme must stay at the high temperature for significant enzymatic activation to occur. A hot-start enzyme is important for some PCRs as most, if not all, the false priming occurs within the first cycle while the components are being heated past the annealing temperature for the first time. Until the reaction mixture reaches the lowest temperature of the thermocycling program, usually 55 to 60C, primers can bind and prime new strand synthesis more easily from the incorrect sequence causing false priming. Once a primer has initiated false priming, the sequence of the primer is incorporated into the rogue sequence. Should the other primer also manage to false prime from the rogue sequence, an unintended PCR product will result as both primer sequences have been incorporated into the PCR product. However, if the DNA polymerase is not active when the reaction is first heating, there will be no opportunity for false priming to occur at that time. This does not mean that there will never be false priming with hot-start enzymes but it will greatly reduce the number of falsely amplified products. Hot-start enzymes are important when the complexity of the DNA or cDNA population being amplified is very high. Examples are genomic DNA or cDNA made from oligo-dT or random primers.

(v) Detection and Quantification of Amplification Products

The question which might be the 'best RT-PCR quantification strategy' to express the exact mRNA content in a sample has still not been answered to universal satisfaction. Numerous papers have been published, proposing various terms, like 'absolute', 'relative', or 'comparative' quantification. The levels of expressed genes may be measured by an 'absolute' quantification or by a relative or comparative qRT-PCR (Pfaffl *et al.*, 2004) (Figure 12.3).

Absolute Quantification

The 'absolute' quantification approach relates the PCR signal to input copy number using a calibration curve (Bustin, 2000; Pfaffl and Hageleit, 2001; Fronhoffs *et al.*, 2002). Absolute quantification uses serially diluted standards of known concentrations to generate a standard curve. The standard curve produces a linear relationship between Ct and initial amounts of total RNA or cDNA, allowing the determination of the concentration of unknowns based on their Ct values (Heid *et al.*, 1996). This method assumes all standards and samples have approximately equal amplification efficiencies (Souaze *et al.*, 1996). Plasmid DNA and *in vitro* transcribed RNA are commonly used to prepare absolute standards. Concentration is measured by A_{260} and converted to the number of copies using the molecular weight of the DNA or RNA. DNA standards have been shown to have a larger quantification range and greater sensitivity, reproducibility, and stability than RNA standards (Pfaffl *et al.,* 2004). However, a DNA standard cannot be used as a standard for a real-time RT-PCR due to the absence of a control for the reverse transcription efficiency (Giulietti, 2001).

Relative Quantification

Relative quantification is based on internal reference genes to determine fold-differences in expression of the target gene. A relative quantification can be deduced considering Ct differences between samples and standards as nicely illustrated by Bovy *et al.* (2002), and improved by Pfaffl (2001). Relative quantification does not require a calibration curve or standards with known concentrations and the reference can be any transcript, as long as its sequence is known (Bustin, 2002). The units used to express relative quantities are irrelevant, and the relative quantities can be compared across multiple real-time RT-PCR experiments (Orlando *et al.,* 1998; Vandesompele *et al.,* 2002). It is the adequate tool to investigate small physiological changes in gene expression levels. Often constant expressed reference genes are chosen as reference genes, which can be co-amplified in the same tube in a multiplex assay (as endogenous controls) or can be amplified in a separate tube (as exogenous controls) (Wittwer *et al.,* 2001; Morse *et al.,* 2005). There are numerous mathematical models available to calculate the mean normalized gene expression from relative quantitation assays. Depending on the method employed, these can yield different results and thus discrepant measures of standard error. A major challenge is the development of exact and reliable gene expression analysis and quantification software. At present, various relative quantification data analysis and software applications *i.e.,* LightCycler® Relative Quantification Software, relative expression software tool (REST), Q-Gene, qBASE, SoFAR, DART-PCR etc are available. An advantage of most described software applications (except SoFAR) is that they are freely available and scientists can use them for their academic research.

Detection Chemistries

A variety of detection chemistries available today can be categorized mainly into two major types: 1) DNA-binding dyes (SYBR Green I), and 2) Dye-labeled, sequence-specific oligonucleotide primers or probes (molecular beacons and TaqMan, hybridization, and Eclipse probes, and Amplifluor, Scorpions, LUX, and BD QZyme primers). The most commonly used detection chemistries are DNA-binding dye SYBR Green I and TaqMan hydrolysis probes. Table 12.1 illustrates the comparison of the characteristic features of various detection chemistries available for real time PCR.

In general, for low-throughput, DNA-binding dyes may be preferable because these assays are easier to design, are faster to set up, and are initially more cost-effective. For high-throughput experiments, however, a fluorescent primer- or probe-based assay (singleplex or multiplex) may be more desirable because the initial cost can be spread over many experiments and the multiplex capability can reduce assay time. Multiplex assays require the use of a fluorescent primer or probe based chemistry,

Table 12.1: Validation of reference genes for plant studies.

Plant	Reference Genes Selected	Suggested Reference Gene	Reference
Rice	18S rRNA, glyceraldehyde-3-phosphate dehydrogenase, actin, and tubulin	18S rRNA	Kim *et al.*, 2003
Potato	Beta-tubulin, cyclophilin, actin, elongation factor 1-alpha (ef1alpha), 18SrRNA, adenine phosphoribosyl transferase (aprt), and cytoplasmic ribosomal protein L2	elongation factor 1-alpha (ef1alpha)	Nicot *et al.*, 2005
Vitis vinifera cv. Cabernet Sauvignon	Actin, AP47 (clathrin-associated protein), cyclophilin, EF1-alpha (elongation factor 1-alpha), GAPDH (glyceraldehyde 3-phosphate dehydrogenase), MDH (malate dehydrogenase), PP2A (protein phosphatase), SAND, TIP41, alpha-tubulin, beta-tubulin, UBC (ubiquitin conjugating enzyme), UBQ-L40 (ubiquitin L40) and UBQ10 (polyubiquitin)	GAPDH, actin, EF1-alpha and SAND	Reid *et al.*, 2006
Coffea arabica	Alcohol dehydrogenase (adh), 14-3-3, polyubiquitin (poly), beta-actin (actin) and glyceraldehyde-3-phosphate dehydrogenase (gapdh)	Gapdh	Barsalobres-Cavallari *et al.*, 2009
Tomato	Actin (ACT), beta-tubulin, elongation factor 1alpha (EF1), glyceraldehyde-3-phosphate dehydrogenase (GAPDH), phosphoglycerate kinase (PGK), ribosomal protein L2 (RPL2), ubiquitin (UBI), and a catalytic subunit of protein phosphatase 2A(PP2Acs)	RPL2PP2Acs, ACT and UBI	Lovdal and Lillo, 2009
Brachiaria brizantha	EF1 (elongation factor 1 alpha),E1F4A (eukaryotic initiation factor 4A), GAPDH (glucose-6-phosphatedehydrogenase), GDP(glyceroldehyde-3-phosphate dehydrogenase), SUCOA(succinyl-CoA ligase), TUB (tubulin), UBCE (ubiquitin conjugating enzyme), UBI(ubiquitin)	E1F4A and EF1	Silveira *et al.*, 2009
Peach [*Prunus persica* (L.) Batsch]	Actin 2/7 (ACT), cyclophilin (CYP2), RNA polymerase II (RP II), phospholipase A2 (PLA2), ribosomal protein L13 (RPL13), glyceraldehyde-3-phosphate dehydrogenase (GAPDH), 18S ribosomal RNA (18S rRNA), tubulin beta (TUB), tubulin alpha (TUA), translation elongation factor 2 (TEF2) and ubiquitin 10 (UBQ10)	TEF2, UBQ10 and RP II	Tong *et al.*, 2009
Salvia miltiorrhiza	18S, EF1alpha, alpha-Tubulin, Ubiquitin and Actin	Actin andUbiquitin	Yang *et al.*, 2009
Soybean	ACT2, ACT11, TUB4, TUA5, CYP,UBQ10, EF1b, SKIP16, MTP, PEPKR1, HDC, TIP41, UKN1, UKN2	ACT11, UKN1 and UKN2	Hu *et al.*, 2009

Contd...

Table 1.1– Contd...

Plant	Reference Genes Selected	Suggested Reference Gene	Reference
Tobacco (*Nicotiana tabacum*)	18S rRNA, EF-1alpha, Ntubc2, alpha- and beta-tubulin, PP2A, L25 and actin	L25 and EF-1alpha	Schmidt *et al.*, 2010
Brassica napus	Acetyl-CoA carboxylase gene (BnACCg8), phosphoenolpyruvate carboxylase (PEP), oleoyl hydrolase gene (FatA), high-mobility-group protein I/Y gene (HMG-I/Y) and cruciferin A gene (CruA	BnACCg8	Wu *et al.*, 2010
Cichorium intybus	Nicotinamide adenine dinucleotide dehydrogenase (NADHD), actin (ACT), beta-tubulin (TUB), glyceraldehyde-3-phosphate-dehydrogenase (GADPH), histone H3 (H3), elongation factor 1-alpha (EF) and 18S rRNA (rRNA)	ACT, EF and rRNA	Maroufi *et al.*, 2010
Gossypium hirsutum	GhACT4, GhEF1alpha5, GhFBX6, GhPP2A1, GhMZA, GhPTB, GhGAPC2, GhbetaTUB3 and GhUBQ14	GhUBQ14 and GhPP2A1	Artico *et al.*, 2010
Citrus rootstock "Swingle" citrumelo	Cyclophilin (CYP), cathepsin (CtP), actin (ACT), glyceraldehyde-3-phosphate dehydrogenase (GAPDH), elongation factor 1alpha (EF1alpha), beta-tubulin (TUB), and ADP ribosylation factor (ADP)-	EF1alpha and ADP	Carvalho *et al.*, 2010
Cicer arietinum L.	ACT1, EF1alpha, GAPDH, IF4a, TUB6, UBC, UBQ5, UBQ10, 18SrRNA, 25SrRNA, GRX andHSP90	EF1alpha and HSP90	Garg *et al.*, 2010
Litchi	Actin (ACTIN), glyceraldehyde-3 phosphate-dehydrogenase (GADPH), elongation factor 1-alpha(EF-1α), poly ubiquitin enzyme (UBQ), α-tubulin (TUA), β-tubulin (TUB) and RNApolymerase-II transcription factor (RPII)	GAPDH + EF-1α or GAPDH + ACTIN	Zhong *et al.*, 2011
Chrysanthemum	Tubulin (alpha-2,4 tubulin), actin, EF1 α (elongation factor 1 α), UBC (ubiquitin C), GAPDH (glyceraldehyde-3-phosphate dehydrogenase), psaA (photosynthesis-related plastid gene representing photosystem I), PP2Acs (catalytic subunit of protein phosphatase 2A), and PGK (phosphoglycerate kinase)	EF1 A, PP2Acs	Gu *et al.*, 2011
Citrus	18SrRNA, ACTB, rpII, UBQI, UBQ10, GAPDH and TUB	18SrRNA, ACTB and rpII	Yan *et al.*, 2012
Blueberry	Ubiquitin –Conjugating enzyme (UBC28),RNA helicase - like (RH8), Clathrin adapter complexes medium subunit family protein (CACSa), and Polyubiquitin (UBQ3b).	UBC28, CACSa, RH8, and UBQ3b	Vashisth *et al.*, 2011
Brassica juncea	ACT2, ELFA, GAPDH, TUA, UBQ9, ACP, CAC, SNF, TIPS-41, TMD, TSB and ZNF	TUA, TIPS-41 and CAC	Chandna *et al.*, 2012

because the lack of specificity of DNA-binding dyes makes them incompatible with quantitative multiplex assays.

1. DNA-Binding Dyes

DNA binding dyes emit fluorescence when bound to dsDNA. As the double-stranded PCR product accumulates during cycling, more dye can bind and emit fluorescence. Thus, the fluorescence intensity increases proportionally to dsDNA concentration (Wittwer *et al.*, 1997). This technique is very flexible because one dye can be used for different gene assays. Consequently, multiplexing reactions is not possible. Because DNA binding dyes do not bind in a sequence-specific manner, these assays are prone to false positives (Simpson *et al.*, 2000). Accurate results demand a specific PCR, which can be confirmed via dissociation curve analysis, where the presence of different PCR products is reflected in the number of first derivative melting peaks (Ririe *et al.*, 1997) or gel analysis (Lekanne Deprez *et al.*, 2002). The most commonly used DNA-binding dye for real-time PCR is SYBR Green I, which binds nonspecifically to double-stranded DNA (dsDNA). SYBR Green I exhibits little fluorescence when it is free in solution, but its fluorescence increases up to 1,000-fold when it binds dsDNA. Therefore, the overall fluorescent signal from a reaction is proportional to the amount of dsDNA present, and will increase as the target is amplified. A new dye, EvaGreen™ (Biotium, Hayward, CA) has been presented for use in real-time PCR. EvaGreen™ costs more than SYBR®Green I but it is reported to show little PCR inhibition which can be a problem with high concentrations of SYBR® Green I, substantially higher fluorescence over SYBR® Green I, and greater stability at high temperatures. More recently new dyes, BOXTO (TATAA Biocenter, Gothenburg, Sweden), LCGreen™ I and LCGreen™ PLUS (Idaho Technology, Salt Lake City, UT) have come onto the market. The advantages of using dsDNA-binding dyes include simple assay design (only two primers are needed; probe design is not necessary), ability to test multiple genes quickly without designing multiple probes (*e.g.*, for validation of gene expression data from many genes in a microarray experiment), lower initial cost (probes cost more), and the ability to perform a melt-curve analysis to check the specificity of the amplification reaction.

2. Dye-Labeled, Sequence-Specific Oligonucleotide Primers or Hybridization Probes

2.1 Hybridization Probes

These can be utilized in either a four or three oligonucleotide manner (Bernard and Wittwer, 2000). The four oligonucleotide method consists of two PCR primers and two sequence-specific probes that bind adjacent to each other in a head-to-tail arrangement. The upstream probe is labeled with an acceptor dye on the 32 end, and the downstream probe with a donor dye on the 5' end (Bernard *et al.*, 1998), allowing the donor and acceptor fluorophores to experience an increase in fluorescence resonance energy transfer (FRET) when bound (Bernard and Wittwer, 2000). The three oligonucleotide method is similar to the four oligonucleotide method, except that the upstream PCR primer is labeled with an acceptor dye on the 3' end, and thus replaces the function of one of the probes from the four oligonucleotide method. In both cases, the downstream probe can be designed to cover a mutation site and discriminate between known alleles and detect new alleles simultaneously (Lay and Wittwer, 1997).

2.2 Hydrolysis Probes

Hydrolysis probes exemplified by the TaqMan chemistry, also known as 5' nuclease assay, fluoresce upon probe hydrolysis to detect PCR product accumulation. The sequence-specific probe is labeled with a reporter dye on the 5' end and a quencher dye on the 3' end (Gibson *et al.*, 1996), which allows the quencher to reduce the reporter fluorescence intensity by FRET when the probe is intact (Clegg, 1992). While both hydrolysis and hybrid-ization probes rely on FRET to alter the intensity of fluorescence emission, the energy transfer works in opposite manners in these two chemistries. FRET reduces fluorescence intensity in hydrolysis probes and increases intensity in hybridization probes. When annealed to the target sequence, the bound and quenched probe will be degraded by the DNA polymerase's 5' nuclease ability during the extension step of the PCR. Probe degradation allows for separation of the reporter from the quencher dye, resulting in increased fluorescence emission (Heid *et al.*, 1996; Gibson *et al.*, 1996).

2.3 Molecular Beacons

These probes consisting of a sequence-specific region (loop region) flanked by two inverted repeats, molecular beacons are the simplest hairpin probe (Tyagi and Kramer, 1996). Reporter and quencher dyes are attached to each end of the molecule, causing a reduction in fluorescence emission via contact quenching (FRET) when the beacon is in hairpin formation (free in solution). When bound to the target, the quencher and reporter are separated, allowing reporter emission. Hairpin probes tend to have greater specificity than linear probes because the probe-target complex is thermodynamically more stable than the hairpin structure itself (Bonnet *et al.*, 1999), a property often exploited for allele discrimination (Marras *et al.*, 1999). To increase fluorescence emission, "wavelength shifting molecular beacons" have been developed, which fluoresce in a number of colors from a single monochromatic light source (Tyagi *et al.*, 2000).

2.4 Scorpions

Scorpions combine the detection probe with the upstream PCR primer (Whitcombe *et al.*, 1999) and consist of a fluorophore on the 5' end, followed by a complementary stem-loop structure (also containing the specific probe sequence), quencher dye, DNA polymerase blocker (a nonamplifiable monomer that prevents DNA polymerase extension), and finally a PCR primer on the 3' end. The probe sequence contained within the hairpin allows the scorpion to anneal to the template strand, which separates the quencher for the fluorophore and results in increased fluorescence. Because sequence-specific priming and probing is a unimolecular event, scorpions perform better than bimolecular methods under conditions of rapid cycling such as the LightCycler (Thelwell *et al.*, 2000). Cycling is performed at a temperature optimal for DNA polymerase activity instead of the reduced temperature necessary for the 5' nuclease assay. Scorpions are specific enough for allele discrimination and may be multiplexed easily (Whitcombe *et al.,* 1999). The scorpion chemistry has been improved with the creation of duplex scorpions in which the reporter dye/probe and quencher fragment are on separate, complementary molecules (Solinas *et al.,* 2001). The duplex scorpions still bind in a unimolecular event, but because the reporter and quenchers are on separate molecules, they yield greater signal intensity because the reporter and quencher can separate completely.

2.5 Sunrise™ Primers

These are created by Oncor (Gaithersburg, MD, USA), Sunrise primers are similar to scorpions in that they combine both the PCR primer and detection mechanism in the same molecule. These probes consist of a dual-labeled (reporter and quencher fluorophores) hairpin loop on the 5' end, with the 3' end acting as the PCR primer. When unbound, the hairpin is intact, causing reporter quenching via FRET. Upon integration into the newly formed PCR product, the reporter and quencher are held far enough apart to allow reporter emission.

2.6 LUX™ Fluorogenic Primers

Light upon extension (LUX) primers (Invitrogen, Carlsbad, CA, USA) are self-quenched single-fluorophore labeled primers almost identical to Sunrise primers. However, rather than using a quencher fluorophore, the secondary structure of the 32 end reduces initial fluorescence to a minimal amount (Nazarenko *et al.,* 2002). Because this chemistry does not require a quencher dye, it is much less expensive than dual-labeled probes. While this system relies on only two oligonucleotides for specificity, unlike the SYBR Green I platform in which a dissociation curve is used to detect erroneous amplification, no such convenient detection exists for the LUX platform. Agarose gels must be run to ensure the presence of a single PCR product, a step that is extremely important not only for the LUX primers but also for the Sunrise primers and scorpions because PCR priming and probe binding are not independent in these chemistries.

Selection of Reference Genes for Plant Studies

The use of RT-PCR for gene expression studies in model species, such as *Arabidopsis* (Czechowski *et al.,* 2005), rice (Kim *et al.,* 2003), and *Medicago* (Kakar *et al.,* 2005) has been widespread. The technique RT-PCR is associated with several problems. These include the inherent quantity and quality variation of RNA, the variability of extraction protocols that may copurify inhibitors, and the variability in RT reactions and PCR efficiency, which have a major impact on the accuracy of the calculated expression result and are critically influenced by PCR reaction components. Therefore, normalization is needed to correct these sources of variability in qPCR analysis. Several strategies have been suggested for normalizing qPCR data, including normalization to sample size, total RNA, or genomic DNA and the use of an artificial molecule incorporated into the sample (Huggett *et al.,* 2005). By far, the most common method is the measurement of an internal reference gene (housekeeping genes) because they potentially account for all of the above-mentioned sources of variability. The expression of this gene should be constant in the tissues or cells under investigation, and should not change in response to experimental treatments. 'Housekeeping' genes

Table 1.2: Studies which have used Real-Time PCR to measure the Abiotic stress responsive gene expression in plants.

Plant	Genes	Result	References
Indica rice cultivar Minghui 86 and Nipponbare	miRNAs miR6254, miR6250, miR169i-3p	Quantitative real-time RT-PCR analysis revealed that miR6254 was predominantly expressed in roots relative to other tissues, whereas miR6250 and miR169i-3p were constitutively expressed in all tissues with high abundance in roots (miR6250) or leaves (miR169i-3p). Also, identified novel miRNAs, six were up-regulated and two were down-regulated under arsenite stress.	Liu et al., 2012
Cynanchum komarovii Al Iljinski	CkChn134 (homologous to chitinase)	The real-time PCR showed that the transcription level of CkChn134 had a significant increase under the stress of ethylene, NaCl, low temperature, drought, and pathogen infection, which indicates that CkChn134 may play an important role in response to abiotic and biotic stresses.	Wang Q et al., 2012
Soy bean	Dicer-like (DCL) genes, GmDCL2a-GmDCL3a and GmDCL1b-GmDCL2b	Among homoeologous or paralogous DCL genes, the Dicer-like 2 (DCL2) gene pair exhibited the strongest response to stress and most conserved co-expression pattern.	Curtin et al., 2012
Malus	Dehydration responsive element binding (DREB) transcription factor	Quantitative real-time PCR showed that transcript levels of some putative MdDREB genes were up-regulated significantly under various abiotic-stress treatments, which indicated their vital roles during stress adaptation.	Zhao et al., 2012
Tamarix hispida	Four Peroxiredoxin genes (ThPrxII, ThPrxIIE, ThPrxIIF, and Th2CysPrx)	The results showed that the four ThPrxs were all expressed in roots, stems and leaves under abiotic stress and ABA. Furthermore, the transcript levels of ThPrxIIE and ThPrxII were the lowest and the highest, respectively, in all tissue types.	Gao et al., 2012
Amur grape (Vitis amurensis Rupr.)	Micro RNAs	The expression levels of va-miRNAs in flowers and berries were found to consistent, a number of regulatory miRNAs exist in Amur grape play an important role in growth, development, and response to abiotic or biotic stress.	Wang C et al., 2012
Salicornia brachiata	Metallothionein (SbMT-2)	Expression of SbMT-2 gene was up-regulated concurrently with zinc, copper, salt, heat and drought stress, down regulated by cold stress while unaffected under cadmium stress. Heterologous expression of SbMT-2 gene enhances metal accumulation and tolerance in E. coli.	Chaturvedi, 2012
Tamarix hispida	MYC genes,	Real-time PCR showed that all nine MYC genes were expressed in root, stem and leaf tissues, but that the levels of the transcripts of these genes in the various tissues differed notably.	Ji et al., 2012

Contd...

Table 12.2–*Contd...*

Plant	Genes	Result	References
Triticum aestivum L.	Expansin gene, TaEXPB23	Real-time PCR analysis revealed that TaEXPB23 transcript expression was upregulated by exogenous methyl jasmonate (MeJA) and salt stress, but downregulated by exogenous gibberellins (GA), ethylene (ET), indole-3-acetic acid (IAA) and a-naphthicetic acid (NAA).	Han *et al.*, 2012
Brassica napus	Modified AP2/ERF Transcription Factor	Real-time PCR analyses also revealed that the expression levels of several stress-regulated genes were altered in the over-expressed BnaERF-B3-hy15-mu3 transgenic lines. The BnaERF-B3-hy15 responded to exogenous ABA. Using RT-PCR analysis, the expression of BnaERF-B3-hy15 at different stages and stress treatments were also analyzed.	Xiong *et al.*, 2012
Cassava	Transcriptome profiling of stress responsive genes under cold conditions	Real-time PCR showed the expression changes of 32 genes under cold and other abiotic stress conditions. Importantly, most of the tested stress-responsive genes were primarily expressed in mature leaves, stem cambia, and fibrous roots rather than apical buds and young leaves.	An *et al.*, 2012
Tomato	AUX/IAAs genes	The transcript abundance of 17 SlIAA genes were increased within a few hours when the seedlings were treated with exogenous IAA. Most of the SlIAA family genes exhibited diverse expression levels under different abiotic stress conditions in tomato seedlings.	Wu, 2012
Solanum lycopersicum	Mitogen-activated protein kinases (MAPKs)	Expression analysis of SlMAPK genes using real-time quantitative PCR demonstrated that all SlMAPK transcripts were able to be detected in at least one investigated tissue, and some of them exhibited tissue-specific expression patterns. The transcript abundance of nearly all SlMAPK genes was increased in response to heat stress treatment.	Kong *et al.*, 2012
Banana fruit	MaCOL1, like CONSTANS (CO)	Real-time PCR analysis showed that MaCOL1 was differentially expressed among various banana plant organs, with higher expression in flower. Accumulation of MaCOL1 transcripts in pulp obviously increased during natural or ethylene-induced fruit ripening, and also enhanced by abiotic and biotic stresses.	Chen *et al.*, 2012

Contd...

Table 1.2– Contd...

Plant	Genes	Result	References
Styela plicata	heat shock protein (hsp70) gene	Quantitative real time PCR showed hsp70 expression varied over 2 years cycle, with higher stress levels recorded in summer and winter. Periodic conditions of high temperatures, particularly when coupled with low salinities, increased hsp70 gene expression.	Pineda *et al.*, 2012
Oryza sativa	OsPFA-DSP1, a protein tyrosine phosphatase	Quantitative real-time PCR showed that OsPFA-DSP1 mRNA was induced by drought stress.	Liu B *et al.*, 2012
Oryza sativa	OsbZIP52/RISBZ5, a basic leucine zipper (bZIP) transcriptionfactor	Real-time PCR analysis revealed that some abiotic stress-related genes, such as OsLEA3, OsTPP1, Rab25, gp1 precursor, b-gal, LOC_Os05g11910 and LOC_Os05g39250, were down-regulated in OsbZIP52 overexpression lines. These results suggest that OsbZIP52/RISBZ5 could function as a negative regulator in cold and drought stress environments.	Liu C *et al.*, 2012
Oryza sativa	Members of annexin gene	Expression analysis by real-time RT-PCR revealed differential temporal and spatial regulation of these genes. The rice annexin genes are also found to be regulated in seedling stage by various abiotic stressors including salinity, drought, heat and cold.	Jami *et al.*, 2012
Triticum aestivum	Pathogenesis-Related 10 (PR-10)	Real-time PCR results showed that the abundance of BQ752893 was generally greater than the abundance of CV778999, particularly when measured in roots across four wheat genotypes. However, CV778999 transcripts were more abundant than BQ752893 in flag leaves.	Mohammadi *et al.*, 2012
Potato	WRKYtranscription factor	The expression profiling clearly demonstrated a widetranscriptional change during the pre-, early and late stages of root colonization.	Gallou *et al.*, 2012
Brassica rapa	miRNAs that are responsive to heat stress	Real-time PCR showed that the conserved miRNAs bra-miR398a and bra-miR398b were heat-inhibitive and guided heat response of their target gene, BracCSD1; and bra-miR156h and bra-miR156g were heat-induced and its putative target BracSPL2 was down-regulated.	Yu *et al.*, 2012
Malus hupehensis	MhGR-RBP1, glycine-rich RNA-binding protein (GR-RBP)	qRT-PCR, indicating that MhGR-RBP1 is expressed abundantly in young leaves but weakly in roots and shoots. Transcript levels in the leaves were increased markedly by drought, hydrogen peroxide (H_2O_2), and mechanical wounding, slightly by salt stress.	Wang S *et al.*, 2012

Contd...

Table 12.2—*Contd...*

Plant	Genes	Result	References
Ceratoides arborescens	Ethylene responsive factor (CeERF)	Expression level of CeERF was highest in leaves and lowest in roots. CeERF expression was induced by 20 per cent PEG. Overexpression of CeERF in transgenic tobacco plants resulted in higher tolerance to abiotic stresses than in control plants.	Dong *et al.,* 2012
Maize	microRNAs in response to low nitrate availability	Nine miRNA families (miR164, miR169, miR172, miR397, miR398, miR399, miR408, miR528, and miR827) were identified in leaves, and nine miRNA families (miR160, miR167, miR168, miR169, miR319, miR395, miR399, miR408, and miR528) identified in roots.	Xu *et al.,* 2011
Zea mays L.	Stress related genes resistant to aflatoxin contamination	Based on the relative-expression levels, the 7 maize inbred lines clustered into two different groups. One group included B73, Lo1016 and Mo17, which had higher levels of aflatoxin contamination and lower levels of overall gene expression. The second group which included Tex6, Mp313E, Lo964 and A638 had lower levels of aflatoxin contamination and higher overall levels of gene expressions. A total of six "cross-talking" genes were identified between the two groups.	Jiang *et al.,* 2011
Sweet sorghum (*Sorghum bicolor* L.).	Alkaline stress-responsive genes of CBL family (Calcineurin B-like proteins)	Real-time PCR analysis showed that SbCBL genes have different tissue-specific expression patterns under normal growth conditions in sweet sorghum.	Zhang *et al.,* 2011
Oryza sativa L.	OsPC, the phytocyanin gene family	The expression profiles showed that a number of OsPC genes abundantly expressed in the various stages of development. Moreover, 17 genes were regulated under the treatments of abiotic stresses.	Ma *et al.,* 2011
Alfalfa	Acetoacetyl-CoA thiolase (MsAACT1)	Real-time RT-PCR analysis indicated that MsAACT1 expression is highly increased in roots and leaves under cold and salinity stress.	Soto *et al.,* 2011
Oryza sativa L.	Heat shock factors	Expression profiling through microarray and quantitative real-time PCR showed that eight OsHsfs express at a higher level during seed development, while six HSFs are up-regulated in all the abiotic stresses studied.	Chauhan *et al.,* 2011
Oryza sativa L.	Jasmonate pathway genes	Real-time RT-PCR data revealed that Exogenous ET (ethephon) application onto the shoots strongly activates JA biosynthesis and signaling genes in the roots.	Nahar *et al.,* 2011

Contd...

Table 12– *Contd...*

Plant	Genes	Result	References
Vitis vinifera L.	Abiotic stress responsive genes	Identification of tissue-specific, abiotic stress-responsive gene expression patterns in wine grape (*Vitis vinifera L.*) based on curation and mining of large-scale EST data sets.	Tillett *et al.*, 2011
Gossypium hirsutum ...	Transcriptome analysis of salt-stress-regulated genes	Key regulatory gene families involved in abiotic and biotic sources of stress such as WRKY, ERF, and JAZ were differentially expressed.	Yao *et al.*, 2011
Cucumis sativus	epibrassinolide (EBR)	The expression of BR biosynthetic genes was repressed in EBR-treated leaves, but elevated significantly in untreated systemic leaves.	Xia, 2011
Gossypium arboreum	biotic and abiotic stress up-regulated ESTs	Total 39 abiotic and biotic up-regulated ESTs were identified.	Barozai and Hussain, 2011
Leymus chinensis	LcDREB3a transcription factor(DREB)	Real-time PCR-based expression analysis shows the transcript accumulates in response to a variety of stress treatments. These results indicate that LcDREB3a is involved in both ABA-dependent and -independent signal transduction in the stress-responsive process of *L. chinensis*.	Xianjun *et al.*, 2011
Vitis vinifera	transcriptomic profiling of genes responsive to light stress (LSCA)	When gene-expression profiles were compared, 'protein metabolism and modification', 'signaling', and 'anti-oxidative' genes were more represented in LSCA.	Carvalho *et al.*, 2011
Arabidopsis thaliana	FTL1/DDF1, AP2 transcription factor of the CBF/DREB1 subfamily	Real-time PCR indicated that FTL1/DDF1 was up-regulated by drought, heat, low temperatures and salt stress in wild-type *Arabidopsis*. Its increased expression in the mutant induced various stress-responsive genes under normal growing conditions, resulting in improved tolerances.	Kang *et al.*, 2011
Setaria italica L.	SiDREB2, DREB2-like gene	Quantitative real-time PCR analysis showed significant up-regulation of SiDREB2 by dehydration (polyethylene glycol)and salinity (NaCl), while its expression was less affected by other stresses.	Lata *et al.*, 2011
Helianthus	Superoxide dismutases, mitochondrialMn-SOD.	Expression of the Mn-SOD I was approximately 12-fold higher than that of Mn-SOD II. The Mn-SOD II showed a significant modulation in response to the assayed biotic and abiotic stresses even when it had no apparent oxidative stress, such as low temperature.	Fernández-Ocaña *et al.*, 2011

Contd...

Table 12.2–*Contd...*

Plant	Genes	Result	References
Cynanchum komarovii Al Iljinski	CkTLP, thaumatin-like proteins (TLPs)	The real-time PCR showed the transcription level of CkTLP had a significant increase under the stress of abscisic acid (ABA), salicylic acid (SA), methyl jasmonate (MeJA), NaCl and drought, which indicates that CkTLP may play an important role in response to abiotic stresses.	Wang *et al.*, 2011
Oryza sativa L.	OsMSR2 (*Oryza sativa* L. Multi-Stress-Responsive gene 2), a novel calmodulin-like protein gene	Expression of OsMSR2 conferred enhanced tolerance to high salt and drought in Arabidopsis (*Arabidopsis thaliana*) accompanied by altered expression of stress/ABA-responsive genes.	Xu *et al.*, 2011
Hordeum vulgare L.	intrinsic proteins	Analysed the expression levels of five tonoplast intrinsic protein isoforms (HvTIP1;2, HvTIP2;1, HvTIP2;2, HvTIP2;3 and HvTIP4;1), a NOD26-like intrinsic protein (HvNIP2;1) and a plasma membrane intrinsic protein (HvPIP2;1) by using the quantitative real-time RT-PCR. Heavy metals (0.2 mM each of Cd, Cu, Zn and Cr) down regulated the transcript levels by 60-80 per cent in roots, whereas 0.2 mM Hg upregulated expressions of most genes in roots.	Ligaba *et al.*, 2011
Arabidopsis thaliana	calcium-dependent lipid-binding protein (AtCLB)	Expression of the AtClb gene was documented in all analysed tissues of *Arabidopsis* (leaf, root, stem, flower, and silique) by real-time PCR analysis. Atclb gene is in negatively regulating responses to abiotic stress in *Arabidopsis thaliana* was identified.	de Silva *et al.*, 2011
Solanum lycopersicum	Ribonucleases, RNase LER, first class II gene of the RNase T2 family	Quantitative real-time RT-PCR analysis showed that treatments with abscisic acid, ethylene or other abiotic and biotic stress factors did not affect RNaseLER expression significantly. Unlike tomato class I genes, RNaseLER represents a constitutively expressed gene with a cell-specific role in stomata and trichomes and no involvement in stress responses.	Köthke and Köck, 2011
Cicer microphyllum	CmMet-2, Metallo-thionein-like gene	Low level of CmMet-2 transcript in the aerial parts than the roots. Quantitative analysis using real-time PCR assay revealed induction of transcript in all parts of plants in response to cold stress at 4°C. Further, regulation of transcript accumulation in response to ABA application, PEG (100 µM)-induced osmotic stress, suggests the involvement of CmMet-2 in multiple stress response.	Singh *et al.*, 2011

Contd...

Table 1: 2– *Contd...*

Plant	Genes	Result	References
Pyrus pyrifolia	four ferritin genes (PpFer1, PpFer2, PpFer3, and PpFer4)	These ferritin genes were differentially expressed in response to various abiotic stresses and hormones treatments. The expression of ferritin wasn't affected by Fe(III)-citrate treatment. ABA significantly enhanced the expression of all four ferritin genes, especially PpFer2, followed by BAP, GA and IAA.	Xi *et al.*, 2011
Triticum aestivum L.	TaABC1 (*Triticum aestivum* L. activity of bc(1) complex) protein kinase	Over expression of TaABC1 increased the expression of stress-responsive genes, such as DREB1A, DREB2A, RD29A, ABF3, KIN1, CBF1, LEA, and P5CS, detected by real-time PCR analysis. TaABC1 overexpression enhances drought, salt, and cold stress tolerance in *Arabidopsis*, and implies that TaABC1 may act as a regulatory factor involved in a multiple stress response pathways.	Wang *et al.*, 2011
Wheat leaves (cv. Suwan 11)	TaCab1 (*Triticum aestivum* calcium binding EF-hand protein 1)	qRT-PCR analyses revealed that TaCab1 was highly expressed in leaves than roots and stems. The transcription of TaCab1 was also up-regulated with phyto-hormones and stress stimuli suggesting tolerance to biotic and abiotic stresses through the SA signaling pathway.	Feng *et al.*, 2011
Panax ginseng C.A. Meyer	Transcript profiling of antioxidant genes during biotic and abiotic stresses.	The transcriptome result under abiotic stresses showed differential expression and elevated up-regulation of PgSOD, PgGPX, PgGS, and PgAPX, thus it may prove the generation of ROS in ginseng.	Sathiyaraj, 2011
Triticum aestivum cv.CPAN 1676	high temperature stress responsive genes	Ten selected genes were analyzed in further detail including one unknown protein and a new heat shock factor, by quantitative real-time PCR in an array of 35 different wheat tissues representing major developmental stages as well as different abiotic stresses.	Chauhan *et al.*, 2011
Triticum aestivum L.	TaMBD2 homoeologous genes encoding methyl CpG-bindingdomain proteins	TaMBD2-5B and TaMBD2-5D were highly responsive to salt stress and TaMBD2-5B was specifically upregulated by low temperature in the seedling leaves.	Hu *et al.*, 2011
Salicornia brachiata	Na(+)/H(+) antiporter-gene (SbNHX1)	Real time PCR analysis revealed that SbNHX1 transcript expresses maximum at 0.5 M NaCl. Overexpression of SbNHX1 gene in tobacco plant showed NaCl tolerance showing that SbNHX1 is a potential gene for salt tolerance, and can be used in future for developing salt tolerant crops.	Jha *et al.*, 2011

Contd...

Table 12.2–*Contd...*

Plant	Genes	Result	References
Setaria italica	transcriptome analysis of genes related to salinity stress	Quantitative real-time PCR of 21 highly up-regulated (e" 2.5-fold) transcripts showed temporal variation in expression under salinity.	Puranik *et al.*, 2011
Morus indica cv. K2	osmotin	Real-time PCR analysis provided evidence for the expression of osmotin in the tobacco transgenic plants under both the constitutive and stress-inducible promoters.	Das *et al.*, 2011
Tamarix hispida	H+-ATPase c subunit gene (ThVHAc1)	Real-time RT-PCR demonstrated that ThVHAc1 gene expression was induced by NaCl, NaHCO$_3$, PEG and CdCl$_2$ stress in *T. hispida* roots, stems and leaves. Exogenous application of abscisic acid (ABA) also stimulated ThVHAc1 transcript levels in the absence of stress, suggesting that ThVHAc1 is involved in ABA-dependent stress signaling pathway in *S.cerevisiae.*	Gao *et al.*, 2011
Tomato	Glutaredoxins (SlGRX1)	Quantitative real-time RT-PCR analysis revealed that SlGRX1 was expressed ubiquitously in tomato including leaf, root, stem and flower, and its expression could be induced by oxidative, drought, and salt stresses. Expression levels of oxidative, drought and salt stress related genes Apx2, Apx6, and RD22 were up-regulated in SlGRX1-overexpressed *Arabidopsis* plants.	Guo *et al.*, 2010
Maize lines Tex6 and B73	genes in developing kernels under drought stress	Several defense-related genes had been downregulated, even though some defense-related or drought responsive genes were upregulated at the later stages. The comparison between Tex6 and B73 revealed that there were significant differences in specific gene expression, patterns and levels.	Luo *et al.*, 2010
Vitis vinifera	transcriptomic analysis of the AP2/ERF superfamily	Expression profile was analyzed by Real-Time quantitative PCR (qRT-PCR) in different vegetative and reproductive tissues and under two different ripening stages.149 sequences,containing at least one ERF domain, were identified.	Licausi *et al.*, 2010
Oryza sativa L.	serine/threonine phosphatases	Expression pattern of the 13 selected genes was validated employing real time PCR. Expression profiling and analysis indicate the involvement of this large gene family in a number of signaling pathways triggered by abiotic stresses and their possible role in plant development.	Singh *et al.*, 2010

Contd...

Table 12.2– Contd...

Plant	Genes	Result	References
Sorghum bicolor	SbPIN, SbLAX and SbPGP	Real-time PCR analysis demonstrated that most of these genes were differently expressed in the organs of sorghum. SbPIN3 and SbPIN9 were highly expressed in flowers, SbLAX2 and SbPGP17 were mainly expressed in stems, and SbPGP7 was strongly expressed in roots. This suggests that individual genes might participate in specific organ development.	Shen *et al.*, 2010
Sorghum bicolor	Auxin-related gene families: auxin-response genes, auxin/indole-3-acetic acid (Aux/ IAA), auxin-response factor (ARF), Gretchen Hagen3 (GH3), small auxin-up RNAs, and lateral organ boundaries (LBD)	Real-time PCR analysis demonstrated these genes are differently expressed in leaf/root of sorghum under IAA, brassinosteroid (BR), salt, and drought treatments. The SbGH3 and SbLBD genes, expressed in low level under natural condition, were highly induced by salt and drought stress consistent with their products being involved in both abiotic stresses. Three genes, SbIAA1, SbGH3-13, and SbLBD32, were highly induced under all the four treatments, IAA, BR, salt, and drought.	Wang S *et al.*, 2010
Oryza sativa L.	Arabinogalactan proteins (AGPs)	AGP-encoding genes are predominantly expressed in anthers and display differential expression patterns in response to abscisic acid, gibberellic acid, and abiotic stresses.	Ma and Zhao, 2010
Wheat	aci-reductone-dioxygenase (TaARD)	qRT-PCR analyses revealed that the TaARD transcript was induced in wheat leaves infected with a compatible stripe rust strain. However, its expression was reduced or suppressed in incompatible interactions and by ABA, ethephon (ET), or salicylic acid (SA) treatments. The expression of TaARD also was inhibited by wounding and environmental stimuli, including high salinity and low temperature.	Xu *et al.*, 2010
Brassica napus	G protein α subunit	BnGA1 expressed was high in root, cotyledon and shoot apex. The expression of BnGA1 was also induced by low gibberellins acid 3 (GA(3)) concentrations and higher GA(3) concentrations inhibit the expression of BnGA1. The BnGA1 was up-regulated in salt and drought stress and down regulated in heat and cold stress.	Gao *et al.*, 2010

Contd...

Table 12.2—*Contd...*

Plant	Genes	Result	References
Chloris virgata Swartz.	expressed sequence tags (ESTs) during NaHCO₃ treatment	The expression profile of 24 genes affected by NaHCO₃ stress was analyzed. NaHCO₃ treatment up-regulated the expressions of pathogenesis-related gene (DC998527), Win1 precursor gene (DC998617), catalase gene (DC999385), ribosomeinactivating protein 1 (DC999555), Na(+)/H(+) antiporter gene (DC998043), and two-component regulator gene (DC998236).	Nishiuchi *et al.*, 2010
Setaria italica L.	Transcriptome analysis of differentially expressed genes during dehydration stress	Real-time PCR showed the comparative expression profiling of randomly chosen 9 up-regulated transcripts (> or =2.5 fold induction) between cv. Prasad (tolerant) and cv. Lepakshi (sensitive) upon dehydration stress. These transcripts showed a differential expression pattern in both cultivars at different time points of stress treatment as analyzed by qRT-PCR.	Lata *et al.*, 2010
Brassica napus	genes in response to high-salinity and drought stresses	The experimental results revealed that some genes may function in ABA-dependent signaling pathway related to drought or salinity stress.	Chen *et al.*, 2010
Gossypium barbadense L.	thaumatin-like protein gene (GbTLP1)	Tobacco transgenic plants with constitutively higher expression of the GbTLP1 showed enhanced resistance against different stress agents.	Munis *et al.*, 2010
Wheat	pathogenesis-related thaumatin-like protein gene TaPR5	qRT-PCR analyses revealed that TaPR5 transcript is significantly induced and upregulated in the incompatible interaction while in the compatible interaction a relative low level of the transcript was detected. TaPR5 was also induced by phyto-hormones (SA, JA and ABA and stress stimuli (wounding, cold temperature and high salinity).	Wang *et al.*, 2010
Pennisetum glaucum	DREB transcription factor	The quantitative Real time PCR results showed higher expression of downstream genes NtERD10B, HSP70-3, Hsp18p, PLC3, AP2 domain TF, THT1, LTP1 and heat shock (NtHSF2) and pathogen-regulated (NtERF5) factors with different stress treatments.	Agarwal *et al.*, 2010
Sugarcane	water-deficit stress- and red-rot-related genes	Real-time reverse transcription-PCR profiling several sugarcane EST clusters showed differential expression in response to biotic and abiotic stress conditions. 25 stress-related clusters showed >2-fold relative expression during water-deficit stress in sugarcane.	Gupta *et al.*, 2010

Contd...

Table 122–*Contd...*

Plant	Genes	Result	References
Buckwheat	The aspartic protease (FeAP9)	Real-time PCR results showed FeAP9 expression is upregulated in buckwheat leaves under the influence of different abiotic stresses, including dark, drought and UV-B light, as well as wounding and salicylic acid.	Timotijevia *et al.,* 2010
Caragana korshinskii (Peashrub)	9-cis-epoxycarotenoid dioxygenase (NCED)	Quantitative real-time PCR analysis showed that salt stress rapidly induced the strong expression of CkNCED1 in leaves and roots of *C. korshinskii*, as well as ABA accumulation. The expression of CkNCED1 and ABA accumulation was also induced by cold stress and the application of exogenous ABA.	Wang *et al.,* 2009
Chorispora bungeana	Polygalacturonase-inhibiting proteins (PGIPs)	Real-time PCR results showed that CbPGIP1 expression was induced by *Stemphylium solani*, salicylic acid (SA), 4° C, -4 °C and NO.	Di *et al.,* 2009
Panax ginseng	alcohol dehydrogenase (ADH)	The expression of PgADH under various environmental stresses was analyzed at different time points using real-time PCR. ABA, SA and especially JA (80-fold) significantly induced PgADH expression within 24 h of treatment. The positive responses of PgADH to abiotic stimuli suggest that ginseng ADH may protect against hormone-related environmental stresses.	Kim *et al.,* 2009
Tamarix hispida Willd.	Plant lipid transfer proteins (LTPs)	The expression profiles showed that all 14 ThLTPs were expressed in root, stem and leaf tissues under normal growth conditions. However, under normal growth conditions, ThLTP abundance varied in each organ, with expression differences of 9000-fold in leaves, 540-fold in stems and 3700-fold in roots. Differential expression of the 14 ThLTPs was observed (> 2-fold) for NaCl, PEG, NaHCO$_3$ and CdCl$_2$ in at least one tissue indicating that they were all involved in abiotic stress responses. All ThLTP genes were highly induced (> 2-fold) under ABA treatment in roots, stems and/or leaves, and particularly in roots, suggesting that ABA-dependent signaling pathways regulated ThLTPs.	Wang *et al.,* 2009
Oryza sativa	C3HC4-type RING finger family genes	Real-time PCR analysis confirmed that five C3HC4-type RING finger genes are preferentially expressed in reproductive tissues or organs such as stamen, panicle, and endosperm. Expression analysis of C3HC4-type RING finger genes under abiotic stresses suggests that twelve genes are differentially regulated by hormones or stress in rice seedlings.	Ma *et al.,* 2009

Contd...

Table 12.2–*Contd...*

Plant	Genes	Result	References
Oryza sativa	Transcript profiling auxin-responsive genes during reproductive development and abiotic stress	The expression profiles of some of the representative genes were confirmed by real-time PCR. The differential expression of auxin-responsive genes during various stages of panicle and seed development implies their involvement in diverse developmental processes.	Jain and Khurana., 2009
Thellungiella halophila	Inositol polyphosphate kinases (ThIPK2)	Real time PCR revealed ubiquitous expression of ThIPK2 in various tissues, including roots, rosette leaves, cauline leaves, stem, flowers and siliques, and shoot ThIPK2 transcript was strongly induced by NaCl or mannitol in *T. halophila*. Transgenic expression of ThIPK2 in *Brassica napus* led to significantly improved salt, dehydration- and oxidative stress resistance.	Zhu *et al*., 2009
Citrullus colocynthis (L.) Shrad	Gene in response to drought stress	18 genes which show similarity to known function genes were confirmed by real-time RT-PCR. These genes are involved in various abiotic and biotic stress and developmental responses.	Si *et al*, 2009
Cynara cardunculus L.	Hydroxycinnamoyl transferase(HQT)	Real time PCR experiments demonstrated an increase in the expression level of HQT in UV-C treated leaves, and established a correlation between the synthesis of phenolic acids and protection against damage due to abiotic stress.	Comino *et al*., 2009
Medicago truncatula	Ozone responsive genes	Temporal profiles of select genes using real-time PCR analysis showed that most of the genes were induced at the later time points.	Puckette *et al*., 2009
Tremula L. x *Populus alba* L.	PtaZFP2, encoding a member of Cys2/His2 zinc finger protein (ZFP)	The real-time PCR experiments showed that PtaZFP2 was rapidly up-regulated in poplar stems in response to gravitropism suggesting that PtaZFP2 is induced by different mechanical signals. Abundance of PtaZFP2 transcripts also increased highly in response to wounding and to a weaker extent to salt treatment and cold.	Martin *et al*, 2009
Alfalfa (*Medicago sativa*)	helicase (MH1)	Expression of MH1 was detected in roots, stems and leaves of alfalfa. Real-time PCR analysis revealed that mannitol, NaCl, methyl viologen and ABA induced the expression of MH1.	Luo *et al*, 2009
Oryza sativa L.	expression profiling analysis of OsIAA gene family	Real-time PCR results show that many genes in these families were responsive to various abiotic stresses, indicating an interaction between plant growth and abiotic stress.	Song *et al*, 2009

Contd...

Table 1.2– *Contd...*

Plant	Genes	Result	References
Chrysanthemum (*Dendranthema x Morifolium*)	DREB genes DmDREBa and DmDREBb	qRT- PCR analysis revealed that both genes were accumulated more in leaves and stems than in roots and flowers. Moreover, DmDREBb reacted earlier and accumulated with higher levels than DmDREBa under cold treatment. Expression of DmDREBa and DmDREBb declined dramatically within 0.5 h of exposure to 100 μM ABA. Besides, DmDREBb expression was variable and recovered to pre-ABA levels at 2, 6 and 12 h, while DmDREBa expression remained low during the 24 h exposure period. Furthermore, both genes expression was totally inhibited at 40°C.	Yang *et al.*, 2009
Salvia miltiorrhiza Bung.	4-hydroxyphenyl-pyruvate dioxygenase gene (Smhppd)	Real-time PCR analysis indicated that Smhppd was constitutively expressed in roots, stems and leaves of *S. miltiorrhiza*, with the high expression in roots.	Xiao *et al.*, 2008
Medicago truncatula	transcription factors involved in root apex responses to salt stress	46 salt-regulated TF genes were identified using massive quantitative real-time PCR TF profiling in whole roots. Analysis of salt-stress regulation in root apexes versus whole roots showed that several TF genes have more than 30-fold expression differences including specific members of AP2/EREBP, HD-ZIP, and MYB TF families.	Gruber *et al.*, 2008
Brassica napus L.	pyridoxal kinase (BnPKL)	Real-time PCR revealed that the relative abundance of two transcripts are modulated by development and environmental stresses. Abscisic acid and NaCl were inclined to decrease PKL expression, but H_2O_2 and cold temperatures induced the PKL expression.	Yu and Luo, 2008
Oryza sativa L.	Receptor-like cytoplasmic kinases (RLCKs)	Real-time PCR-based expression profiling for a selected few genes showed differential expression patterns for majority of OsRLCKs during development and stress suggest their involvement in diverse functions in rice.	Vij *et al.*, 2008
Oryza sativa	leucine zipper (bZIP) transcription factor family	GeneChip and real-time PCR analyses revealed that hundreds of genes were up- or down-regulated in the rice plants overexpressing OsbZIP23.	Xiang *et al.*, 2008
Medicago truncatula	trehalase gene (MTTRE1)	MTTRE1 expression is induced in nodules compared with leaves and roots, indicating atranscriptional regulation of trehalase in the presence of the microsymbiont. Under salt stress conditions, trehalase activity is downregulated at the transcriptional level, allowing trehalose accumulation.	López *et al.*, 2008

Contd...

Table 12.2– *Contd...*

Plant	Genes	Result	References
Oryza sativa L.	NAC transcription factors	Real-time polymerase chain reaction (PCR) analysis revealed 12 genes with different tissue-specific (such as callus, root, stamen, or immature endosperm) expression patterns, suggesting that these genes may play crucial regulatory roles during growth and development of rice. The expression levels of this family were also checked under various abiotic stresses including drought, salinity, and low temperature. Most of these stress-responsive genes belonged to the group III (SNAC).	Fang *et al.*, 2008
Arabidopsis thaliana	Transcription factors regulating leaf senescence	41 TF genes that were gradually up-regulated as leaves progressed through these developmental stages and 144 TF genes were down-regulated during senescence.	Balazadeh *et al.*, 2008
Pea	Pathogenesis-related proteins (ABR17)	Transcriptional analysis in ABR17 transgenic Arabidopsis plants, both under normal and saline conditions, revealed significant changes in abundance of transcripts for many stress responsive genes, as well as those related to plant growth and development.	Krishnaswamy *et al.*, 2008
Ginkgo biloba	anthocyanidin synthase	The expression analysis by real-time PCR showed that GbANS expressed in a tissue-specific manner in *G. biloba*. GbANS was also found to be up-regulated by all of the six tested abiotic stresses, UV-B, abscisic acid, sucrose, salicylic acid, cold and ethylene, consistent with the promoter region analysis of GbANS.	Xu *et al.*, 2008
Oryza sativa L.	tonoplast intrinsic proteins (TIPs)	Rice TIP expression patterns under various abiotic stress conditions including dehydration, high salinity, ABA and during seed germination were investigated by real-time PCR. OsTIP1s (OsTIP1;1 and OsTIP1;2) were highly expressed during seed germination, whereas OsTIP3s (OsTIP3;1 and OsTIP3;2) were specifically expressed in mature seeds with a decrease in expression levels upon germination.	Li *et al.*, 2008
Zea mays	aldehyde dehydrogenase (ALDHs)	Real-time PCR analysis indicates that ZmALDH22A1 is expressed differentially in different tissues. Various elevated levels of ZmALDH22A1 expression have been detected when the seedling roots exposed to abiotic stresses including dehydration, high salinity and abscisic acid (ABA).	Huang *et al.*, 2008
Oryza sativa L.	Genes in response to copper stress	Expression profile showed that a large proportion of general and defence stress response genes are up-regulated under excess Cu conditions, whereas photosynthesis and transport-related genes are down-regulated.	Sudo *et al.*, 2008

Contd...

Table 12.2–*Contd...*

Plant	Genes	Result	References
Limonium bicolor	expressed sequence tags (ESTs) analysis under NaHCO₃ stress	The expression of 18 putative stress-related genes were further analyzed in roots and leaves of *L. bicolor* using real-time RT-PCR, and 14 genes were differentially expressed by more than 2-fold as a result of the NaHCO₃ stress.	Wang *et al.*, 2008
Oryza sativa indica cultivar var. Pusa Basmati	Mitogen activated protein kinase kinase (MKK)	Analysis of transcript regulation by quantitative real time PCR revealed that these five members 1, 3, 4, 6 and 10-2 are differentially regulated by cold, heat, salinity and drought stresses. MAP kinase kinases 4 and 6 are strongly regulated by cold and salt stresses while MAP kinase kinase 1 is regulated by salt and drought stresses. MAP kinase kinase 10-2 is regulated only by cold stress.	Kumar *et al.*, 2008
Seteria italica	Genes in response to salt stress	The expression patterns of 7 out of 9 genes showed a significant increase of differential expression in tolerant variety after 1 h of salt stress in comparison to salt-sensitive variety as analyzed by qRT-PCR.	Jayaraman *et al.*, 2008
Phaseolus vulgaris	Delta(1)-pyrroline-5-carboxylatesynthetase (P5CS)	The expression patterns of PvP5CS in common bean treated with drought, cold (4 degrees C), and salt (200 mM NaCl) stresses were examined using real-time quantitative PCR. These abiotic stresses caused significant up-regulation of the expression of PvP5CS in leaves. The PvP5CS mRNA transcript increased to 2.5 times after 4d drought stress. A rapid up-regulation of PvP5CS, to about 16.3 times under salt stress. A significant increase in PvP5CS expression (11.7-fold) was detected after 2h of cold stress.	Chen *et al.*, 2008
Prunus incisa x serrula.	GUS-marker gene (uidA)	Transcription levels were assessed by real-time PCR in leaves of plantlets under various abiotic stresses like low- and high temperature, salicylic acid and wounding.	Maghuly *et al.*, 2008
Nicotiana tabacum	NADP-malic enzyme	Real-time reverse transcription-PCR studies show that the three nadp-me tobacco transcripts respond differently to several biotic and abiotic stress stimuli.	Müller *et al.*, 2008
Arabidopsis thaliana	The receptor-like cytoplasmic protein kinases (RLCKs)	qRT-PCR was used to determine the relative transcript levels in the various organs of the *Arabidopsis* plant as well as under a series of abiotic stress/hormone treatments in seedlings.	Jurca *et al.*, 2008
Theobroma cacao (Cacao)	genes involved in polyamine biosynthesis	Expression of all five PA associated transcripts was enhanced (1.5-3-fold) in response to treatment with abscisic acid. TcODC and TcADC, were also responsive to mechanical wounding. Constitutive expression of PA biosynthesis genes was generally highest in mature leaves and open flowers.	Bae *et al.*, 2008

Contd...

Table 12.2–*Contd...*

Plant	Genes	Result	References
Zea mays	putative H+-translocating pyrophosphatase (H+-PPase)	The Real-time RT-PCR assays showed that the expression of ZmGPP was up-regulated both in shoots and roots of maize seedlings under dehydration, cold and high salt stresses.	Yue *et al.*, 2008
Rice cv IR651	salt responsive genes	To investigate the correlation between mRNA and protein level in response to salinity, quantitative Real-Time PCR analysis of three genes that were salt responsive at the protein level, including 1,4 Benzoquinone reductase, a putative remorin and a hyper-sensitive induced response protein were done. No concordance was detected between the changes in levels of gene and protein expression.	Nohzadeh Malakshah *et al.*, 2007
Pennisetum glaucum	ESTs inresponse to abiotic (salinity, drought and cold) stresses	The relative mRNA abundance of 38 selected genes, quantified using real-time PCR, demonstrated the existence of a complex gene regulatory network that differentially modulates gene expression in a kinetics-specific manner in response to different abiotic stresses.	Mishra *et al.*, 2007
Solanum lycopersicum	14-3-3 proteins (phosphoserine-binding proteins)	By using real-time RT-PCR, it was found that (a) under normal growth conditions, there were significant differences in the mRNA levels of 14-3-3 gene family members in young tomato roots and (b) 14-3-3 proteins exhibited diverse patterns of gene expression in response to salt stress and potassium and iron deficiencies in tomato roots.	Xu WF and Shi, 2006
Arabidopsis thaliana	annexin gene family	Real-time RT-PCR results showed the expression of most of the Arabidopsis annexin genes is differentially regulated by exposure to salt, drought, and high- and low-temperature conditions, indicating a likely role for members of this gene family in stress responses.	Cantero *et al.*, 2006
Oryza sativa	cytokinin-responsive type-A response regulators	The transcripts of OsRR genes detected by real-time PCR in all organs of the light- and dark-grown rice seedlings/plants, although there were quantitative differences. The steady-state transcript levels of most of the OsRR genes increased rapidly (within 15 min) on exogenous cytokinin application even in the presence of cycloheximide. Moreover, the expression of the OsRR6 gene was enhanced in rice seedlings exposed to salinity, dehydration and low temperature stress.	Jain *et al*, 2006

Contd...

Table 12.2– *Contd...*

Plant	Genes	Result	References
Pennisetum glaucum	vacuolar ATPase subunit c (VHA-c)	Real-time PCR approach showed three isoforms are regulated in a tissue-specific manner under salinity stress. While isoform III is constitutively expressed in roots and shoots and does not respond to stress, isoform I is upregulated under stress. Isoform II is expressed mainly in roots; however, under salinity stress its expression is down-regulated in roots and upregulated in shoots. Tissue specific expression under salinity stress of isoform II was also seen after exogenous application of calcium.	Tyagi *et al.*, 2006
Oryza sativa L. cv. Nipponbare	Proteomic analysis of genes in response to chilling stress	Gene expression analysis of 44 different proteins by quantitative real time PCR showed that the mRNA level was not correlated well with the protein level.	Yan SP, 2006
Triticum aestivum	Lipocalins	Expression analyses by quantitative real-time PCR showed that expression of the wheat lipocalins and lipocalin-like proteins is associated with abiotic stress response and is correlated with the plant's capacity to develop freezing tolerance.	Charron *et al.*, 2005
Triticum durum and *T. aestivum*	group 2 (dehydrins) and group 4 Late embryo-genesis abundant (Lea) genes	The five genes exhibited clear differences in their accumulation pattern in wheat seed and in response to dehydration, low temperature, salinity and ABA. Td29b, Td16 and Td27e gene transcripts accumulate late in embryogenesis as expected for Lea genes, Td11 gene transcripts were present throughout seed development whereas no Td25a gene transcripts were detected in seeds. Drastic changes in the relative levels of Td29b, Td16, Td27e and Td11 transcripts occurred at the shift between the cell expansion and desiccation phases.	Ali-Benali *et al.*, 2005
Rumex palustris	alpha-expansin genes	Several genes were expressed differentially in response to low O$_2$. Expression of some expansin mRNAs increased in 'mature zones' of roots; these expansins might be involved in root hair formation or in formation of lateral root primordia.	Colmer *et al.*, 2004
Arabidopsis thaliana	aquaporin genes, plasma membrane intrinsic protein (PIP)	Several PIP genes were predominantly expressed either in the roots or in the flowers. The expressions of both the highly expressed aquaporins including PIP1;1, PIP1;2, and PIP2;7 and the weakly expressed aquaporins such as PIP1;4, PIP2;1, PIP2;4, and PIP2;5 were modulated by external stimuli. Only the PIP2;5 was up-regulated by cold treatment, and most of the PIP genes were down-regulated by cold stress. Marked up- or down-regulation in PIP expression was observed by drought stress, whereas PIP genes were less-severely modulated by high salinity.	Jang *et al.*, 2004

Contd...

Table 12.2–*Contd...*

Plant	Genes	Result	References
Cucumis sativus	protein kinase, serine/threonine and tyrosine kinase	A quantitative real-time PCR analysis revealed that, among the abiotic stresses tested, drought treatment markedly decreased the transcript level of the kinase, whereas the expression of the kinase gene significantly increased by cold treatment. High salinity did not influence its expression.	Jang *et al.*, 2004
Arabidopsis thaliana	Pyruvate decarboxylase (PDC)	By using real-time PCR, the expression levels of each individual gene in different tissues, under normal growth conditions, and when the plants were subjected to anoxia or other environmental stress conditions were determined. PDC1 is the only gene induced under oxygen limitation among the PDC1 gene family and that a pdc1 null mutant is involved in anoxia tolerance but not other environmental stresses.	Kürsteiner *et al.*, 2003

involved in basic cellular functions (*e.g.*, actin and ubiquitin genes) are often assumed to have a uniform expression pattern, and have been used extensively as reference genes (Bustin, 2002). However, many studies have found considerable variation in house-keeping gene expression (Suzuki *et al.,* 2000; Lee *et al.,* 2002; Dheda *et al.,* 2004; Czechowski *et al.,* 2005), and the stability of potential reference genes must be systematically determined prior to their use in RT-PCR normalization (Guenin *et al.,* 2009). Several reports describing proper selection and evaluation of multiple housekeeping genes as internal control genes for accurate normalization of real-time quantitative RT-PCR have been published (Table 12.1). These studies have validated the necessity of utilizing multiple reference genes when evaluating quantitative RT-PCR data.

Real-Time PCR for Abiotic Stress Studies

Abiotic stresses such as drought, cold and high salt significantly limit crop productivity thereby reducing average yields of most of the major crops (Bray *et al.,* 2000). Plants adapt to stress via induction of morphological, physiological and biochemical changes which are brought about by altered expression of several stress responsive genes which leads to plant induction to stress, mainly by reduction in growth and development. Transcriptome studies have helped to provide a better understanding of plant stress responses. Through these studies, numerous novel stress-responsive genes have been discovered. Overlaps of genes induced by various stress conditions suggest extensive cross-talk between the signalling pathways. Transcriptome studies with multiple time points suggest that plant responses progress from general to specific responses (Sung *et al.,* 2003). The analysis of gene expression requires sensitive, precise, and reproducible measurements for specific mRNA sequences. Gene expression levels were commonly determined using northern blot analysis. However, this technique is time consuming and requires a large quantity of RNA (Dean *et al.,* 2002). RT-PCR is, at present, the most sensitive method for the detection of low abundance mRNAs (Bustin, 2000), and could be used for the analysis of tissue-specific gene expression in plants (Gachon *et al.,* 2000) (Table 12.2).

Acknowledgements

Corresponding author would like to sincerely thank Dr.M.K.Reddy, Senior Research Scientist, Plant Molecular Biology Group, ICGEB, New Delhi for introducing the topic of real-time PCR. This work in C.O.R.P laboratory was supported by the DBT, DST and C.S.I.R, Government of India.

References

Agarwal, P., Agarwal, P.K., Joshi, A.J., Sopory, S.K. and Reddy, M.K. 2010. Overexpression of PgDREB2A transcription factor enhances abiotic stress tolerance and activates downstream stress-responsive genes. Mol. Biol. Rep. **37(2):** 1125-35.

Ali-Benali, M.A., Alary, R., Joudrier, P. and Gautier, M.F. 2005. Comparative expression of five Lea Genes during wheat seed development and in response to abiotic stresses by real-time quantitative RT-PCR. Biochim. Biophys. Acta. **1730(1):** 56-65.

Alwine, J.C., Kemp, D.J. and Stark G.R. 1977. Method for detection of specific RNAs in agarose gels by transfer to diazobenzyloxymethyl-paper and hybridization with DNA probes. Proc. Natl. Acad. Sci. USA **74**: 5350-5354.

An, D., Yang, J. and Zhang, P. 2012. Transcriptome profiling of low temperature-treated cassava apical shoots showed dynamic responses of tropical plant to cold stress. BMC Genomics. **13**: 64.

Artico, S., Nardeli, S.M., Brilhante, O., Grossi-de-Sa, M.F. and Alves-Ferreira, M. 2010. Identification and evaluation of new reference genes in *Gossypium hirsutum* for accurate normalization of real-time quantitative RT-PCR data.BMC Plant Biol.**10**: 49.

Bae, H., Kim, S.H, Kim, M.S., Sicher, R.C., Lary, D., Strem, M.D., Natarajan, S. and Bailey, B.A. 2008. The drought response of *Theobroma cacao* (cacao) and the regulation of genes involved in polyamine biosynthesis by drought and other stresses. Plant Physiol. Biochem. **46(2)**: 174-88.

Balazadeh, S., Riaño-Pachón, D.M. and Mueller-Roeber, B. 2008. Transcription factors regulating leaf senescence in *Arabidopsis thaliana.* Plant Biol (Stuttg). Sep;**10** Suppl 1: 63-75.

Barozai, M.Y. and Husnain, T. 2011. Identification of biotic and abiotic stress up-regulated ESTs in *Gossypium arboreum*. Mol. Biol. Rep. **39(2)**: 1011-8.

Barsalobres-Cavallari, C.F., Severino, F.E., Maluf, M.P. and Maia, I.G. 2009. Identification of suitable internal control genes for expression studies in *Coffea arabica* under different experimental conditions. BMC Mol. Biol.**10**: 1. doi: 10: 1186/1471-2199-10-1.

Bernard, P.S., Ajioka, R.S., Kushner, J.P. and Wittwer, C.T. 1998. Homogeneous multiplex genotyping of hemochromatosis mutations with fluorescent hybridization probes. Am. J. Pathol. **153**: 1055-1061.

Bernard, P.S. and Wittwer, C.T. 2000. Homogeneous amplification and variant detection by fluorescent hybridization probes. Clin. Chem. **46**: 147-148.

Bonnet, G., Tyagi, S., Libchaber, A. and Kramer, F.R. 1999. Thermodynamic basis of the enhanced specificity of structured DNA probes. Proc. Natl. Acad. Sci. USA **96**: 6171-6176.

Bovy, A, de Vos, R., Kemper M., *et al.* 2002. High-flavonol tomatoes resulting from the heterologous expression of the maize transcription factor genes LC and C1. The Plant Cell. **14**: 2509–2526.

Bray, E.A., Bailey-Serres, J. and Weretilnyk, E. 2000. Responses to abiotic stresses. In Gruissem, W., Buchannan, B., Jones, R. (eds.) Biochemistry and Molecular Biology of Plants. American Society of Plant Physiologists, Rockville, MD, pp 1158–1249.

Brooks, E.M., Sheflin, L.G. and Spaulding, S.W. 1995. Secondary structure in the 32 UTR of EGF and the choice of reverse transcriptases affect the detection of message diversity by RT-PCR. BioTechniques **19**: 806-815.

Buell, G.N., Wickens, M.P., Payvar, F., and Schimke, R.T 1978. Synthesis of full length cDNAs from four partially purified oviduct mRNAs. J. Biol. Chem. **253:** 2471-2482.

Bustin, S.A. 2000. Absolute quantification of mRNA using real-time reverse transcription polymerase chain reaction assays. J. Mol. Endocrinol. **25:** 169-193.

Bustin, S.A. 2002. Quantification of mRNA using real-time reverse transcription PCR (RT-PCR): trends and problems. J. Mol. Endocrinol. **29:** 23–29.

Cantero, A., Barthakur, S., Bushart, T.J., Chou, S., Morgan, R.O., Fernandez, M.P., Clark, G.B. and Roux, S.J. 2006. Expression profiling of the Arabidopsis annexin gene family during germination, de-etiolation and abiotic stress. Plant Physiol. Biochem. **44(1):**13-24.

Carvalho, K., de Campos, M.K., Pereira, L.F. and Vieira, L.G. 2010. Reference gene selection for real-time quantitative polymerase chain reaction normalization in "Swingle" citrumelo under drought stress.Anal. Biochem. **402(2):** 197-9.

Carvalho, L.C., Vilela, B.J., Mullineaux, P.M. and Amâncio, S. 2011. Comparative transcriptomic profiling of *Vitis vinifera* under high light using a custom-made array and the Affymetrix GeneChip. Mol. Plant **4(6):** 1038-51.

Chandna, R., Augustine, R. and, Bisht, N.C. 2012. Evaluation of candidate reference genes for gene expression normalization in *Brassica juncea* using Real Time Quantitative RT-PCR. PLoS One. 7(5):e36918.

Charron, J.B., Ouellet, F., Pelletier, M., Danyluk, J., Chauve, C. and Sarhan, F. 2005. Identification, expression, and evolutionary analyses of plant lipocalins. Plant Physiol. **139(4):** 2017-28.

Chaturvedi, A.K., Mishra, A., Tiwari, V. and Jha, B. 2012. Cloning and transcript analysis of type 2 metallothionein gene (SbMT-2) from extreme halophyte *Salicornia brachiata* and its heterologous expression in *E.coli.* Gene. **499(2):** 280-7.

Chauhan, H., Khurana, N., Tyagi, A.K., Khurana, J.P. and Khurana, P. 2011. Identification and characterization of high temperature stress responsive genes in bread wheat (*Triticum aestivum* L.) and their regulation at various stages of development. Plant Mol. Biol. **75(1-2):** 35-51.

Chen, C., Bai, L.H., Qiao, D.R., Xu, H., Dong, G.L., Ruan, K., Huang. F. and Cao, Y. 2007. Cloning and expression study of a putative carotene biosynthesis related (cbr)gene from the halotolerant green alga *Dunaliella salina.* Mol. Biol. Rep. **35(3):** 321-7.

Chen, J., Chen, J.Y., Wang, J.N., Kuang, J.F., Shan, W. and Lu, W.J. 2012. Molecular characterization and expression profiles of MaCOL1, a CONSTANS-like gene in banana fruit. Gene. **496(2):** 110-7.

Chen, J.B., Wang, S.M., Jing, R.L. and Mao, X.G. 2008. Cloning the PvP5CS gene from common bean (*Phaseolus vulgaris*) and its expression patterns under abiotic stresses. J. Plant Physiol, **166(1):** 12-9.

Chen, L., Ren, F., Zhong, H., Feng, Y., Jiang, W. and Li, X. 2010. Identification and expression analysis of genes in response to high-salinity and drought stresses in *Brassica napus*. Acta Biochim Biophys Sin (Shanghai). **42(2)**: 154-64.

Clegg, R.M 1992. Fluorescence resonance energy transfer and nucleic acids. Methods Enzymol. **211**: 353-388.

Colmer, T.D., Peeters, A.J., Wagemaker, C.A., Vriezen, W.H., Ammerlaan, A., Voesenek, L.A. 2004. Expression of alpha-expansin genes during root acclimations to O$_2$ deficiency in *Rumex palustris*. Plant Mol. Biol. **56(3)**: 423-37.

Comino, C., Hehn, A., Moglia, A., Menin, B., Bourgaud, F., Lanteri, S. and Portis, E. 2009. The isolation and mapping of a novel hydroxycinnamoyltransferase in the globe artichoke chlorogenic acid pathway. BMC Plant Biol. **9**: 30.

Curtin, S.J., Kantar, M.B., Yoon, H.W., Whaley, A.M., Schlueter, J.A. and Stupar, R.M. 2012. Co-expression of soybean Dicer-like genes in response to stress and development. Funct Integr Genomics. Apr 15. [Epub ahead of print]

Czechowski, T., Stitt, M., Altmann, T., Udvardi, M.K., and Scheible, W.R. 2005. Genome-wide identification and testing of superior reference genes for transcript normalization in *Arabidopsis*. Plant Physiol. **139**: 5–17.

Das, M., Chauhan, H., Chhibbar, A., Rizwanul Haq, Q.M. and Khurana, P. 2011. High-efficiency transformation and selective tolerance against biotic and abiotic stress in mulberry, *Morus indica* cv. K2, by constitutive and inducible expression of tobacco osmotin. Transgenic Res. **20(2)**: 231-46.

de Silva, K., Laska, B., Brown, C., Sederoff, H.W. and Khodakovskaya, M. 2011. *Arabidopsis thaliana* calcium-dependent lipid-binding protein (AtCLB): a novel repressor of abiotic stress response. J. Exp. Bot. **62(8)**: 2679-89.

DeStefano, J.J. 1995. The orientation of binding of human immunodeficiency virus reverse transcriptase on nucleic acid hybrids. Nucleic Acids Res. **23**: 3901–3908.

DeStefano, J.J., Buiser, R.G., Mallaber, L.M., Myers, T.W., Bambara, R.A. and Fay, P.J. 1991. Polymerization and RNase H activities of the reverse transcriptases from avian myeloblastosis, human immunodeficiency, and Moloney murine leukemia viruses are functionally uncoupled. J. Biol. Chem. **266**: 7423–7431.

Dheda. K., Huggett, J.F., Bustin, S.A., Johnson, M.A., Rook, G. and Zumla, A. 2004. Validation of housekeeping genes for normalizing RNA expression in real-time PCR. Biotechniques **37**: 112–114, 116,118–119.

Di, C., Li, M., Long, F., Bai, M., Liu, Y., Zheng, X., Xu, S., Xiang, Y., Sun, Z. and An, L. 2009. Molecular cloning, functional analysis and localization of a novel gene encoding polygalacturonase-inhibiting protein in *Chorispora bungeana*. Planta. **231(1)**: 169-78.

Dong, J., Wang, X., Wang, K., Wang, Z. and Gao, H. 2012. Isolation and characterization of a gene encoding an ethylene responsive factor protein from *Ceratoides arborescens*. Mol. Biol. Rep. **39(2)**:1349-57.

Fang, Y., You, J., Xie, K., Xie, W. and Xiong, L. 2008. Systematic sequence analysis and identification of tissue-specific or stress-responsive genes of NAC transcription factor family in rice. Mol. Genet. Genomics. **280(6):** 547-63.

Feng, H., Wang, X., Sun, Y., Wang, X., Chen, X., Guo, J., Duan, Y., Huang, L. and Kang, Z. 2011. Cloning and characterization of a calcium binding EF-hand protein gene TaCab1 from wheat and its expression in response to *Puccinia striiformis* f. sp. *tritici* and abiotic stresses. Mol. Biol. Rep. **38(6):** 3857-66.

Fernández-Ocaña, A., Chaki, M., Luque, F., Gómez-Rodríguez, M.V., Carreras, A., Valderrama, R., Begara-Morales, J.C., Hernández, L.E., Corpas, F.J. and Barroso, J.B. 2011. Functional analysis of superoxide dismutases (SODs) in sunflower under biotic and abiotic stress conditions. Identification of two new genes of mitochondrial Mn-SOD. J. Plant. Physiol. **168(11):** 303-8.

Freeman, W.M., Walker, S.J. and Vrana, K.E. 1999. Quantitative RT-PCR: pitfalls and potential. BioTechniques **26:** 112-115.

Fronhoffs, S., Totzke, G., Stier, S., Wernert, N., Rothe, M., Bruning, T., Koch, B., Sachinidis, A., Vetter, H. and Ko, Y. 2002. A method for the rapid construction of cRNA standard curves in quantitative real-time reverse transcription polymerase chain reaction. Mol. Cell. Probes **16(2):** 99–110.

Gachon, C., Mingam, A., and Charrier, B. 2004. Real-time PCR: what relevance to plant studies? J. Expt. Bot. **55:** 1445–1454.

Gallou, A., Declerck, S. and Cranenbrouck, S. 2011. Transcriptional regulation of defence genes and involvement of the WRKY transcription factor in arbuscular mycorrhizal potato root colonization. Funct. Integr. Genomics. **12(1):** 183-98.

Gao, C., Wang, Y., Jiang, B., Liu, G., Yu, L., Wei, Z. and Yang, C. 2011. A novel vacuolar membrane H+-ATPase c subunit gene (ThVHAc1) from *Tamarix hispida* confers tolerance to several abiotic stresses in *Saccharomyces cerevisiae*. Mol. Biol. Rep. **38(2):** 957-63.

Gao, C., Zhang, K., Yang, G. and Wang, Y. 2012. Expression analysis of four peroxiredoxin genes from *Tamarix hispida* in response to different abiotic stresses and exogenous abscisic acid (ABA). Int. J. Mol. Sci. **13(3):** 3751-64.

Gao, Y., Li, T., Liu, Y., Ren, C., Zhao, Y. and Wang, M. 2010. Isolation and characterization of gene encoding G protein α subunit protein responsive to plant hormones and abiotic stresses in *Brassica napus*. Mol. Biol. Rep. **37(8):** 3957-65.

Garg, R., Sahoo, A., Tyagi, A.K. and Jain, M. 2010. Validation of internal control genes for quantitative gene expression studies in chickpea (*Cicer arietinum* L.). Biochem. Biophys. Res. Commun. **396(2):** 283-8.

Gentle, A., Anastasopoulos, F. and Mc-Brien, N.A. 2001. High-resolution semi-quantita-tive real-time PCR without the use of a stan-dard curve. BioTechniques **31:** 502-508.

Gerard, G.F., Fox, D.K., Nathan, M. and D'Alessio, J.M. 1997. Reverse transcriptase. The use of cloned Moloney murine leukemia virus reverse transcriptase to synthesize DNA from RNA. Mol. Biotechnol. **8:** 61–77.

Gibson, U.E., Heid, C.A. and Williams, P.M. 1996. A novel method for real time quantitative RT-PCR. Genome Res. **6:** 995-1001.

Giulietti, A., Overbergh, L., Valckx, D., Decallonne, B., Bouillon, R. and Mathieu, C. 2001. An overview of real-time quantitative PCR: applications to quantify cytokine gene expression. Methods **25:** 386-401.

Gruber, V., Blanchet, S., Diet, A., Zahaf, O., Boualem, A., Kakar, K., Alunni, B., Udvardi, M., Frugier, F. and Crespi, M. 2009. Identification of transcription factors involved in root apex responses to salt stress in *Medicago truncatula*. Mol. Genet. Genomics. **281(1):** 55-66.

Gu, C., Chen, S., Liu, Z., Shan, H., Luo, H., Guan, Z. and Chen, F. 2011. Reference gene selection for quantitative real-time PCR in *Chrysanthemum* subjected to biotic and abiotic stress. Mol. Biotechnol. **49(2):** 192-7.

Guenin, S., Mauriat, M., Pelloux, J., Van Wuytswinkel, O., Bellini, C. and Gutierrez, L. 2009. Normalization of qRT-PCR data: The necessity of adopting a systematic, experimental conditionsspecific, validation of references. J. Exp. Bot. **60:** 487–493.

Guo, Y., Huang, C., Xie, Y., Song, F. and Zhou, X. 2010. A tomato glutaredoxin gene SlGRX1 regulates plant responses to oxidative, drought and salt stresses. Planta. **232(6):** 1499-509.

Gupta, V., Raghuvanshi, S., Gupta, A., Saini, N., Gaur, A., Khan, M.S., Gupta, R.S., Singh, J., Duttamajumder, S.K., Srivastava, S., Suman, A., Khurana, J.P., Kapur, R. and Tyagi, A.K. 2010. The water-deficit stress- and red-rot-related genes in sugarcane. Funct Integr Genomics. **10(2):** 207-14.

Han, Yy., Li, Ax., Li, F., Zhao, Mr. and Wang, W. 2012. Characterization of a wheat (*Triticum aestivum* L.) expansin gene, TaEXPB23, involved in the abiotic stress response and phytohormone regulation. Plant Physiol. Biochem. **54:** 49-58.

Heid, C.A., Stevens, J., Livak, K.J. and Williams, P.M. 1996. Real time quantitative PCR. Genome Res. **6:** 986-994.

Higuchi, R., Fockler, C., Dollinger, G. and Watson, R. 1993. Kinetic PCR analysis: real-time monitoring of DNA amplification reac-tions. Biotechnology (NY), **11:** 1026-1030.

Hu, R., Fan, C., Li, H., Zhang, Q. and Fu, Y.F. 2009. Evaluation of putative reference genes for gene expression normalization in soybean by quantitative real-time RT-PCR. BMC Mol. Biol. **10:** 93.

Hu, Z., Yu, Y., Wang, R., Yao, Y., Peng, H., Ni, Z. and Sun, Q. 2011. Expression divergence of TaMBD2 homoeologous genes encoding methyl CpG-binding domain proteins in wheat (*Triticum aestivum* L.). Gene. **471(1-2):**13-8.

Huang, W., Ma, X., Wang, Q., Gao, Y., Xue, Y., Niu, X., Yu, G. and Liu, Y. 2008. Significant improvement of stress tolerance in tobacco plants by overexpressing a stress-responsive aldehyde dehydrogenase gene from maize (*Zea mays*). Plant Mol. Biol. **68(4-5):** 451-63.

Huggett, J., Dheda, K., Bustin, S. and Zumla, A. 2005. Real-time RT-PCR normalisation; strategies and considerations. Genes and Immunity. **6:** 279-284.

Jain, M. and Khurana, J.P. 2009. Transcript profiling reveals diverse roles of auxin-responsive genes during reproductive development and abiotic stress in rice. FEBS J. **276(11):** 3148-62.

Jain, M., Tyagi, A.K. and Khurana, J.P. 2006. Molecular characterization and differential expression of cytokinin-responsive type-A response regulators in rice (*Oryza sativa*). BMC Plant Biol. **13:** 6:1.

Jami, S.K., Clark, G.B., Ayele, B.T., Roux, S.J. and Kirti, P.B. 2012. Identification and characterization of annexin gene family in rice. Plant Cell Rep. **31(5):** 813-25.

Jang, J.Y., Kim, D.G., Kim, Y.O., Kim, J.S. and Kang, H. 2004. An expression analysis of a gene family encoding plasma membrane aquaporins in response to abiotic stresses in *Arabidopsis thaliana*. Plant Mol. Biol. **54(5):** 713-25.

Jang, J.Y., Kwak, K.J. and Kang, H. 2004. Molecular cloning and characterization of a cDNA encoding a kinase in *Cucumis sativus* and its expression by abiotic stress treatments. Biochim. Biophys. Acta. **1679(1):**74-9.

Jayaraman, A., Puranik, S., Rai, N.K., Vidapu, S., Sahu, P.P., Lata, C. and Prasad, M. 2008. cDNA-AFLP analysis reveals differential gene expression in response to salt stress in foxtail millet (*Setaria italica* L.). Mol. Biotechnol. **40(3):** 241-51.

Jha, A., Joshi, M., Yadav, N.S., Agarwal, P.K. and Jha, B. 2011. Cloning and characterization of the *Salicornia brachiata* Na(+)/H(+) antiporter gene SbNHX1 and its expression by abiotic stress. Mol. Biol. Rep. **38(3):**1965-73.

Ji, X., Wang, Y., and Liu, G. 2012. Expression Analysis of MYC Genes from *Tamarix hispida* in Response to Different Abiotic Stresses. Int. J. Mol. Sci. **13(2):** 1300-13.

Jiang, T., Zhou, B., Luo, M., Abbas, H.K., Kemerait, R., Lee, R.D., Scully, B.T. and Guo, B. 2011. Expression analysis of stress-related genes in kernels of different maize (*Zea mays* L.) inbred lines with different resistance to aflatoxin contamination. Toxins (Basel). **3(6):** 538-50.

Jurca, M.E., Bottka, S. and Fehér, A. 2008. Characterization of a family of *Arabidopsis* receptor-like cytoplasmic kinases (RLCK class VI). Plant Cell Rep. **27(4):**739-48.

Kakar, K., Wandrey, M., Czechowski, T., *et al.* 2008. A community resource for high-throughput quantitative RT-PCR analysis of transcription factor gene expression in *Medicago truncatula*. Plant Methods **4:** 18.

Kang, H.G., Kim, J., Kim, B., Jeong, H., Choi, S.H., Kim, E.K., Lee, H.Y. and Lim, P.O. 2011. Overexpression of FTL1/DDF1, an AP2 transcription factor, enhances tolerance to cold, drought, and heat stresses in *Arabidopsis thaliana*. Plant Sci. **180(4):** 634-41.

Kim, B.R., Nam, H.Y., Kim, S.U., Kim, S.I. and Chang, Y.J. 2003. Normalization of reverse transcription quantitative-PCR with housekeeping genes in rice. Biotechnology Letters 25, 1869–1872.

Kim, Y.J., Shim, J.S., Lee, J.H., Jung, D.Y., Sun, H., In, J.G. and Yang, D.C. 2009. Isolation and characterization of a novel short-chain alcohol dehydrogenase gene from *Panax ginseng*. BMB Rep. **42(10)**: 673-8.

Kong, F., Wang, J., Cheng, L., Liu, S., Wu, J., Peng, Z. and Lu, G. 2012. Genome-wide analysis of the mitogen-activated protein kinase gene family in *Solanum lycopersicum*. Gene. **499(1)**:108-20.

Kotewicz, M.L., Sampson, C.M., D'Alessio, J.M., and Gerard, G.F 1988. Isolation of cloned Moloney murine leukemia virus reverse transcriptase lacking ribonuclease H activity. Nucleic Acids Res. **16**: 265-277.

Köthke, S. and Köck, M. 2011. The *Solanum lycopersicum* RNaseLER is a class II enzyme of the RNase T2 family and shows preferential expression in guard cells. J. Plant Physiol. **168(8)**: 840-7.

Krishnaswamy, S.S., Srivastava, S., Mohammadi, M., Rahman, M.H., Deyholos, M.K. and Kav, N.N. 2008. Transcriptional profiling of pea ABR17 mediated changes in gene expression in *Arabidopsis thaliana*. BMC Plant Biol. **10**: 8:91.

Kumar, K., Rao, K.P., Sharma, P. and Sinha, A.K.2008. Differential regulation of rice mitogen activated protein kinase kinase (MKK) by abiotic stress. Plant Physiol. Biochem. **46(10)**: 891-7.

Kürsteiner, O., Dupuis, I. and Kuhlemeier, C. 2003. The pyruvate decarboxylase1 gene of *Arabidopsis* is required during anoxia but not other environmental stresses. Plant Physiol. **132(2)**: 968-78.

Lata, C., Bhutty, S., Bahadur, R.P., Majee, M. and Prasad, M. 2011. Association of an SNP in a novel DREB2-like gene SiDREB2 with stress tolerance in foxtail millet (*Setaria italica* L.). J. Exp. Bot. **62(10)**: 3387-401.

Lata, C., Sahu, P.P. and Prasad, M. 2010. Comparative transcriptome analysis of differentially expressed genes in foxtail millet (*Setaria italica* L.) during dehydration stress. Biochem. Biophys. Res. Commun. **393(4)**: 720-7.

Lay, M.J. and Wittwer, C.T. 1997. Real-time fluorescence genotyping of factor V Leiden during rapid-cycle PCR. Clin. Chem. **43**: 2262-2267.

Lee, P.D., Sladek, R., Greenwood, C.M.T. and Hudson, T.J. 2002. Control genes and variability: Absence of ubiquitous reference transcripts in diverse mammalian expression studies. Genome Res **12**: 292–297.

Lee, J., J. Sayegh, J., Daniel, S. Clarke, and M.T. Bedford 2005. PRMT8, a new membrane bound tissue-specific member of the protein arginine methyltransferase family. J. Biol. Chem. **280**: 32890-32896.

Lekanne Deprez, R.H., Fijnvandraat, A.C., Ruijter, J.M. and Moorman, A.F. 2002. Sensitivity and accuracy of quantitative real-time polymerase chain reaction using SYBR green I depends on cDNA synthesis conditions. Anal. Biochem. **307**: 63-69.

Li, G.W., Peng, Y.H., Yu, X., Zhang, M.H., Cai, W.M., Sun, W.N. and Su, W.A. 2008. Transport functions and expression analysis of vacuolar membrane aquaporins in response to various stresses in rice. J. Plant Physiol. **165(18):** 1879-88.

Liao, Z., Chen, M., Guo, L., Gong, Y., Tang, F., Sun, X. and Tang, K. 2004. Rapid isolation of high-quality total RNA from *Taxus* and *Ginkgo*. Preparative Biochem. Biotechnol. **34:** 209-214.

Licausi, F., Giorgi, F.M., Zenoni, S., Osti, F., Pezzotti, M. and Perata, P. 2010. Genomic and transcriptomic analysis of the AP2/ERF superfamily in *Vitis vinifera*. BMC Genomics, **11:** 719.

Ligaba, A., Katsuhara, M., Shibasaka, M. and Djira, G. 2011. Abiotic stresses modulate expression of major intrinsic proteins in barley (*Hordeum vulgare*). C R Biology, **334(2):** 127-39.

Liu, B., Fan, J., Zhang, Y., Mu, P., Wang, P., Su, J., Lai, H., Li, S., Feng, D., Wang, J. and Wang, H. 2012. OsPFA-DSP1, negatively regulates drought stress responses in transgenic tobacco and rice plants. Plant Cell Rep. **31(6):** 1021-32.

Liu, C., Wu, Y. and Wang, X, 2012. bZIP transcription factor OsbZIP52/RISBZ5: a potential negative regulator of cold and drought stress response in rice. Planta, Dec 22. [Epub ahead of print].

Liu, Q. 2012. Novel miRNAs in the control of arsenite levels in rice. Funct Integr Genomics. [Epub ahead of print].

López, M., Tejera, N.A., Iribarne, C., Lluch, C. and Herrera-Cervera, J.A. 2008. Trehalose and trehalase in root nodules of *Medicago truncatula* and *Phaseolus vulgaris* in response to salt stress. Physiol. Plant. **134(4):** 575-82.

Løvdal, T. and Lillo, C. 2009. Reference gene selection for quantitative real-time PCR normalization in tomato subjected to nitrogen, cold, and light stress. Anal Biochem. **387(2):** 238-42.

Luo, M., Liu, J., Lee, R.D., Scully, B.T. and Guo, B. 2010. Monitoring the expression of maize genes in developing kernels under drought stress using oligo-microarray. J. Integr. Plant Biol. **52(12):** 1059-74.

Luo, Y., Liu, Y.B., Dong, Y.X., Gao, X.Q., Zhang, X.S. 2009. Expression of a putative alfalfa helicase increases tolerance to abiotic stress in *Arabidopsis* by enhancing the capacities for ROS scavenging and osmotic adjustment. J. Plant Physio. **166(4):** 385-94.

Ma, H., Zhao, H., Liu, Z. and Zhao, J. 2011. The phytocyanin gene family in rice (*Oryza sativa* L.): genome-wide identification, classification and transcriptional analysis. PLoS One, **6(10):** e25184.

Ma, H. and Zhao, J. 2010. Genome-wide identification, classification, and expression analysis of the arabinogalactan protein gene family in rice (*Oryza sativa* L.). J. Exp. Bot. **61(10):**2647-68.

Ma, K., Xiao, J., Li, X., Zhang, Q. and Lian, X. 2009. Sequence and expression analysis of the C3HC4-type RING finger gene family in rice. Gene. **444(1-2):** 33-45.

Maderazo, A.B., Belk, J.P., He, F., Jacobson, A. 2003. Nonsense-containing mRNAs that accumulate in the absence of a functional nonsense-mediated mRNA decay pathway are destabilized rapidly upon its restitution. Mol. Cell. Biol. **23**: 842-851.

Maghuly, F., Khan, M.A., Fernandez, E.B., Druart, P., Watillon, B. and Laimer, M. 2008. Stress regulated expression of the GUS-marker gene (uidA) under the control of plant calmodulin and viral 35S promoters in a model fruit tree rootstock: *Prunus incisa x serrula.* J. Biotech. **135(1)**: 105-16.

Malinen, E., Kassinen, A., Rinttila, T. and Palva, A. 2003. Comparison of real-time PCR with SYBR Green I or 52-nuclease assays and dot-blot hybridization with rDNA-targeted oligonucleotide probes in quantification of se-lected faecal bacteria. Microbiology **149**: 269-277.

Maroufi, A., Van Bockstaele, E. and De Loose, M. 2010. Validation of reference genes for gene expression analysis in chicory (*Cichorium intybus*) using quantitative real-time PCR. BMC Mol Biol.11:15.

Marras, S.A., Kramer, F.R. and Tyagi, S. 1999. Multiplex detection of single-nucleotide variations using molecular beacons. Genet. Anal. **14**: 151-156.

Martin, L., Leblanc-Fournier, N., Azri, W., Lenne, C., Henry, C., Coutand, C. and Julien, J.L. 2009. Characterization and expression analysis under bending and other abiotic factors of PtaZFP2, a poplar gene encoding a Cys2/His2 zinc finger protein. Tree Physiol. **29(1)**: 125-36.

Melton, D.A., Krieg, P.A. Rebagliati, M.R. Maniatis, T. Zinn, K. and Green, M.R. 1984. Efficient *in vitro* synthesis of biologically active RNA and RNA hybridization probes from plasmids containing a bacteriophage SP6 promoter. Nucleic Acids Res. **12**: 7035-7056.

Mishra, R.N., Reddy, P.S., Nair, S., Markandeya, G., Reddy, A.R., Sopory, S.K. and Reddy, M.K. 2007. Isolation and characterization of expressed sequence tags (ESTs) from subtracted cDNA libraries of *Pennisetum glaucum* seedlings. Plant Mol. Biol. **64(6)**: 713-32.

Mohammadi, M., Srivastava, S., Hall, J.C., Kav, N.N. and Deyholos, M.K. 2012. Two Wheat (*Triticum aestivum*) Pathogenesis-related 10 (PR-10) transcripts with distinct patterns of abundance in different organs. Mol. Biot. **51(2)**: 103-8.

Morse, D.L., Carroll, D., Weberg, L., Borgstrom, M.C., Ranger-Moore, J. and Gillies, R.J. 2005. Determining suitable internal standards for mRNA quantification of increasing cancer progression in human breast cells by real-time reverse transcriptase polymerase chain reaction. Anal. Biochem. **342(1)**: 69–77.

Müller, G.L., Drincovich, M.F., Andreo, C.S. and Lara, M.V. 2008. *Nicotiana tabacum* NADP-malic enzyme: cloning, characterization and analysis of biological role. Plant Cell Physiol. **49(3)**: 469-80.

Mullis, K.B. 1990. Target amplification for DNA analysis by the polymerase chain reaction. Ann. Biol. Clin. (Paris) **48**: 579–582.

Munis, M.F., Tu, L., Deng, F., Tan, J., Xu, L., Xu, S., Long, L. and Zhang, X. 2010. A thaumatin-like protein gene involved in cotton fiber secondary cell wall development enhances resistance against *Verticillium dahliae* and other stresses in transgenic tobacco. Biochem. Biophys. Res. Commun. **393(1):** 38-44.

Nahar, K., Kyndt, T., De Vleesschauwer, D., Höfte, M. and Gheysen, G. 2011. The jasmonate pathway is a key player in systemically induced defense against root knot nematodes in rice. Plant Physiol. **157(1):** 305-16.

Nazarenko, I., Lowe, B., Darfler, M., Ikonomi, P., Schuster, D. and Rashtchian, A. 2002. Multiplex quantitative PCR using self-quenched primers labeled with a single fluorophore. Nucleic Acids Res. 30:e37.

Nicot, N., Hausman, J.F., Hoffmann, L. and Evers, D. 2005. Housekeeping gene selection for real-time RT-PCR normalization in potato during biotic and abiotic stress. J. Exp. Bot. **56(421):** 2907-14.

Nishiuchi, S., Fujihara, K., Liu, S. and Takano, T. 2010. Analysis of expressed sequence tags from a NaHCO(3)-treated alkali-tolerant plant, *Chloris virgata*. Plant Physiol. and Biochem. **48(4):** 247-55.

Nohzadeh Malakshah, S., Habibi Rezaei, M., Heidari, M. and Salekdeh, G.H. 2007. Proteomics reveals new salt responsive proteins associated with rice plasma membrane. Biosci. Biotechnol. Biochem. 71(9):2144-54.

Orlando, C., Pinzani, P. and Pazzagli, M. 1998. Developments in quantitative PCR. Clin. Chem. Lab. Med. **36(5):** 255–269.

Palmer, S., Wiegand, A.P., Maldarelli, F., Bazmi, H., Mican, J.M., Polis, M., Dewar, R.L. and Planta, A. 2003. New real-time reverse tran-scriptase-initiated PCR assay with single-copy sensitivity for human immunodeficiency virus type 1 RNA in plasma. J. Clin. Microbiol. **41:** 4531-4536.

Pfaffl, M.W. 2001. A new mathematical model for relative quantification in real-time RT-PCR. Nucleic Acids Research 29 E45.

Pfaffl, M.W. and Hageleit, M. 2001. Validities of mRNA quantification using recombinant RNA and recombinant DNA external calibration curves in real-time RT-PCR. Biotechn. Lett. **23:** 275–282.

Pfaffl, M.W., Tichopad, A., Prgomet, C. and Neuvians, T.P. 2004. Determination of stable housekeeping genes, differentially regulated target genes and sample integrity: BestKeep-er-Excel-based tool using pair-wise correla-tions. Biotechnol. Lett. **26:** 509-515.

Pineda, M.C., Turon, X. and López-Legentil, S. 2012. Stress levels over time in the introduced ascidian *Styela plicata*: the effects of temperature and salinity variations on hsp70 gene expression. Cell Stress Chaperones. Jan 17. [Epub ahead of print].

Puckette, M., Peal, L., Steele, J., Tang, Y. and Mahalingam, R. 2009. Ozone responsive genes in *Medicago truncatula*: analysis by suppression subtraction hybridization. J. Plant Physiol. 15; **166(12):** 1284-95.

Puranik, S., Jha, S., Srivastava, P.S., Sreenivasulu, N. and Prasad, M. 2011. Comparative transcriptome analysis of contrasting foxtail millet cultivars in response to short-term salinity stress. J. Plant Physiol. **168(3):** 280-7.

Reid, K.E., Olsson, N., Schlosser, J., Peng, F. and Lund, S.T. 2006. An optimized grapevine RNA isolation procedure and statistical determination of reference genes for real-time RT-PCR during berry development.BMC Plant Biol. **6:** 27.

Ririe, K.M., Rasmussen, R.P. and Wittwer, C.T. 1997. Product differentiation by analysis of DNA melting curves during the polymerase chain reaction. Anal. Biochem. **245:**154-160.

Sathiyaraj, G., Lee, O.R., Parvin, S., Khorolragchaa, A., Kim, Y.J. and Yang, D.C, 2011. Transcript profiling of antioxidant genes during biotic and abiotic stresses in *Panax ginseng* C. A. Meyer. Mol. Biol. Rep. **38(4):** 2761-9.

Schmidt, G.W. and Delaney, S.K. 2010. Stable internal reference genes for normalization of real-time RT-PCR in tobacco (*Nicotiana tabacum*) during development and abiotic stress. Mol. Genet. Genomics. **283(3):** 233-41.

Shen, C., Bai, Y., Wang, S., Zhang, S., Wu, Y., Chen, M., Jiang, D. and Qi, Y. 2010. Expression profile of PIN, AUX/LAX and PGP auxin transporter gene families in *Sorghum bicolor* under phytohormone and abiotic stress. FEBS J. **277(14):** 2954-69.

Shimomaye, E. and Salvato, M. 1989. Use of avian myeloblastosis virus reverse transcriptase at high temperature for sequence analysis of highly structured RNA. Gene Anal. Tech. **6:** 25-28.

Si, Y., Zhang, C., Meng, S. and Dane, F. 2009. Gene expression changes in response to drought stress in *Citrullus colocynthis*. Plant Cell Report. **28(6):** 997-1009.

Silveira, E.D., Alves-Ferreira, M., Guimarães, L.A., da Silva, F.R. and Carneiro, V.T. 2009. Selection of reference genes for quantitative real-time PCR expression studies in the apomictic and sexual grass *Brachiaria brizantha*. BMC Plant Biol.**9:** 84.

Simpson, D.A., Feeney, S., Boyle, C. and Stitt, A.W. 2000. Retinal VEGF mRNA measured by SYBR green I fluorescence: A versatile approach to quantitative PCR. Mol. Vis. **6:**178-183.

Singh, A., Giri, J., Kapoor, S., Tyagi, A.K. and Pandey, G.K. 2010. Protein phosphatase complement in rice: genome-wide identification and transcriptional analysis under abiotic stress conditions and reproductive development. BMC Genomics **16:** 11:435.

Singh, R.K., Anandhan, S., Singh, S., Patade, V.Y., Ahmed, Z. and Pande, V. 2011. Metallothionein-like gene from *Cicer microphyllum* is regulated by multiple abiotic stresses. Protoplasma. **248(4):** 839-47.

Solinas, A., Brown, L.J., McKeen, C., Mellor, J.M., Nicol, J., Thelwell, N. and Brown, T. 2001. Duplex Scorpion primers in SNP analysis and FRET applications. Nucleic Acids Res. 29:E96.

Song, Y., Wang, L. and Xiong, L. 2009. Comprehensive expression profiling analysis of OsIAA gene family in developmental processes and in response to phytohormone and stress treatments. Planta, **229(3):** 577-91.

Soto, G., Stritzler, M., Lisi, C., Alleva, K., Pagano, M.E., Ardila, F., Mozzicafreddo, M., Cuccioloni, M., Angeletti, M. and Ayub, N.D. 2011. Acetoacetyl-CoA thiolase regulates the mevalonate pathway during abiotic stress adaptation. J. Exp. Bot. **62(15):** 5699-711.

Souaze, F., Ntodou-Thome, A., Tran, C.Y., Rostene, W. and Forgez P. 1996. Quantitative RT-PCR: limits and accuracy. BioTechniques **21:** 280-285.

Southern, E.M. 1975. Detection of specific sequences among DNA fragments separated by gel electrophoresis. J. Mol. Biol. **98:** 503–517.

Sudo, E., Itouga, M., Yoshida-Hatanaka, K., Ono, Y. and Sakakibara, H. 2008. Gene expression and sensitivity in response to copper stress in rice leaves. J. Exp. Bot. **59(12):** 3465-74.

Sung, D.Y., Kaplan, F., Lee, K.J., and Guy, C.L. 2003. Acquired tolerance to temperature extremes. Trends Plant Sci. **8:** 179–187.

Suzuki, T., Higgins, P.J. and Crawford, D.R. 2000. Control selection for RNA quantitation. Biotechniques **29:** 332–337.

Thelwell, N., Millington, S., Solinas, A., Booth, J. and Brown, T. 2000. Mode of action and application of Scorpion primers to mutation detection. Nucleic Acids Res. **28:**3752-3761.

Tillett, R.L., Ergül, A., Albion, R.L., Schlauch, K.A., Cramer, G.R., Cushman, J.C. 2011. Identification of tissue-specific, abiotic stress-responsive gene expression patterns in wine grape (*Vitis vinifera* L.) based on curation and mining of large-scale EST data sets. BMC Plant Biol. **11:** 86.

Timotijeviæ, G.S., Milisavljeviæ, M.D.J., Radoviæ, S.R., Konstantinoviæ, M.M. and Maksimoviæ, V.R. 2010. Ubiquitous aspartic proteinase as an actor in the stress response in buckwheat. J. Plant Physiol. **167(1):** 61-8.

Tong, Z., Gao, Z., Wang, F., Zhou, J. and Zhang, Z. 2009. Selection of reliable reference genes for gene expression studies in peach using real-time PCR.BMC Mol. Biol.**10:** 71.

Tyagi, S. and Kramer, F.R. 1996. Molecular beacons: probes that fluoresce upon hybridization. Nat. Biotechnol. **14:** 303-308.

Tyagi, S., Marras, S.A. and Kramer, F.R. 2000. Wavelength-shifting molecular beacons. Nat. Biotechnol. **18:**1191-1196.

Tyagi, W., Singla-Pareek, S., Nair, S., Reddy, M.K. and Sopory, S.K. 2006. A novel isoform of ATPase c subunit from pearl millet that is differentially regulated in response to salinity and calcium. Plant Cell Rep. **25(2):** 156-63.

Vandesompele, J., De Preter, K., Pattyn, F., Poppe, B., Van Roy, N., De Paepe, A., and Speleman, F. 2002. Accurate normalization of real-time quantitative RT-PCR data by geometric averaging of multiple internal control genes. Genome Biol **3(7):** 0034.1–0034.11.

Vashisth, T., Johnson, L.K. and Malladi, A. 2011. An efficient RNA isolation procedure and identification of reference genes for normalization of gene expression in blueberry. Plant Cell Rep. **(12):** 2167-76.

Vij, S., Giri, J., Dansana, P.K., Kapoor, S. and Tyagi, A.K. 2008. The receptor-like cytoplasmic kinase (OsRLCK) gene family in rice: organization, phylogenetic relationship, and expression during development and stress. Mol. Plant. **1(5):** 732-50.

von Ahsen, N., Schutz, E., Armstrong, V.W. and Oellerich, M. 1999. Rapid detection of prothrombotic mutations of prothrombin (G20210A), factor V (G1691A), and methylenetetrahydrofolate reductase (C677T) by real-time fluorescence PCR with the LightCycler. Clin. Chem. **45:** 694-696.

Wang, C., Han, J., Liu, C., Kibet, K.N., Kayesh, E., Shangguan, L., Li. X. and Fang, J. 2012. Identification of microRNAs from Amur grape (*Vitis amurensis* Rupr.) by deep sequencing and analysis of microRNA variations with bioinformatics. BMC Genomics. **29:** 13:122.

Wang, C., Jing, R., Mao, X., Chang, X., and Li, A. 2011. TaABC1, a member of the activity of bc1 complex protein kinase family from common wheat, confers enhanced tolerance to abiotic stresses in *Arabidopsis*. J. Exp. Bot. *62(3):* 1299-311.

Wang, C., Yang, C., Gao, C., and Wang, Y. 2009. Cloning and expression analysis of 14 lipid transfer protein genes from *Tamarix hispida* responding to different abiotic stresses. Tree Physiol. **29(12):** 1607-19.

Wang, L., Yu, X., Wang, H., Lu, Y.Z., de Ruiter, M., Prins, M,. and He, Y.K. 2011. A novel class of heat-responsive small RNAs derived from the chloroplast genome of Chinese cabbage (*Brassica rapa*). BMC Genomics. **3:** 12:289.

Wang, Q., Li, F., Zhang, X., Zhang, Y., Hou, Y., Zhang, S., and Wu, Z. 2011. Purification and characterization of a CkTLP protein from *Cynanchum komarovii* seeds that confers antifungal activity. PLoS One, **6(2):** e16930.

Wang, Q., Zhang, Y., Hou, Y., Wang, P., Zhou, S., Ma, X., and Zhang, N. 2012. Purification, characterization of a CkChn134 protein from *Cynanchum komarovii* seeds and synergistic effect with CkTLP against *Verticillium dahliae*. Protein Sci. **21(6):** 865-75.

Wang, S., Bai, Y., Shen, C., Wu, Y., Zhang, S., Jiang, D., Guilfoyle, T.J., Chen, M., and Qi, Y. 2010. Auxin-related gene families in abiotic stress response in *Sorghum bicolor*. Funct. Integr. Genomics. **10(4):** 533-46.

Wang, S., Wang, R., Liang, D., Ma, F. and Shu, H. 2012. Molecular characterization and expression analysis of a glycine-rich RNA-binding protein gene from *Malus hupehensis* Rehd. Mol. Biol. Rep. **39(4):** 4145-53.

Wang, X., Tang, C., Deng, L., Cai, G., Liu, X., Liu, B,. Han, Q., Buchenauer, H., Wei, G., Han, D, Huang, L. and Kang, Z. 2010. Characterization of a pathogenesis-related thaumatin-like protein gene TaPR5 from wheat induced by stripe rust fungus. Physiol. Plant. **139(1):** 27-38.

Wang, X., Wang, Z., Dong, J., Wang, M. and Gao, H. 2009. Cloning of a 9-cis epoxycarotenoid dioxygenase gene and the responses of *Caragana korshinskii* to a variety of abiotic stresses. Genes and Genetic Sys. **84(6):** 397-405.

Wang, Y., Ma, H., Liu, G., Zhang, D., Ban, Q., Zhang, G., Xu, C. and Yang, C. 2008. Generation and analysis of expressed sequence tags from a NaHCO3-treated *Limonium bicolor* cDNA library. Plant Physiol. Biochem. **46(11):** 977-86.

Wang, T. and Brown, M.J. 1999. mRNA quantification by real time TaqMan polymerase chain reaction: validation and compar compar-ison with RNase protection. Anal. Biochem. **269:**198-201.

Wei, A., He, C., Li, B., Li, N. and Zhang, J. 2011. The pyramid of transgenes TsVP and BetA effectively enhances the drought tolerance of maize plants. Plant Biotech. J. **9(2):** 216-29.

Whitcombe, D., Theaker, J., Guy, S.P., Brown, T. and Little, S. 1999. Detection of PCR products using self-probing amplicons and fluorescence. Nat. Biotechnol. **17:** 804-807.

Wittwer, C.T., Herrmann, M.G., Moss, A.A. and Rasmussen, R.P. 1997. Continuous fluorescence monitoring of rapid cycle DNA amplification. BioTechniques **22:** 130-138.

Wittwer, C.T., Herrmann, M.G., Gundry, C.N. and Elenitoba-Johnson, K.S.J. 1991. Real-Time Multiplex PCR Assays. Methods. 25, 430–442.

Wu, G., Zhang, L., Wu, Y., Cao, Y., and Lu, C. 2010. Comparison of five endogenous reference genes for specific PCR detection and quantification of *Brassica napus*. J. Agric. Food Chem. **58(5):** 2812-7.

Wu, J., Peng, Z., Liu, S., He, Y., Cheng, L., Kong, F., Wang, J., and Lu, G. 2012. Genomewide analysis of Aux/IAA gene family in Solanaceae species using tomato as a model. Mol. Genet. Genomics. **287(4):** 295-11.

Xi, L., Xu, K., Qiao, Y., Qu, S., Zhang, Z. and Dai, W. 2011. Differential expression of ferritin genes in response to abiotic stresses and hormones in pear (*Pyrus pyrifolia*). Mol. Biol. Rep. **38(7):** 4405-13.

Xia, X.J., Zhou, Y.H., Ding, J., Shi, K., Asami, T., Chen, Z. and Yu, J.Q. 2011. Induction of systemic stress tolerance by brassinosteroid in *Cucumis sativus*. New Phytol. **191(3):** 706-20.

Xiang, Y., Tang, N., Du, H., Ye, H., and Xiong, L. 2008. Characterization of OsbZIP23 as a key player of the basic leucine zipper transcription factor family for conferring abscisic acid sensitivity and salinity and drought tolerance in rice. Plant Physiol. **148(4):** 1938-52.

Xianjun, P., Xingyong, M., Weihong, F., Man, S., Liqin, C., Alam, I., Lee, B.H., Dongmei, Q., Shihua, S., and Gongshe, L. 2011. Improved drought and salt tolerance of *Arabidopsis thaliana* by transgenic expression of a novel DREB gene from *Leymus chinensis*. Plant Cell Rep. 30(8):1493-502.

Xiao, Y., Di, P., Chen, J., Liu, Y., Chen, W. and Zhang, L. 2009. Characterization and expression profiling of 4-hydroxyphenylpyruvate dioxygenase gene (Smhppd) from *Salvia miltiorrhiza* hairy root cultures. Mol. Biol. Rep. **36(7)**: 2019-29.

Xiong, A.S., Jiang, H.H., Zhuang, J., Peng, R.H., Jin, X.F., Zhu, B., Wang, F., Zhang, J. and Yao, Q.H. 2012. Expression and function of a modified AP2/ERF transcription factor from *Brassica napus* enhances cold tolerance in transgenic *Arabidopsis*. Mol. Biotech. Feb 20. [Epub ahead of print].

Xu, F., Cheng, H., Cai, R., Li, L.L., Chang, J., Zhu, J., Zhang, F.X., Chen, L.J., Wang, Y., Cheng, S.H. and Cheng, S.Y. 2008. Molecular cloning and function analysis of an anthocyanidin synthase gene from *Ginkgo biloba,* and its expression in abiotic stress responses. Mol. Cells. **26(6)**: 536-47.

Xu, G.Y., Rocha, P.S., Wang, M.L., Xu, M.L., Cui, Y.C., Li, L.Y., Zhu, Y.X. and Xia, X. 2011. A novel rice calmodulin-like gene, OsMSR2, enhances drought and salt tolerance and increases ABA sensitivity in *Arabidopsis*. Planta, **234(1)**: 47-59.

Xu, W.F. and Shi, W.M. 2006. Expression profiling of the 14-3-3 gene family in response to salt stress and potassium and iron deficiencies in young tomato (*Solanum lycopersicum*) roots: analysis by real-time RT-PCR. Annal. Bot. **98(5)**: 965-74.

Xu, Z., Zhong, S., Li, X., Li, W., Rothstein, S.J., Zhang, S., Bi, Y., and Xie, C. 2011. Genome-wide identification of microRNAs in response to low nitrate availability in maize leaves and roots. PLoS One; **6(11)**: e28009.

Yan, J., Yuan. F., Long, G., Qin, L. and Deng, Z. 2012. Selection of reference genes for quantitative real-time RT-PCR analysis in citrus. Mol. Biol. Rep. **(2)**: 1831-8.

Yan, S.P., Zhang, Q.Y., Tang, Z.C., Su, W.A. and Sun, W.N. 2006. Comparative proteomic analysis provides new insights into chilling stress responses in rice. Mol. Cell Proteom. **5(3)**: 484-96.

Yang, Y., Hou, S., Cui, G., Chen, S., Wei, J. and Huang, L. 2009. Characterization of reference genes for quantitative real-time PCR analysis in various tissues of *Salvia miltiorrhiza*. Mol. Biol. Rep. **(1)**: 507-13.

Yang, Y., Wu, J., Zhu, K., Liu, L., Chen, F. and Yu, D. 2009. Identification and characterization of two chrysanthemum (*Dendronthema x morifolium*) DREB genes, belonging to the AP2/EREBP family. Mol. Biol. Rep. **36(1)**: 71-81.

Yao, D., Zhang, X., Zhao, X., Liu, C., Wang, C., Zhang, Z., Zhang, C., Wei, Q., Wang, Q., Yan, H., Li, F., and Su, Z. 2011. Transcriptome analysis reveals salt-stress-regulated biological processes and key pathways in roots of cotton (*Gossypium hirsutum* L.). Genomics, **98(1)**: 47-55.

Yu, S. and Luo, L. 2008. Expression analysis of a novel pyridoxal kinase messenger RNA splice variant, PKL, in oil rape suffering abiotic stress and phytohormones. Acta Biochim. Biophys. Sin (Shanghai). **40(12)**: 1005-14.

Yu, X., Wang, H., Lu, Y., de Ruiter, M., Cariaso, M., Prins, M., van Tunen, A., and He, Y. 2012. Identification of conserved and novel microRNAs that are responsive to heat stress in *Brassica rapa*. J. Exp. Bot. **63(2)**: 1025-38.

Yue, G., Sui, Z., Gao, Q. and Zhang, J. 2008. Molecular cloning and characterization of a novel H+-translocating pyrophosphatase gene in *Zea mays*. DNA Seq. **19(2):** 79-86.

Zhang, C., Bian, M., Yu, H., Liu, Q., Yang, Z. 2011. Identification of alkaline stress-responsive genes of CBL family in sweet sorghum (*Sorghum bicolor* L.). Plant Physiol. Biochem. **49(11):**1306-12.

Zhang., J. and Byrne, C.D. 1999. Differential priming of RNA templates during cDNA synthesis markedly affects both accuracy and reproducibility of quantitative competitive re-verse transcriptase PCR. Biochem. J. **337(Pt 2):** 231-241.

Zhao, T., Liang, D., Wang, P., Liu, J. and Ma, F. 2012 Genome-wide analysis and expression profiling of the DREB transcription factor gene family in *Malus* under abiotic stress. Mol. Genet. Genomics. **287(5):** 423-36.

Zhong, H.Y., Chen, J.W., Li, C.Q., Chen, L., Wu, J.Y., Chen, J.Y., Lu, W.J. and Li, J.G. 2011. Selection of reliable reference genes for expression studies by reverse transcription quantitative real-time PCR in litchi under different experimental conditions. Plant Cell Rep. **(4):** 641-53.

Zhu, J.Q., Zhang, J.T., Tang, R.J., Lv, Q.D., Wang, Q.Q., Yang, L. and Zhang, H.X. 2009. Molecular characterization of ThIPK2, an inositol polyphosphate kinase gene homolog from *Thellungiella halophila*, and its heterologous expression to improve abiotic stress tolerance in *Brassica napus*. Physiol. Plant. **136(4):** 407-25.

Zinn, K., DiMaio, D. and Maniatis, T. 1983. Identification of two distinct regulatory regions adjacent to the human beta-interferon gene. Cell. **34:** 865-879.

2013, **Abiotic Stress and Biotechnology**
Editor: T. Pullaiah
Published by: REGENCY PUBLICATIONS, NEW DELHI

Pages 285–309

Chapter 13

Subtractive Hybridization: A Powerful Method to Isolate Differentially Expressed Genes

Chandra Obul Reddy Puli[1]*, Sravani Konduru[1], Jayanna Naik Banavath[1], Varakumar Pandit[1], Krishnaveni Thiruveedhi[2], Chandra Sekhar Akila[2]

[1]*Department of Plant Sciences,* [2]*Department of Biotechnology, School of Life Sciences, Yogi Vemana University, Vemanapuram, Kadapa – 516 003, A.P., India*

ABSTRACT

Here we have described the cDNA subtractive hybridization method for high-throughput cloning of differentially expressed genes from plants that can be applied to any experimental system. The comparison of two RNA populations that differ from the effects of a single independent variable, such as stress exposed or a specific genetic defect can identify differences in the abundance of specific transcripts that vary in a population dependent manner. There are a variety of methods for identifying differentially expressed genes, including microarray, SAGE, qRT-PCR and DDGE. This protocol describes a easy and cost effective alternative that does not require prior knowledge of the transcriptomes under investigation and is particularly applicable when minimal levels of starting material (RNA) are available. This protocol describes the use of biotin-streptavidin

* Corresponding Author: E-mail: coreddy@yogivemanaauniversity.ac.in

interaction and magnetic bead technology for efficient removal of common cDNA population during subtractive hybridization to reduce the non-target cDNA population prior to subtractive cDNA library construction for the enrichment of differentially expressed genes. The final products are cDNA populations enriched for significantly over-represented transcripts in either of the two input RNA preparations. These cDNA populations may then be cloned to make subtracted cDNA libraries and/or used as probes to screen subtracted cDNA, global cDNA or genomic DNA libraries. Since understanding of specific gene expression patterns that regulate developmental and stress responses is a major concern of molecular biology, Subtractive Hybridization has become a very popular molecular technique during the past decade.

Keywords: Subtractive hybridization, Differentially expressed genes, Driver, Tester.

Introduction

Cells respond to environmental changes by altering their gene expression patterns to produce the proteins required for cell adaptation to the new environments. Such alterations are generally mediated through signal transduction pathways switched on by master regulators or receptors which initiate cascades of gene induction processes and thus evoke the developmental progression and/or the responses to the environmental situation. The conception that, obtaining information on transcriptional changes will yield information on biologically relevant metabolism pathways lies at the heart of RNA transcript profiling. By comparing the concentrations of individual mRNAs present in samples originating from different genotypes, developmental stages or growth conditions, genes can be identified that are differentially expressed and hence may have specific metabolic or morphogenetic functions. A pressing problem is to identify and to characterize those genes that are differentially expressed in order to understand the molecular nature.

To find these differentially expressed genes, several methods have been designed to detect and isolate different DNA sequences that are present in one expressed gene library but absent in the other. Isolation of DNA fragments responsible for the phenotypic differences is one of the most important problems of molecular biology. The identification and characterization of genes that are differentially expressed have been severely hampered by technical limitations. A variety of methods were developed and improved over time to identify genes that express differentially; these include simple differential display (Liang and Pardee, 1992) and were later improved to multicolor fluorescent-labeled differential display (Cho *et al.*, 2000). However, these techniques suffered from a lack of reproducibility and had a high rate of false-positive signals. To overcome the problems associated with differential display, a cDNA–amplification fragment-length polymorphism analysis (cDNA–AFLP) protocol, using stringent polymerase chain reaction (PCR) conditions, was developed (Vos *et al.*, 1995) and later improved to high-throughput analysis (Buntjer *et al.*, 2001). Later, numerous PCR-based differentially expressed gene-cloning methods, such as cDNA–representational difference analysis (cDNA–RDA) (Hubank and Schatz 1994; Chang and Denny, 1998; Welford *et al.*, 1998) and serial analysis of gene expression (SAGE)

(Dale *et al.,* 2001) have been used to identify differentially expressed genes. These methods differ in their convenience, sensitivity, automation, and throughput.

Recently, microarray technology has become a useful tool for the analysis of gene expression at the global level. It uses either cDNA spotted chips or the array of synthetic oligonucleotides (Schena *et al.,* 1995; Lockhart *et al.,* 1996). These two methods have been reviewed extensively (Duggan *et al.,* 1999; Lipshutz *et al.,* 1999). This technology is possible only in model systems where the complete genome sequences are known or where large numbers of characterized cDNA clones are available, and its use is further limited to a few research laboratories due to the high cost of capital equipment, operational expenses, the requirement of high technical expertise, and problems coping with the massive amount of data generated and the subsequent interpretation of data in the context of biological knowledge. This technology, therefore, is largely becoming a cross-disciplinary endeavor requiring the collaboration of biologists, engineers, database designers, physicists, and mathematicians. As an alternative, a number of laboratories have prepared subtractive cDNA libraries to eliminate common genes and access the differentially expressed genes.

Subtractive library production and screening are powerful methods to isolate transcripts that are exclusively or selectively expressed either in certain tissues, or during development or in disease states. The original method was developed by Davis and colleagues in classical work on the isolation of T-lymphocyte-specific transcripts, including a third type of murine T-cell receptor gene (Hedrick *et al.,* 1984; Chien *et al.,* 1984). Over the years numerous subtractive hybridization techniques have been developed and used to isolate significant genes in many systems (Sargent and Dawid, 1983; Hedrick *et al.,* 1984; Hara *et al.,* 1991; Wang and Brown, 1991; Hubank and Schatz, 1994). However, the principle behind the method is essentially the same, and only a few modifications have improved on the remarkable sensitivity and efficiency of the original method.

Principle of Subtractive Hybridization

Subtractive cloning uses a process called driver excess hybridization. Nucleic acid from which one wants to isolate differentially expressed sequences (the tracer) is hybridized to complementary nucleic acid that is believed to lack sequences of interest (the driver). Driver nucleic acid is present at much higher concentration (at least five-fold) than is tracer, and it dictates the speed of the reannealing reaction. The driver and tracer nucleic acid populations are allowed to hybridize, and only sequences common to the two populations can form hybrids. After hybridization, driver-tracer hybrids and unhybridized driver are removed. This is the subtraction step. The tracer that remains behind is enriched for sequences specific to the tracer tissue source [often called the plus (+) source] and depleted for sequences common to tracer and driver [often called the minus (–) source] (Figure 13.1). Usually, the process must be performed reiteratively in order to remove all the sequences common to both the driver and the tracer. After subtraction, remaining nucleic acid can be used to prepare a library enriched in tracer-specific clones or to make a probe that can be used to screen a library for tracer-specific clones. Subtractive hybridization may be also used for isolation of sequences that are present both in driver and in tracer but in different abundance. Such a problem often arises during subtraction of cDNA libraries for the determination of the differences in levels of gene expression in closely related cell types. This type of the target will be called the "copied" target. Subtractive hybridization may be used also for the isolation of homologous sequences present both in driver and tracer. This type of target will be called "homologous" target.

The procedure consists of primarily a subtractive hybridization between two complimentary single stranded nucleic acid populations. The mRNA populations from tester and approximately five-fold excess of complimentary 1st strand cDNA from corresponding driver were heat denatured and subjected to limited renaturation with empirically determined Cot value. The hybridization of mRNA strands belonging to driver sample with the complementary cDNA strands belonging to tester sample obeys the ideal pseudo-first order kinetics as described in the equation-1. The hybridization kinetics is sharper as compared to the second order kinetics, where two different double stranded nucleic acids (from tester and control) were together subjected to hybridization after denaturation.

$$\frac{dC}{dt} = -kC^2 \tag{1}$$

where, 'C' is molar concentration of a single-stranded target nucleic acid, t' is time and 'k' is the rate constant. Equation-1, can be integrated and solved yielding equation-2 as

$$C_t = \frac{C_0}{C_0 kt + 1} \tag{2}$$

where, 'C$_0$' is the starting concentration of the single-strand nucleic acid molecules and 'C$_t$' is the concentration of the remaining single-strand nucleic acid molecules, at time 't'. The concentration of remaining single stranded nucleic acid molecules is determined mainly by its hybridization rate constant 'k' and hybridization time 't' and is independent of its starting concentration 'C$_0$', where 'k' depends for any given nucleic acid molecule on the relative proportion (ratio 'R') of its presence in tester (mRNA) to its complementary nucleic acid molecules present in control (cDNA) population. During shorter hybridizations the complementary single-strand nucleic acid molecules present in tester and control fail to drive the hybridization to completion, however, the abundant transcripts will form mRNA-cDNA hybrids quickly whereas the low-abundant mRNA species will anneal less rapidly and the remaining mRNA fraction will become progressively normalized during the course of hybridization. Biotin tag was introduced into the 5' ends of all control cDNA molecules during cDNA synthesis. By exploiting the biotin-streptavidin interaction using streptavidin coated paramagnetic beads, mRNA-cDNA hybrids as well as excess cDNA molecules were physically separated with the help of magnetic separator. The enrichment is far more efficient for differentially expressed genes with a large value for 'R', but at the same time not eliminating the genes with lesser value for 'R' in this subtraction, even though it appears all-or-non-type after subtraction.

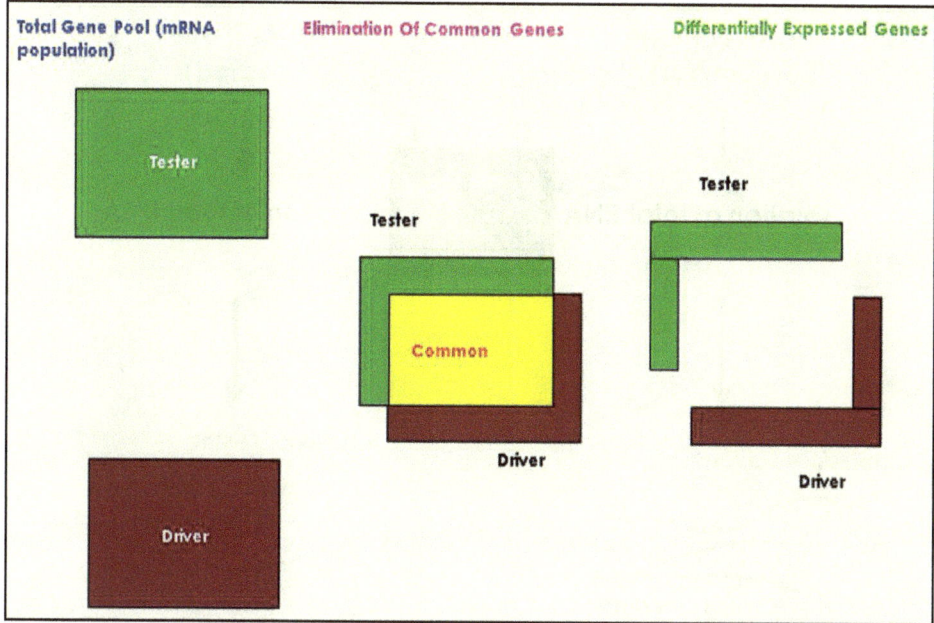

Figure 13.1: Principle of Subtractive hybridization.

Steps in the Subtractive cDNA Library Construction

Construction and screening of a subtractive cDNA library can be divided into several steps (Figure 13.2).

1. Preparation of driver and tester nucleic acids.
2. Subtractive hybridization between driver and tester.
3. Cloning of the remaining cDNA to construct a subtractive cDNA library.
4. Preparation and labelling of probes for screening of the subtractive cDNA library.
5. Analysis of clones.

For many of these steps, a number of options exist, and the choice of procedures should meet the following conditions:

1. *Sensitivity.* It should be possible to isolate cDNA clones representing mRNAs with abundance as low as 0.0001-0.001 per cent.
2. *Efficiency.* The method should require modest amounts of starting RNA and allow the isolation of as many specific transcripts as possible of wide size range.
3. *Convenience.* The procedures should not be too cumbersome, and should avoid the isolation of any artefactual clones.

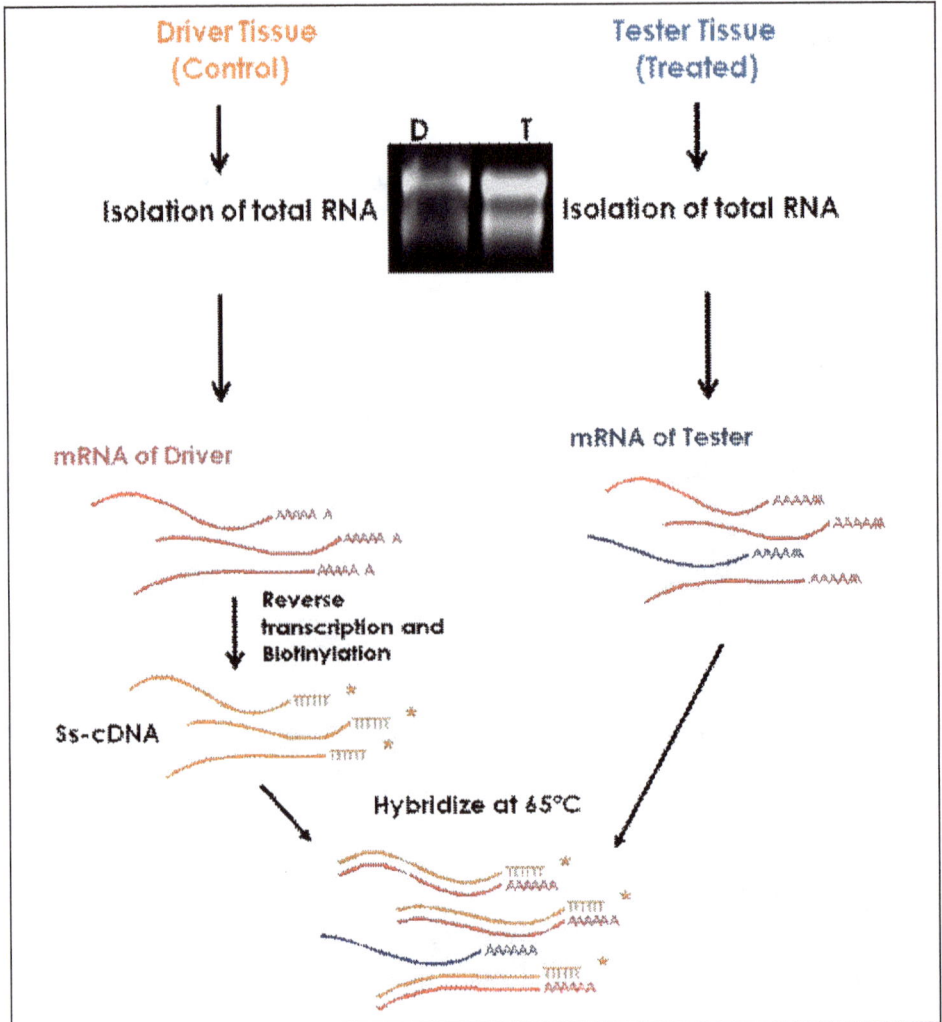

Figure 13.2: Schematic representation of subtractive hybridization described in the chapter.

Subtractive hybridization typically involves annealing of nucleic acids (poly(A)⁺ RNA or single-strand cDNA (ss-cDNA) or double-strand cDNA (ds-cDNA)) extracted or/and synthesized from *driver* tissues with at least 20-fold less ss- or ds-cDNA synthesized from another source *(target)* for which specific transcripts have to be isolated. To achieve a maximum of efficiency and sensitivity, contamination of target cDNA by driver nucleic acids after hybridization has to be completely avoided. Furthermore, one round of subtractive hybridization should result in as much specific subtraction as possible of common abundant molecules. Whatever the separation method of target cDNA from both annealed and free driver nucleic acids which coexist after subtractive hybridization, slight cross-contamination is always found. Therefore, either driver nucleic acids need to be eliminated after hybridization while target cDNA remains intact, or unseparated annealed and free driver nucleic acids must not interfere with cloning of the subtracted target sequences.

The Choice of Target and Driver Tissues

One of the major deciding factors that determines the efficiency and sensitivity of the subtractive cDNA library is the choice of target and driver tissues. Both the abundance and range of transcripts are influenced by the target cDNA and driver nucleic acids with the following reasons

1. Target tissues do not always comprise a homogenous population of cells. Therefore, it may be necessary to select the part of the target tissue which contains desired specific transcripts.

2. The relative proportion of various target-specific transcripts is determined by the presence of common transcripts in target and driver tissues.

Preparation of Poly (A)⁺ RNA from Driver and Target Tissues

Target and driver poly(A)⁺RNA can be purified from either total RNA or cytoplasmic RNA, or RNA extracted from ribosomes, which are associated with the endoplasmic reticulum. The majority of workers use total RNA to purify poly(A)⁺RNA.

Protocol 1. Isolation of Total RNA from Target and Driver Tissues

Reagents and Solutions Required
☆ Trizol reagent (Invitrogen Catalogue number 15596018)
☆ Chloroform
☆ Isopropanol
☆ 75 per cent Ethanol
☆ DEPC treated autoclaved water

Method
1. Grind approximately 3 gm of plant tissue in liquid nitrogen to a fine powder using a mortar and pestle.
2. Transfer powdered tissue to Oakridge tube and let the liquid nitrogen evaporate.
3. Add approximately 30 ml (10 ml of Trizol reagent per g of tissue). Start adding the reagent before the tissues start to thaw.
4. Mix vigorously.
5. Keep on a rocker to shake gently for 5 min.
6. Add 6 ml of chloroform. (2 ml chloroform per g of tissue).
7. Shake vigorously for about 15 seconds.
8. Incubate at room temperature for 5 min.
9. Centrifuge at 12, 000 g for 15 min at 4°C. After centrifugation you should see a red phenol chloroform layer at the bottom of the tube, an interphase and a colourless aqueous phase. Transfer upper, colourless, phase to a

fresh tube. Sacrifice some of the sample rather than risk contamination with the interphase.

10. Add 15 ml of iso-propanol (0.5 ml per ml of Trizol). Mix well. The mixture should appear a little cloudy at this stage.
11. Incubate at room temperature for 10 min.
12. Spin at 12,000 g, 10 min.
13. Decant supernatant.
14. Wash pellet with 5 ml of 75 per cent ethanol.
15. Spin at 12,000 g for 10 min at 4°C.
16. Decant supernatant.
17. Air dry horizontally on clean paper towels for about 15 min.
18. Dissolve the pellet in DEPC-treated autoclaved water (0.5-1 ml/g starting tissue).
19. Store at –80°C until use.

Concentration and Quality Determination

Use a spectrophotometer to measure absorption at (260 nm) and ratio (A_{260} vs. A_{280} nm) of total RNA. 1 OD_{260} = approx 40 µg/ml. Typical yield is between 500-700 µg of total RNA per gram of leaf tissue. The $A_{260} : A_{280}$ value should fall in the range of 1.7-2.1.

A critical aspect of poly(A) RNA isolation is that RNAs from both target and driver tissues have to be- completely DNA-free. Even a minor contamination (few nanograms) of RNA by genomic DNA will lead to the generation of a great number of artefactual clones in a subtractive cDNA library, which contains inserts of genomic DNA. Total RNA always contains the remains of genomic DNA, regardless of what method has been used to extract RNA. Affinity chromatography on oligo(dT) cellulose significantly reduces contamination of RNA by genomic DNA. None the less, treatment of poly(A) RNA with DNAse I to eliminate genomic DNA completely is a critical requirement.

Protocol 2. Purification of Poly (A)⁺ from Target and Driver Total RNA

Reagents and Solutions required

- ☆ Biotinylated oligo-dT primer (T_{20})
- ☆ Streptavidin linked paramagnetic beads (Roche; Catalogue number 11641786001)
- ☆ 10 mM Tris-Cl pH 7.5 (made in DEPC treated autoclaved water)
- ☆ 0.5 M KCl (made in DEPC treated autoclaved water)
- ☆ 10 x TM (500 mM Tris, pH7.5, 70 mM $MgCl_2$)

Methods

1. Dissolve approximately 500 µg of total RNA completely in 10 mM Tris-Cl pH 7.5 then increase the salt concentration to 0.5 M KCl. [NOTE: make sure that the RNA is completely dissolved before adding KCl or else it is very difficult to dissolve the RNA after the addition of KCl]

2. Add 100 to 200 ng of biotinylated oligo-dT primer to the RNA solution.

3. Heat the contents to 65°C for 5 min and then cool to room temperature.

4. Meanwhile wash and equilibrate streptavidin para-magnetic beads in 10 mM Tris-Cl pH 7.5 containing 0.5 M KCl.

5. Re-suspend the washed beads in the RNA solution. Mix gently. Leave at room temperature for about 5 min.

6. Immobilize the mRNA onto the beads and capture with magnet allowing the remaining RNA to be removed and wash the beads with 10 mM Tris-Cl pH 7.5 containing 0.5 M KCl.

7. Elute the captured mRNA into RNAase free water and separate the beads with magnet.

8. Add 8µl water, 2µl 110 x TM and 1µl DNase 1 (5Uµl^{-1}). Incubate for 30 min at 37 °C.

9. Add 15µl 2M sodium acetate, pH4.0, 115 u.1 water, and extract the DNase-treated poly(A)+ RNA with equal volume of water-saturated phenol.

10. Precipitate poly(A)$^+$ RNA with 2.5 volumes ethanol. Keep poly(A)$^+$ RNA under ethanol at least for 1 day at - 70°C.

11. Spin at 15 000g for 1 h at 4°C, wash poly(A)- RNA pellet twice with *70 per cent* ethanol, dry, and dissolve driver poly (A)$^+$RNA at 1 -5 mg ml^{-1}, and target poly(A)$^+$RNA in 10µl (< mg ml$^{-1)}$ store at –80°C till further use.

Protocol 3. Synthesis of Driver ss-cDNA with Biotinylated Oligo dT (T$_{20}$) Primer

cDNA is prepared by oligo(dT) or random priming. The problem with this method is that large amounts of starting material must be obtained, and only two rounds of subtraction can be performed before the amount of remaining tracer becomes too small. With complex starting tissues, subtraction is never complete at this point.

Five micrograms of poly(A)+ RNA from driver samples will be used to synthesize 1st strand cDNA after priming with 5′-biotinylated oligo dT (T$_{20}$) primer using Superscript II or Superscript III reverse transcriptase (Invitrogen; Catalogue number 18080044) following the protocol mentioned below:

Annealing 5′-biotinylated oligo dT (T$_{20}$) primer to mRNA:

mRNA (5 µg)	10.0 µl
Biotinylated oligo dT primer (1-2 µg)	1.0 µl

Mix mRNA and 5′-biotinylated oligo dT (T$_{20}$) primer, heat to 70°C for 5 min and cool on ice for about 5 mins. Add the following chemicals in the order given below:

5X first strand buffer	10.0 µl
10 mM DTT solution	5.0 µl
20 mM dNTP mix	2.0 µl
RNaseOUT (Invitrogen)	1.0 µl
Superscript III (200 U/ml)	1.0 µl
RNase free water	20.0 µl
Total	*50.0 µl*

Mix contents gently and centrifuge briefly. Incubate at 42°C for 1 h.

Heat inactivate the Superscript III by incubating the tube at 70°C for 10 min.

Allow the sample to come to room temperature.

Add RNaseH (5 units/µl)	1.0 µl

Incubate at 37°C for 15 min

Heat inactivate by heating the reaction to 65°C for 20 min.

The 1st strand cDNA is purified and excess oligo dT primers removed using Qiagen purification column (catalogue number 28104). The samples are eluted in 25 µl of TE (10 mM Tris-HCl, pH 8.0; EDTA 1 mM).

Subtractive Hybridization

An important parameter controlling the success of the hybridization is the tester: driver ratio, which should be at least 1:10 to allow the driver to govern the subtraction. Other factors include the absolute concentration of driver and the time allowed for hybridization, both of which should be as large as practical because both factors affect the Rot or Cot (for DNA driver) that is achieved and therefore determine the extent of hybridization. In general, a Rot of at least 1000 should be obtained. The hybridization time is limited by the degradation of driver and tester during hybridization, particularly when the driver is RNA.

It is also important to consider whether the driver and tester will be single or double stranded. Single-stranded driver is the most efficient choice, since the concentration of driver decreases only slightly as driver hybridizes to tracer, which allows high Rot or Cot values to be reached. In contrast, when double stranded driver is used, driver-driver and driver-tracer duplexes form during hybridization. Driver-driver duplex formation competes with the desired reaction, and over time, decreases the concentration of driver available, reducing the efficiency of subtraction. As a result, more rounds of subtraction must be performed with a double-stranded driver than with a single-stranded driver to obtain equivalent subtraction.

Protocol 4. Subtractive Hybridization

Preparation of differentially up-regulated mRNA molecules by cDNA subtraction.

In a 1.5 ml silanised Eppendorf tube add:

Tester mRNA (1 µg/µl)	5.0 µl
5′-biotinylated 1st strand cDNA from driver sample (5 µg) (Protocol.3)	20.0 µl
2X RNase free hybridization buffer (20 mM Tris-Cl pH 7.5, 1.0 M KCl, 10 mM EDTA)	25.0 µl
DEPC treated water (if required, to make volume up to 50 µl)	
	_____
Total	*50.0 µl*

Incubate (Hybridize) at 68°C for 2 h.

Protocol 5. Separation of Tester Specific Nucleic Acids using Streptavidin-Coated Magnetic Beads

There are two different approaches to separating target mRNA with driver nucleic acids.

1. Chromatography on a hydroxyapatite column provides a way to separate ss- DNA and/or RNA from ds-DNA or/and RNA-DNA duplex (Hedrick *et al.,* 1984; Travis and Sutcliffe 1988).

2. Biotinylated driver nucleic acids can be easily separated from as target driver nucleic acid duplex and removed by phenol-chloroform extraction when complexed with streptavidin (Sive and St John. 1988).

The use of biotinylated driver nucleic acids offers several advantages.

1. Chromatography on a hydroxyapatitc column is much more difficult to perform than the removal of driver nucleic acids with the streptavidin/ phenol-chloroform method (Sive and St John. 1988).

2. Chromatography on a hydroxyapatite column is not highly effective, and 30-50 per cent ssDNA or RNA may irreversibly be attached to the hydroxyapatite.

3. Hydroxyapatite does not effectively separate ss-DNA from ds-DNA or RNADNA duplex, and about 1-2 per cent of ds-DNA or RNA-DNA may contaminate single stranded nucleic acids.

Altogether, the employment of biotinylated driver nucleic acids to separate target mRNA effectively from driver nucleic acids is favoured over the use of Chromatography on a hydroxyapatite column.

Preparation of Streptavidin-Coated Magnetic Beads

Take 50 µl of streptavidin-coated magnetic beads (Roche; Catalogue number 11641786001) after vortexing the stock briefly. Wash the beads twice with 0.5 ml of RNase free 1X hybridization buffer. [To remove the wash solution place the Eppendorf tube on the magnetic stand and allow to stand for 30-60 sec. Carefully remove the solution. Remove the tube from the stand and add a fresh volume of RNase free 1X hybridization buffer volume. Mix beads gently or vortex and remove the buffer, as above, by keeping the tube on the magnetic stand]. Re-suspended the beads in 50 µl of RNase free 1X hybridization buffer. Pre-incubate for 20 min at room temperature with 40 µg of yeast tRNA to block all possible non-specific binding sites. After blocking, wash the magnetic beads thrice with RNase free 1X hybridization buffer.

Purification of Differentially Up-Regulated mRNA

Mix pre-washed streptavidin-coated magnetic beads with hybridization samples (1st strand cDNA from control sample mixed with mRNA of stressed sample) and incubate further for 20 min at 68°C.

The poly(A) RNA-cDNA hybrids and unhybridized cDNA population get immobilized on to the beads. The beads are then separated on the magnetic stand and the unbound differentially up-regulated poly(A) RNAs in the solution is transferred to a fresh silanized tube and precipitated with 2 volumes of ethanol.

Protocol 6. Construction of subtractive cDNA library using ZAP-cDNA Gigapack III Gold Cloning Kit.

Here we have described the protocol for construction of Subtracted cDNA library using ZAP-cDNA Gigapack III Gold Cloning Kit (Stratagene catalog number 200450) according to the manufacturer's protocol.

The template is tester mRNA and the primer is a 50-base oligonucleotide with the following sequence:

5′GAGAGAGAGAGAGAGAGAGAGAACTAGT**CTCGAG**TTTTTTTTTTTTTTTTTT3′ "GAGA" Sequence *Xho* poly(dT). This oligonucleotide was designed with a "GAGA" sequence to protect the *Xho* I restriction enzyme recognition site and an 18-base poly(dT) sequence. The restriction site allows the finished cDNA to be inserted into the Uni-ZAP XR vector1 in a sense orientation (*EcoR* I-*Xho* I) with respect to the *lacZ* promoter. The poly(dT) region binds to the 3′ poly(A) region of the mRNA template, and AccuScript RT begins synthesis of first-strand cDNA. The nucleotide mixture for the first strand contains normal dATP, dGTP, and dTTP plus the analog 5-methyl dCTP. The complete first strand will have a methyl group on each cytosine base, which will protect the cDNA from restriction enzymes used in subsequent cloning steps.

During second-strand synthesis, RNase H nicks the RNA bound to the firststrand cDNA to produce a multitude of fragments, which serve as primers for DNA polymerase I. DNA polymerase I "nick-translates" these RNA fragments into second-strand cDNA. The second-strand nucleotide mixture has been supplemented with dCTP to reduce the probability of 5-methyl dCTP incorporation into the second strand. This ensures that the restriction sites in the linker–primer will be susceptible to restriction digestion. The uneven termini of the double-stranded cDNA are nibbled back or filled in with cloned *Pfu* DNA polymerase, and *EcoR* I adapters, with the sequence shown below, are ligated to the blunt ends.

5'-OH-AATTCGGCACGAGG-3'

3'-GCCGTGCTCCp-5'

These adapters are composed of 10- and 14-mer oligonucleotides, which are complementary to each other with an *Eco*R I cohesive end. The 10-mer oligonucleotide is phosphorylated, which allows it to ligate to other blunt termini available in the form of cDNA and other adapters. The 14-mer oligonucleotide is kept dephosphorylated to prevent it from ligating to other cohesive ends. After adapter ligation is complete and the ligase has been heat inactivated, the 14-mer oligonucleotide is phosphorylated to enable its ligation to the dephosphorylated vector arms. The *Xho* I digestion releases the *Eco*R I adapter and residual linker–primer from the 3´ end of the cDNA. These two fragments are separated on a drip column containing Sepharose® CL-2B gel filtration medium. The size fractionated cDNA is then precipitated and ligated to the Uni-ZAP XR vector.

Synthesis of ss-cDNA from Tester Specific mRNA

Two choices need to be made for the preparation of target ss-cDNA: the type of primer to be used for ss-cDNA synthesis and the method of removing primers from the cDNA, A number of different primers—oligo(dT), adapter-primer, random-primer—can be used to synthesize ss-cDNA. However, random primer-synthesized cDNA is relatively difficult to clone, because ligation of ds-cDNA with adapters and attendant processes, like phospnorylation of cDNA and separating cDNA from free adapters, have to be carried out. Moreover, an entire coding region is more likely to be cloned from an oligo(dT)- primed cDNA subtractive library rather than from a random-primed cDNA library. A further advantage of oligo(dT)-primed ss-cDNA is that the heterogeneous cDNA pool can be amplified by the polymerase chain reaction (PCR) using non-specific primers, such as oligo(dT) and oligo(dC). To simplify the cloning of" target ss-cDNA after subtractive hybridization, the use of adapter-primers is preferable to oligo(dT), The best plan to be followed is to include sequences of oligo(dT) with a XhoI site in an adapter-primer (see Protocol *3).* Such adapter-primers anneal to the poly(A)-tail of mRNA. Therefore, the resulting subtracted target ds-cDNA is unlikely to be cut with XhoI during the process of generating XhoI sites for ligation into appropriate vectors.

Synthesizing First-Strand cDNA from Tester Specific (Subtracted) mRNA

1. Preheat a 42°C water bath.
2. The final volume of the first-strand synthesis reaction is 50 µl.
3. In an RNase-free microcentrifuge tube, add the following reagents in order:

   5 µl of 10× first-strand buffer

   3 µl of first-strand methyl nucleotide mixture

   2 µl of linker–primer (1.4 µg/µl)

   11 µl of DEPC-treated water

   1 µl of RNase Block Ribonuclease Inhibitor (40 U/µl) Mix the reaction and then add

   25 µl of tester specific poly(A)+ RNA (subtracted) (5 µg) (Protocol).

4. Mix gently. Allow the primer to anneal to the template for 10 minutes at room temperature.

5. Add 3 μl of AccuScript RT to the first-strand synthesis reaction. The final volume of the first-strand synthesis reaction should now be 50 μl.

6. Mix the sample gently and spin down the contents in a microcentrifuge.

7. Incubate the first-strand synthesis reactions, including the control reaction, at 42°C for one hour.

8. After 1 hour, remove the first-strand synthesis reactions from the 42°C water bath. Place the first-strand synthesis reaction on ice. On this gel, run the 5μl first-strand reaction.

Synthesizing Second-Strand Subtracted Tester Specific cDNA

1. Thaw all nonenzymatic second-strand components. Briefly vortex and spin in a microcentrifuge before placing the tubes on ice.

   Note: It is important that all reagents be <16°C when the DNA polymerase I is added.

2. Add the following components in order to the 45-μl first-strand synthesis reaction on ice:

   20 μl of 10× second-strand buffer

   6 μl of second-strand dNTP mixture

   116 μl of sterile distilled water (DEPC-treated water is not required)

   Add the following enzymes to the second-strand synthesis reaction:

   2 μl of RNase H (1.5 U/μl)

   11 μl of DNA polymerase I (9.0 U/μl)

3. Gently vortex the contents of the tube, spin the reaction in a microcentrifuge, and incubate for 2.5 hours at 16°C. Check the water bath occasionally to ensure that the temperature does not rise above 16°C. Temperatures above 16°C can cause the formation of hairpin structures, which are unclonable and interfere with the efficient insertion of correctly synthesized cDNA into the prepared vector.

4. After second-strand synthesis for 2.5 hours at 16°C, immediately place the reaction tube on ice.

Blunting the cDNA Termini

1. Add the following to the second-strand synthesis reaction:

   23 μl of blunting dNTP mix

   2 μl of cloned *Pfu* DNA polymerase (2.5 U/μl)

2. Quickly vortex the reaction and spin in a microcentrifuge. Incubate the reaction at 72°C for 30 minutes. **Do not exceed 30 minutes!!**

3. Thaw the 3 M sodium acetate.

4. Remove the reaction from incubation at 72°C, then add 200 μl of phenol–chloroform [1:1 (v/v)] and vortex the mixture.

   Note: The phenol must be equilibrated to pH 7–8.

5. Spin the reaction in a microcentrifuge at maximum speed for 2 minutes at room temperature and transfer the upper aqueous layer, containing the cDNA, to a new tube. Be careful to avoid removing any interface that may be present.

6. Add an equal volume of chloroform and vortex the mixture.

7. Spin the reaction in a microcentrifuge at maximum speed for 2 minutes at room temperature and transfer the upper aqueous layer, containing the cDNA, to a new tube.

8. Precipitate the cDNA by adding the following to the saved aqueous layer:

   20 μl of 3 M sodium acetate

   400 μl of 100 per cent (v/v) ethanol

   Vortex the reaction.

9. Precipitate overnight at –20°C.

10. In order to orient the direction of precipitate accumulation, place a mark on the microcentrifuge tube or point the tube hinge away from the center of the microcentrifuge as an indicator of where the pellet will form.

11. Spin in a microcentrifuge at maximum speed for 60 minutes at 4°C.

12. Avoid disturbing the pellet and carefully remove and discard the supernatant.

    Note: The conditions of synthesis and precipitation produce a large white pellet. The pellet accumulates near the bottom of the microcentrifuge tube and may taper up along the marked side of the tube.

13. Gently wash the pellet by adding 500 μl of 70 per cent (v/v) ethanol to the side of the tube away from the precipitate. **Do not mix or vortex!**

14. Spin in a microcentrifuge at maximum speed for 2 minutes at room temperature with the orientation marked as in step 10.

15. Aspirate the ethanol wash and dry the pellet by vacuum centrifugation.

16. Resuspend the pellet in 9 μl of *Eco*R I adapters and incubate at 4°C for at least 30 minutes to allow the cDNA to resuspend. To ensure that the cDNA is completely in solution, transfer the cDNA to a fresh microcentrifuge tube.

17. Transfer 1 μl of this second-strand synthesis reaction to a separate tube.

    At this point, run the samples of the first- and second-strand synthesis reactions on an alkaline agarose gel.

    Note: The second-strand synthesis reaction can be stored overnight at –20°C.

Ligating the *Eco*R I Adapters

1. Add the following components to the tube containing the blunted cDNA and the *Eco*R I adapters:

   1 µl of 10× ligase buffer

   1 µl of 10 mM rATP

   1 µl of T4 DNA ligase (4 U/µl)

2. Spin down the reaction in a microcentrifuge and incubate overnight at 8°C. Alternatively, the ligation can be incubated at 4°C for 2 days.

3. In the morning, heat-inactivate the ligase by placing the tubes in a 70°C water bath for 30 minutes.

Phosphorylating the *Eco*R I Ends

1. After the ligase is heat inactivated, spin the reaction in a microcentrifuge for 2 seconds. Cool the reaction at room temperature for 5 minutes.

2. Phosphorylate the adapter ends by adding the following components:

   1 µl of 10× ligase buffer

   2 µl of 10 mM rATP

   5 µl of sterile water

   2 µl of T4 polynucleotide kinase (5 U/µl)

3. Incubate the reaction for 30 minutes at 37°C.

4. Heat-inactivate the kinase for 30 minutes at 70°C.

5. Spin down any condensation in a microcentrifuge for 2 seconds and allow the reaction to equilibrate to room temperature for 5 minutes.

Digesting with *Xho* I

1. Add the following components to the reaction:

   28 µl of *Xho* I buffer supplement

   3 µl of *Xho* I (40 U/µl)

2. Incubate the reaction for 1.5 hours at 37°C.

3. Add 5 µl of 10× STE buffer and 125 µl of 100 per cent (v/v) ethanol to the microcentrifuge tube.

4. Precipitate the reaction overnight at −20°C.

5. Following precipitation, spin the reaction in a microcentrifuge at maximum speed for 60 minutes at 4°C.

6. Discard the supernatant, dry the pellet completely, and resuspend the pellet in 14 µl of 1× STE buffer.

7. Add 3.5 µl of the column loading dye to each sample.

The sample is now ready to be run through a drip column containing Sepharose CL-2B gel filtration medium.

Size Fractionating

Before attempting the experimental protocols outlined within this section, please read this section in its entirety in order to become familiar with the procedures. Review of the *Troubleshooting* section may also prove helpful. The drip columns should be prepared and the cDNA should be eluted in a single day. Because a full day is required to complete these procedures, gathering all necessary materials in advance is recommended.

Assembling the Drip Column

1. Perform the following preparatory steps while assembling the drip columns:
    a. Remove the Sepharose CL-2B gel filtration medium and the 10× STE buffer from refrigeration and equilibrate the two components to room temperature.
    b. Prepare 50 ml of 1× STE buffer by diluting 10× STE buffer 1:10 in sterile water.

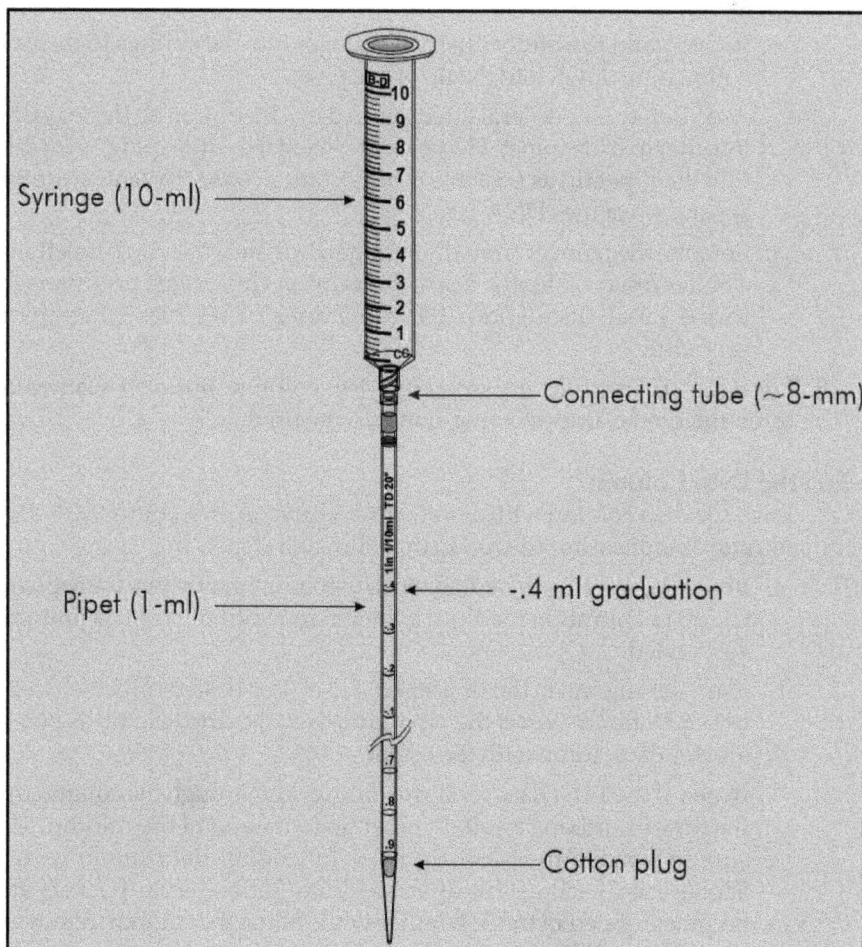

2. Assemble the drip columns as outlined in the following steps (see diagram of the final setup):

   Note Wear gloves while assembling the drip columns.

   a. Remove the plastic wrapper from the top of a sterile 1-ml pipet.

   b. Using a sterile needle or a pair of fine-tipped forceps, **carefully** tease the cotton plug out of each pipet, leaving a piece of the cotton plug measuring ~3–4 mm inside. Cut off the external portion of the cotton plug.

   c. Push the remaining 3- to 4-mm piece of the cotton plug into the top of each pipet with the tip of the needle or forceps.

   d. Cut a small piece of the connecting tubing measuring ~8 mm. Use this small tube to connect the 1-ml pipet to the 10-ml syringe. First attach one end of the connecting tube to the pipet and then connect the other end to the syringe. There should be no gap between the pipet and the syringe when joined by the connecting tube.

   e. Rapidly and forcefully push the plunger into the syringe to thrust the cotton plug down into the tip of the pipet.

      Note: It may take several attempts to drive the cotton all the way down into the tip of the pipet. However, pushing the cotton plug as far down into the pipet tip as possible is important in order to achieve optimal separation of the cDNA fractions.

   f. Remove the plunger from the syringe. Because the syringe functions as a buffer reservoir for the drip column, leave the syringe firmly attached to the pipet throughout the remainder of the size fractionation procedure.

3. Locate a support for the assembled drip column. Butterfly clamps or a three-fingered clamp on a ring stand can be used.

Loading the Drip Column

1. Load the drip column with a uniform suspension of Sepharose CL-2B gel filtration medium as outlined in the following steps:

   a. Immediately prior to loading the drip column, gently mix the Sepharose CL-2B gel filtration medium by inversion until the resin is uniformly suspended.

   b. Place the column in the ring stand. Fill a glass Pasteur pipet with ~2 ml of 1× STE buffer. Insert the pipet as far into the drip column as possible and fill the column with the buffer.

      Notes: If the 1× STE buffer flows too quickly through the column, stem the flow by affixing a yellow pipet tip to the end of the column. Make sure to remove the pipet tip prior to loading the column with the Sepharose CL-2B gel filtration medium. If bubbles or pockets of air become trapped in the STE buffer while filling the column, remove the trapped air prior to packing the column with the resin. To remove the bubbles or air, re-insert the Pasteur pipet into the top of the column and gently pipet the STE buffer in and out of the pipet until the trapped air escapes through the top of the column.

c. Immediately add a uniform suspension of Sepharose CL-2B gel filtration medium to the column with a Pasteur pipet by inserting the pipet as far into the column as possible. As the resin settles, continue adding the Sepharose CL-2B gel filtration medium. Stop adding the resin when the surface of the packed bed is ¼ inch below the "lip of the pipet." The lip of the pipet is defined as the point where the pipet and the syringe are joined.

Notes: If air bubbles form as the resin packs, use a Pasteur pipet as described in step 1b to remove the blockage. Failure to remove bubbles can impede the flow of the column and result in a loss of the cDNA. If the preparation of Sepharose CL-2B gels filtration medium settles and becomes too viscous to transfer from the stock tube to the column, add a small volume (~1–5 ml) of 1× STE buffer to resuspend the resin.

2. Wash the drip column by filling the buffer reservoir (*i.e.*, the syringe) with a minimum of 10 ml of 1× STE buffer. As the column washes, the buffer should flow through the drip column at a steady rate; however, it may take at least 2 hours to complete the entire wash step. After washing, do not allow the drip column to dry out, because the resin could be damaged and cause sample loss. If this occurs, pour another column.

Note: If a free flow of buffer is not observed, then bubbles or pockets of air have become trapped in the drip column. In this case, the column must be repacked. If cDNA is loaded onto a column on which a free flow of buffer is not observed, the sample could become irretrievably lost.

3. When ~50 µl of the STE buffer remains above the surface of the resin, immediately load the cDNA sample using a pipettor. Gently release the sample onto the surface of the column bed, but avoid disturbing the resin as this may affect cDNA separation.

4. Once the sample enters the Sepharose CL-2B gel filtration medium, fill the connecting tube with buffer using a pipettor.

Note: Do not disturb the bed while filling the connecting tube with buffer.

Gently add 3 ml of 1× STE buffer to the buffer reservoir by trickling the buffer down the inside wall of the syringe. Do not squirt the buffer into the reservoir because this will disturb the resin, resulting in loss of the sample.

5. As the cDNA sample elutes through the column, the dye will gradually diffuse as it migrates through the resin. Because the dye is used to gauge when the sample elutes from the column, monitor the progress of the dye, or the cDNA sample could be irretrievably lost.

Collecting the Sample Fractions

The drip column containing the Sepharose CL-2B gel filtration medium separates molecules on the basis of size. Large cDNA molecules elute first followed by smaller cDNA and finally unincorporated nucleotides. Although this material elutes from the column in parallel with the dye, unincorporated nucleotides are usually not collected because the cDNA has already eluted from the column. For standard cDNA size fractionation (>400 bp), collect ~12 fractions using the procedure described in this section. The progression of the leading edge of the dye through the column will be used as a guideline to monitor collection. Until the fractions have been assessed for the presence of cDNA on a thin 0.8 per cent agarose gel electrophoresis do not discard any fractions.

1. Using a fresh microcentrifuge tube to collect each fraction, begin collecting three drops per fraction when the leading edge of the dye reaches the –.4 ml graduation on the pipet.

2. Continue to collect fractions until the trailing edge of the dye reaches the.3 ml graduation. A minimum of 12 fractions, each containing ~100 µl (*i.e.*, three drops), should be collected.

3. Before processing the fractions and recovering the size-fractionated cDNA, remove 8 µl of each collected fraction and save for later analysis. These aliquots will be electrophoresed on a 0.8 per cent agarose gel electrophoresis to assess the effectiveness of the size fractionation and to determine which fractions will be used for ligation.

Processing the cDNA Fractions

In this section of the size fractionation procedure, the fractions collected from the drip column are extracted with phenol–chloroform and are precipitated with ethanol to recover the size-selected cDNA. The purpose of the organic extractions is to remove contaminating proteins; of particular concern is kinase, which can be carried over from previous steps in the synthesis. Because kinase often retains activity following heat treatment, it is necessary to follow the extraction procedures.

1. Begin extracting the remainder of the collected fractions by adding an equal volume of phenol–chloroform [1:1 (v/v)].

2. Vortex and spin in a microcentrifuge at maximum speed for 2 minutes at room temperature. Transfer the upper aqueous layer to a fresh microcentrifuge tube.

3. Add an equal volume of chloroform.

4. Vortex and spin in a microcentrifuge at maximum speed for 2 minutes at room temperature. Transfer the upper aqueous layer to a fresh microcentrifuge tube.

5. To each extracted sample, add a volume of 100 per cent (v/v) ethanol that is equal to twice the individual sample volume.

   Note The 1× STE buffer contains sufficient NaCl for precipitation.

6. Precipitate overnight at –20°C.

7 Spin the sample in the microcentrifuge at maximum speed for 60 minutes at 4°C. Transfer the supernatant to another tube.

8. Carefully wash the pellet with 200 μl of 80 per cent (v/v) ethanol, ensuring that the pellet remains undisturbed. *Do not mix or vortex!* Spin the sample in a microcentrifuge at maximum speed for 2 minutes at room temperature. Remove the ethanol and verify that the pellet has been recovered by visual inspection. Vacuum evaporate the pellet for ~5 minutes or until dry. Do not dry the pellet beyond the point of initial dryness or the cDNA may be difficult to solubilize and resuspend the cDNA in 5 μl of sterile water. Mix by pipetting up and down.

To help ensure ligation success, quantitate the cDNA before proceeding.

Best results are usually obtained by ligating 100 ng of cDNA/1 μg of vector. Place the remaining cDNA at –20°C for short term storage only. The cDNA is most stable after ligation into vector arms and may be damaged during long-term storage.

Ligating the cDNA Insert

The Uni-ZAP XR vector arms are shipped in 10 mM Tris-HCl (pH 7.0) and 0.1 mM EDTA and can be stored up to 1 month at 4°C or frozen in aliquots at –20°C for longer storage. The pBR322 test insert should be stored at –20°C. However, do not put samples through multiple freeze-thaw cycles.

1. Set up a control ligation to ligate the test insert into the Uni-ZAP XR vector as follows:

   Prepare the sample ligation in a separate tube as follows:

   2.5 μl of resuspended cDNA (~100 ng)

   0.5 μl of 10× ligase buffer

   0.5 μl of 10 mM rATP (pH 7.5)

   1.0 μl of the predigested Uni-ZAP XR vector (1 μg)

   Then add

   0.5 μl of T4 DNA ligase (4 U/μl)

2. Incubate the reaction tubes overnight at 12°C or for up to 2 days at 4°C.

In vitro Packing Using Packaging Extraction

After ligation is completed, 4 μl of ligation reaction package in to XL1-Blue MRF ÿlIÿcells using Gigapack III Gold packaging extract according to the packaging instructions. Phage and the bacteria incubate at 37°C for 15 minutes to allow the phage to attach to the bacterial host cells. Add 3 ml of NZY top agar to the packaging reaction and allow to cool up to ~48°C, and immediately plate on to dry prewarmed NZY agar plates. Allow the plates to set for 10 minutes, after that invert the plates and incubated at 42°C. After 6 hours visible plaques will be observed on plates. Randomly selected plaques from the agar plate were transferred in to a sterile microcentrifuge tube containing 500 μl of SM buffer and 20 μl of chloroform and store at -80°C for further analysis.

Protocol 7. Analysis of Subtracted Library

Presence of Insert

Colonies containing insert can be screened by subjecting the isolated plaques directly to polymerase chain reaction (PCR). Colony PCR can be carried out by taking 2µl single individual plaque which was dissolved in SM buffer, as a template and amplify by PCR using M13 universal primers forward (5′-GTAAAACGACGGCCAGT-3′) and reverse primers (5′-GGAAACAGCTATGACCATG-3′) complementary to vector sequences flanking the cDNA inserts, PCR can be performed in 96 well microplate. The 100 ml reaction mixture contains 2µl of each primer (150ng/µl), 10 µl of 10X Taq-polymerase assay buffer, 2 µl (10 mM) dNTPs, and two units of Taq-polymerase enzyme. The inserts can be amplified in programmed thermal cycler using amplification programe of 94⁰C for 5 minutes, 94⁰C for 1 minute, 55⁰C for 1 minute, 72⁰C for 1 minute, with the final extension at 72⁰C for 10 minutes and the cycle can be repeated for 30 times. After the reaction the product can be separated on 0.8 per cent agarose gel to identify the positive clones.

Differential and Confirmatory Screening of Subtractive cDNA Library

To detect clones bearing cDNA inserts, which constitute target-specific transcripts, special screening, named differential screening, should be performed. The principle behind differential screening is the separate hybridization of the cDNA library with a target-derived probe as well as with a driver-derived probe. The cDNA inserts, which are detected by hybridization with a target-derived probe, but not a driver-derived probe, could be considered as target-specific transcripts. The original method of differential screening of a conventional cDNA library has been described by Dworkin and Dawid (1980).

Basic Analysis of Specific Transcripts

To complete the process of isolation of specific transcripts, three different types of analyses should be executed:

1. In order to eliminate redundant clones and select clones bearing the longest insert, the insert probes are random-prime labelled and cross-hybridized with each other. By this means the range of independent clones is isolated.

2. The clones obtained from the subtractive cDNA library after differential screening may be expressed exclusively, preferentially or non-specifically in the target tissue. This question can be answered through Northern blot and/or in situ hybridization analysis of the selected inserts. Northern blot analysis enables us to determine not only the degree of specificity of the selected clones, but also the sizes of transcripts. This in turn allows us to estimate how many of the selected clones are full-length. In situ hybridization offers a way to assess cell-specificity of the selected transcripts. Furthermore, *in situ* hybridization can be used when tissue is limiting for making a Northern blot.

3. In order to determine the nature of transcripts, sequence analysis of the 5'-end of selected inserts should be carried out. Sequence analysis may assign each transcript to one of following groups: known tissue-specific transcripts, known transcripts expressed preferentially in target tissue, known non-specific transcripts, novel tissue-specific transcripts which contain known motifs), novel transcripts with recognizable motifs) expressed preferentially in target tissue and new transcripts with non-recognizable motifs). For new transcripts with non-recognizable motifs), the isolation of full-length cDNA is recommended, because sequencing of the full-length clones may provide fresh insights into the nature of these transcripts.

Conclusion

The gene enrichment strategy used in this study increases the representation of differentially up-regulated cDNA populations. The procedure consists of primarily a subtractive hybridization between two complimentary single stranded nucleic acid populations. The mRNA populations from tester population and approximately five-fold excess of complimentary 1st strand cDNA from driver population were heat denatured and subjected to limited renaturation with empirically determined Cot value. The hybridization of mRNA strands belonging to tester with the complementary cDNA strands belonging to driver obeys the ideal pseudo-first order kinetics. The hybridization kinetics is sharper as compared to the second order kinetics, where two different double stranded nucleic acids (from tester and control) were together subjected to hybridization after denaturation. The post-cloning duplication of recombinant clones in cDNA library construction (amplification) was also avoided by selecting individual recombinant phages from the primary library without amplification. This further reduces the probability of repeat selecting and subsequent sequencing of redundant clones and therefore the efforts can be better directed towards sequencing clones with low redundancy.

Acknowledgements

Corresponding author would like to sincerely thank his Ph.D. and Postdoctoral advisors, Prof.Chinta Sudhakar, Department of Botany, Sri Krishnadevaraya University, Anantapur and Dr.M.K.Reddy, Senior Research Scientist, Plant Molecular Biology Group, ICGEB, New Delhi for introducing the topic of Subtractive Hybridization. Protocol described in this chapter is adopted from Dr.M.K. Reddy laboratory, ICGEB, New Delhi. This work in C.O.R.P. laboratory was supported by the DBT, DST and C.S.I.R., Government of India.

References

Buntjer, J., Ruiter, M.D., Vallins, B., Minarik, M., Mahtani, M., Schaik, R.V., Vos, P. and Eijk, M.V. 2001. High throughput analysis of AFLP fragments on the MegaBASE capillary sequencer. In: Plant and animal genome IX conference. January 13–17, 2001, San Diego, CA, USA.

Chang, D. and Denny, C, 1998, Isolating differentially expressed genes by representational difference analysis, in: P. Siebert, J. Larric (Eds.), Gene Cloning and Analysis by RT–PCR, BioTechniques Books, Natick, MA, pp. 193–202.

Chien, Y.H., Gascoigne, N.R., Kavaler, J., Lee, N.E. and Davis, M.M. 1984. Somatic recombination in a murine T-cell receptor gene. Nature. 24; **309(5966):** 322–326.

Cho, Y.J., Meade, J.D., Walden, J.C., Chen, X., Guo, Z., Liang, P. 2001. Multicolor fluorescent differential display. Biotechniques.30 **(3):** 562-8, 570, 572.

Dale, A.P., Ian, E.K., Selim, N., Dennis, S., Carolyn, M.K., JeVrey, R.M., Gregory, R. and Kornelia, P. 2001. A SAGE (serial analysis of gene expression) view of breast tumor progression, Cancer Res. **61:** 5697–5702.

Duggan, D.J, Bittner, M., Chen, Y., Meltzer, P.and Trent, J.M, 1999. Expression profiling using cDNA microarrays. Nat Genet. 21(1 Suppl):10-4.

Dworkin, M.B. and Dawid, I.B. 1980. Use of a cloned lthrary for the study of ahundant poly$^{(A)}$RNA during Xenopus laevis development. Dev. Biol. **76:** 449-464.

Hara, E., Kato, T., Nakada, S., Sekiya, S. and Oda, K. 1991. Subtractive cDNA cloning using oligo(dT)30-latex and PCR: isolation of cDNA clones specific to undifferentiated human embryonal carcinoma cells. Nucleic Acids Res. **19:**7097–104.

Hedrick, S.M., Cohen, D.I., Nielsen, E.A. and Davis, M.M. 1984. Isolation of cDNA clones encoding T cell-specific membrane-associated proteins. Nature 8-14; **308(5955):** 149-53.

Hubank, M. and Schatz, D.G. 1994. Identifying differences in mRNA expression by representational difference analysis of cDNA. Nucleic Acids Research 22: 5640–5648.

Liang, P. and Pardee, A.B. 1992. Differential display of eukaryotic messenger RNA by means of the polymerase chain reaction. Science.14; **257(5072):** 967-71.

Lipshutz, R.J., Fodor, S,P., Gingeras, T.R. and Lockhart, D.J. 1999. High density synthetic oligonucleotide arrays. Nat. Genet. **21**(1 Suppl):20-4.

Lockhart, D.J., Dong, H., Byrne, M.C., Follettie, M.T., Gallo, M.V., Chee, M.S., Mittmann, M., Wang, C., Kobayashi, M., Horton, H. and Brown, E.L. 1996. Expression monitoring by hybridization to high-density oligonucleotide arrays. Nat. Biotechnol. **14(13):** 1675-80.

Sargent, T.D., Dawid, I.B. 1983. Differential gene expression in the gastrula of *Xenopus laevis*. Science **222(4620):** 135-9.

Schena, M., Shalon, D., Davis, R.W., Brown, P.O. 1995. Quantitative monitoring of gene expression patterns with a complementary DNA microarray. Science **270(5235):** 467-70.

Sive, H.L. and St John, T. 1988. A simple subtractive hybridization technique employing photoactivatable biotin and phenol extraction. Nucleic Adds Res. **16:** 10937.

Travis, G.H. and Sutcliffe, J.G. 1988. Phenol emulsion-enhanced DNA-driven subtractive cDNA cloning: isolation of low-abundance monkey cortex-specific mRNAs. Proc. Natl. Acad. Sci. USA, 85: 1696

Vos, P. and Eijk, M.V. 2001. High throughput analysis of AFLP fragments on the MegaBASE capillary sequencer. In: Plant and Animal Genome IX, January 13-17, 2001, San Diego, CA.

Vos, P., Hogers, R., Bleeker, M., Reijans, M., van der Lee, T., Hornes, M., Frijters, A., Pot, J., Peleman, J., Kuiper, M. and Zabeau, M. 1995. AFLP: A new technique for DNA fingerprinting. Nucleic Acids Research **18**: 7213-7218.

Wang, Z. and Brown, D.D. 1991. A gene expression screen. Proc. Natl. Acad. Sci. USA **88**:11505–9.

Welford, S.M., Gregg, J., Chen, E., Garrison, D., Sorensen, P.H., Denny, C.T. and Nelson, S.F. 1998. Detection of differentially expressed genes in primary tumor tissues using representation differences analysis coupled to microarray hybridization. Nucleic Acids Research. **26**: 3059–3065.

2013, Abiotic Stress and Biotechnology
Editor: T. Pullaiah
Published by: REGENCY PUBLICATIONS, NEW DELHI

Pages 311–320

Chapter 14

In vitro Propagation of Dwarfing Apple Rootstock Malling 26

P. Sharma, M. Modgil* and M. Thakur

*Department of Biotechnology,
Dr. Y.S. Parmar University of Horticulture and Forestry,
Nauni, Solan – 173 230, India*

ABSTRACT

The present studies were undertaken to develop a workable micropropagation technique of clonal apple rootstock Malling 26 using axillary buds harvested from field trees. Cultures were initiated on MS medium supplemented with 1.0 mg l^{-1} BA, 0.1 mg l^{-1} IBA and 0.5 mg l^{-1} GA$_3$. Phenol exudation, contamination and sterilent toxicity affected the survival and growth of explants. Use of initial liquid cultures was found suitable to rinse the phenols. Shoot multiplication was influenced by the type and concentration of cytokinin. 4-8 fold multiplication with 0.5–3.5 cm long shoots were found on MS medium supplemented with 0.5 mg l^{-1} BA and 0.5 mg l^{-1} Kin which on further supplementing with 0.05 mg l^{-1} IBA slightly increased the number and length of shoots. 98 per cent shoots rooted on medium supplemented with 0.3 mg l^{-1} IBA and 162 mg l^{-1} phloroglucinol when shoots exposed for a brief period. Rooted plantlets resulted in 65 percent success in potting mixture during hardening.

Keywords: Apple rootstocks, Micropropagation, Malling 26.

* Corresponding Author: E-mail: manju_modgil@yahoo.com

Abbreviations

BA: Benzyl adenine

IBA: Indolebutyric acid

GA_3: Gibberelic acid

Kin: Kinetin

MS: Murashige and Skoog

NaOCl: Sodium hypochlorite.

Introduction

Apple is the most ubiquitous of temperate fruits and has been cultivated in Europe and Asia from ancient times. It is the most important temperate fruit crop of India. The economic upliftment of rural masses in the hilly states completely depends upon the temperate fruit industry wherein apple dominates the scenario. Apple fruiting varieties are grafted on two types of rootstocks: the seedling rootstock and clonal rootstock. Seedling rootstocks are propagated from seeds and are highly variable in their behaviour, while clonal rootstocks are propagated by vegetative means and are uniform in performance and precocious in fruit bearing. Different clonal rootstocks of Malling and Malling Merton series have been recommended for commercial use in different states of India.

Malling 26 is a clonal rootstock obtained by cross between Malling 9 x Malling16 and produces a dwarf tree which is the most winter hardy and tolerant to any mineral deficiency. It is suitable for high density plantations in soils where M9 is likely to perform poorly and recommended for commercial use in certain areas of Himachal Pradesh (Kanwar, 1987). The conventional methods of propagation of dwarfing rootstocks are slow and have proved inadequate, and therefore adoption of micropropagation is required to produce uniform and quality planting material on large scale throughout the year. Production of defined clonal rootstocks of apple through micropropagation is a routine practice in several countries of the world and M26 has been micropropagated through shoot meristem or axillary bud culture by a number of research workers outside India (Gjamovski *et al.,* 2009, Le and Collet, 1991, Magyar *et al.*2002). Now, it is desirable to undertake *in vitro* clonal multiplication of improved rootstocks suitable for agroclimatic conditions of apple growing areas of India. We had already initiated some work on micropropagation of Malling (M) and Malling Merton (MM) series of apple rootstocks (Modgil *et al.*1999, Soni *et al.* 2011) and achieved some success. In view of the importance of rootstock M26 for mid hill areas with flat and irrigated land, the present investigation was aimed at *in vitro* clonal multiplication through axillary bud culture.

Materials and Methods

Axillary buds were collected from shoots of ten years old plants of M26, growing in the field of Dr. Y. S. Parmar University of Horticulture and Forestry, Nauni, Solan (H.P.) in early spring. The explants were surface sterilised with 0.1 per cent NaOCl for 10 minutes. All traces of NaOCl were washed off with three rinses of autoclaved distilled water. Explants were cultured onto 20 ml of half strength MS (Murashige and Skoog, 1962) basal liquid/solid medium supplemented with 100 mg l⁻¹ ascorbic acid and 1.0 g l⁻¹ PVP (Polyvinylpyrrolidone) and agitated in rotary shaker for diluting the phenols. The explants were then aseptically placed in 18 x 2.5cm culture tubes having 20ml of agar gelled MS medium supplemented with varying concentrations of 0.5-2.0 mg l⁻¹ benzyladenine, 0.1 mg l⁻¹ indole-3-butyric acid or 0.05-1.0 mg l⁻¹ napthalene-1-acetic acid and 0.5 mg l⁻¹ gibberellic acid, 3 per cent sucrose and 0.7 per cent agar for initiation of bud cultures. Each treatment consisted of thirty buds and experiments were repeated thrice.

Axillary shoots from axenic bud cultures were subcultured for multiplication of shoots on MS medium supplemented with different combinations of 0.5-2.0 mg l⁻¹ BA, 0.5- 1.0 mg l⁻¹ kinetin, 1.0 mg l⁻¹ N⁶(2-isopentyl) adenine, 0.5-1.0 mg l⁻¹ GA₃ and 0.01- 1.0 mg l⁻¹ IBA. Each treatment comprised of ten vessels each with five shoots. Shoots were subcultured on same medium for three passages. Growth data pertaining to number of shoots per explant formed and length of shoots in each subculture were recorded after four weeks of culture. Green and healthy shoots, 2-4 cm long were cut off and transferred to 15 ml rooting medium either for entire rooting period or for brief period (4-7 days) which then transferred to auxin free medium. The medium contained 1/2 MS salts, 0.5 mg l⁻¹ of thiamine, 20 g l⁻¹ sucrose, 4 g l⁻¹ agar and a series of IBA, IAA and NAA concentrations. Some of the medium combinations were enriched with 162 mg l⁻¹ phloroglucinol (PG). Observations were recorded with respect to the number of rooted shoots and number of roots per shoot after four weeks.

Plantlets with adequate roots were washed and treated with fungicide (0.5 per cent carbendazim) for half an hour and planted in paper cups containing cocopeat with 5gl⁻¹ bio control agent (*Trichoderma viridae* defence SP) and were remained covered with plastic bags (with holes for air circulation). The cups were placed in the glasshouse in shade for hardening. The plants were irrigated with water and Knop's nutrient solution twice a week or depending upon the requirement of each plant. Percent survival of acclimatized plants was noted after six weeks.

Results and Discussion

The initial explants from field trees of apple rootstock M26 were difficult to establish *in vitro* because of the phenol exudation, contamination and sterilent toxicity which affected the survival and growth of explants. It has been observed that during the establishment of cultures, around 20 per cent explants were contaminated with micro-organisms and 25 per cent failed to grow because of toxicity caused by tissue browning due to oxidation of phenols and sterilent. Some of the explants which remained green and swollen or showed very late bud break were discarded after two months of culture initiation. The main phenolic compound in apple *i.e.* phloridzin seemed to be involved in the browning reaction of explants (Block and Lankes,1995). When explants were cultured on solid initiation medium containing PVP and ascorbic acid, there is accumulation of phenols around the explants and frequent transfer to fresh medium was required. It introduces an extra labour which is economically disadvantageous. Therefore, to control the browning of explants, we followed our previous approach (Modgil *et al.* 1999, Soni *et al.* 2011) of shaking explants in liquid cultures while, dipping of explants in 0.1µM glutathione solution (GSH) prior to culture prevented the browning in shoot tips of M26 (Nomura *et al.* 1998).

The established buds showed bud break in 3rd week of culture initiation (Figure 14.1a). It is evident from Table 14.1 that the growth regulators composition significantly affected the bud break. Highest bud break was found in two combinations *i.e.* 1.0 mg l^{-1} BA with 0.1 mg l^{-1} IBA and 0.5 mg l^{-1} GA$_3$ (38.31 per cent) and other without GA$_3$ (34.11 per cent). Both treatments were significant and at par with each other. A similar response was observed by Simmonds (1983) who used almost similar growth regulators in M26. This is further supported by Miri *et al* (2003) that maximum bud break in various apple rootstocks was obtained on medium supplemented with GA$_3$ along with BA and IBA but the concentration depends upon the type of genotype.

Table 14.1: Influence of growth regulators, added to MS medium, on axillary bud initiation.

Treatment No.	Concentration of Growth Regulators (mg l^{-1})				Per cent Bud Break after 6 Weeks
	BA	IBA	GA$_3$	NAA	
1	1.0	0.1	–	–	34.17 (35.79)
2	1.0	0.1	0.5	–	38.31 (38.26)
3	0.5	0.1	0.5	–	0.00 (0.00)
4	1.0	–	–	0.1	25.11 (30.08)
5	1.5	–	–	0.1	20.75 (27.11)
6	1.0	–	0.5	0.05	0.00 (0.00)
S.E.					0.57
C.D.$_{0.05}$					1.24

Figures in parenthesis are arc sine transformed values.

Explants varied in their initial growth, with approximately half of the explants continued their growth after initial expansion of leaflets while leaflets of others decayed gradually. The vigorous buds were selected for transfer to fresh medium and developed into mother cultures. The surviving cultures either elongated into small shoots or formed rosette types of leaves (Figure 14.1b) after second subculture. The frequency with which axillary buds developed into shoots in newly initiated cultures is quite low but increased with the number of subcultures and when cultures became adapted, growth of shoots enhanced and became uniform after fifth subculture.

Multiplication rate is the major economic parameter for successful large scale plant production. Shoot multiplication was influenced by the type and concentration of cytokinin and a relationship between concentration of BA, number and size of shoots was observed. Among the different concentrations of BA (0.5-2.0 mg l^{-1}) alone,

Figure 14.1(a-e): *In vitro* propagation of apple rootstock Malling 26, (a) axillary buds showing bud break on MS medium with 1mgl⁻¹ BA, 0.1 mgl⁻¹ IBA and 0.5 mgl⁻¹ GA₃, (b) axillary shoot proliferation on the same medium, (c) *In vitro* shoot multiplication on MS medium with 0.5 mgl⁻¹ each of BA and Kin, (d) *In vitro* root induction, (e) acclimatization of M26 plantlets.

Contd...

Figure 14.1(a-e)–*Contd...*

only 1.0 mg l^{-1} showed one to four fold multiplication with 0.5-2.0 cm long shoots (Table 14.2), which was not found reproducible. There was no multiplication when 0.5 mg l^{-1} BA was combined with 0.1 mg l^{-1} IBA. Four times multiplication rate with 2.5 cm long shoots were observed in most of the cultures, when 1.0 mg l^{-1} BA was supplemented with 0.1 mg l^{-1} IBA and 0.5 mg/l GA$_3$. Multiplication rate increased to eight fold with 0.5-3.5 cm long shoots when 0.5 mg l^{-1} BA was combined with 0.5 mg l^{-1} kinetin. Higher growth and multiplication of shoots (4-9 shoots per explant and upto 4.0 cm length) was recorded when BA and kinetin were further supplemented with 0.05 mg l^{-1} IBA (Table 14.2, Figure 14.1c). Arello *et al.*(1991) and Kataoka and

Inove (1993) support the present findings. Considerable decline in the multiplication rate was observed when 0.5 mg/l GA_3 was used in combination with 1.0 mg l^{-1} BA and 0.5 mg l^{-1} Kinetin. Auxin and gibberellic acid have been used by many laboratories in apple micropropagation but have not been shown to be essential and can inhibit shoot growth at some concentrations (Lane, 1978). Nevertheless, they may have beneficial effects on shoot quality. However, auxins were responsible to induce more shoot length which was better suited for root induction as in present case. It may be beneficial when combined with other growth regulators at specific concentrations. Shoots did not multiply with 2-ip. Thus, it has been found that shoot induction potential of low concentrations of BA alongwith kinetin, and even with very low concentrations of IBA seemed to yield the best growth and multiplication, while BA alone failed to induce such a response. While working with a number of rootstocks and cultivars of apple, we have found that response of BA varies with different cultivars and rootstocks (Modgil and Sharma, 2005).

Table 14.2: Influence of growth regulators added to MS medium on multiplication of shoots.

Treatment No.	Concentration of Growth Regulators (mg l^{-1})					Multiplication Ratio	Length of Shoots (cm)
	BA	Kin	IBA	GA_3	2–ip		
1	0.5	–	–	–	–	1 :1	0.5 – 1.0
2	1.0	–	–	–	–	1 :1–4	0.5 – 2.0
3	2.0	–	–	–	–	1 :1	0.5 – 1.2
4	0.5	0.5	–	–	–	1 : 4–8	0.5 – 3.5
5	0.5	0.5	0.05	–	–	1 : 4–9	0.5 – 4.0
6	1.0	0.5	–	0.5	–	1 : 1	0.5 – 0.9
7	0.5	–	0.1	–	–	1 : 1	0.5 – 1.0
8	0.5	–	0.1	0.5	–	1 : 1	0.2 – 0.5
9	1.0	–	0.1	0.5	–	1 : 1–4	0.5 – 2.5
10	–	–	–	–	1.0	1 : 1	0.5 – 0.7

There was no root initiation with 0.1-0.5 mg l^{-1} IBA. Only swelling and frequent stimulation of callus was observed at the base of shoots, which may be due to high endogenous auxin levels in M26. Therefore, a short auxin treatment period was found necessary. Supplementing phloroglucinol to IBA, roots appeared within two weeks in all the combinations tested. Maximum 98.61 per cent root induction was observed in 0.3 mg l^{-1} IBA with root number 4-15 (Table 14.3, Figure 14.1d). While studying the rooting response to various concentrations of IBA in M26, the highest rooting percentage was obtained with lowest IBA level. The results reported here are similar to numerous workers (Magyar *et al.*2002, Welander, 1991). Supplementing PG in rooting medium seemed to have a synergistic effect with IBA at all the concentrations, as it stimulated rooting and have a beneficial effect by reducing or preventing callus formation at the base of shoots. These results are supported by Jones (1976) who reported that phloridzin and phloroglucinol stimulated rooting in Malling 26. The effect of PG on rooting in a number of apple scion cultivars *in vitro* was evaluated by

Zimmermann and Broome (1981). It has been suggested that the enhanced rooting with phenolics results from these compounds inhibiting IAA – oxidase, thus preventing the destruction of auxin. Growth activity of PG varies with the plant species and even with the different growth phases of same plant. In case, when IBA was combined with IAA, more callus was stimulated, and often inhibited the shoot growth. Although low concentration of NAA (0.1- 0.2 mg l⁻¹) promoted rooting but production of callus caused thick, fleshy and malformed roots whereas root development was inhibited.

Table 14.3: Effect of IBA/NAA in the presence and absence of phloroglucinol (PG) on rooting of M26.

Treatment No.	Concentration of IBA/NAA (mg l^{-1}) IBA	Presence (+)/ Absence (–) of PG (162 mg l^{-1})	Rooting Percentage (per cent)	Number of Roots per Shoot
1	0.2	+	81.33 (64.41)	4 – 10
		–	0 (0.00)	
2	0.3	+	98.67 (84.58)	4 – 15
		–	0 (0.00)	
3	0.4	+	60.67 (51.16)	5 – 11
		–	0 (0.00)	
4	0.5	+	40.00 (39.21)	3 – 10
		–	0 (0.00)	
5	0.6	+	60.00 (50.79)	3 – 12
	NAA	–	0 (0.00)	
6	0.1	–	66.67 (54.74)	3 – 8
7	0.2	–	50.00 (45.00)	3 – 5
S.E.			2.16	
C.D.$_{0.05}$			4.63	

Figures in parenthesis are arc sine transformed values.

From all the experiments, we have seen that if shoots remained throughout in auxin enriched medium, callus produced at the base of the cutting, tended to cause malformed roots and often inhibited shoot growth but if short period treatment in IBA was given, better rooting frequency and good quality roots were developed. The reason may be that root emergence and further growth was inhibited if auxin was present throughout the rooting phase. The importance of a number of factors like strength of basal medium, percentage of agar and use of separate media for root initiation and root elongation has been discussed in several reports (Marin and Marin, 1998, Modgil *et al.* 2009). Because too long treatment of auxin leads to root inhibition, the auxin treatment must be appropriately interrupted by transplanting the root induced shoots on hormone free medium. In most of the cases, this allowed root initiation and avoided callogenesis.

Hardening is a crucial step prior to transplantation of plants to the soil. Plants were established in cocopeat, with a survival frequency of 65 per cent (Figure 14.1e). Cocopeat was found better than soil containing mixtures in our previous studies (Kaushal *et.al.* 2005) because of better aeration, less water retention capacity, moderate pH and lack of any organic components. The humidity maintenance around the plantlets during acclimatization by using polybags was found to be very economic and beneficial, because holes in the bags could be increased to decrease the humidity. It was noticed from the present studies that in order to ensure high percentage and healthy plant growth during hardening, the roots must be emerged without callus and rooted plantlets be healthy. In conclusion, the technique presented here has a potential for multiplication of apple Malling 26 shoots on a large scale or commercial basis.

Acknowledgements

Financial assistance in the form of a research project funded by Department of Biotechnology, Govt. of India, New Delhi is gratefully acknowledged.

References

Arello, E.F., Pasqual, M. and Pinto, J.E.B.P. 1991. Benzylaminopurine in the *in vitro* propagation of the apple rootstocks 'MM111'. Revista-Ceres. **38:** 94-100.

Block, R. and Lankes, C. 1995. Reasons for tissue browning of explants of the apple rootstock M9 during *in vitro* establishment. Gartenbauwissenschaft **60:** 276-279.

Gjamovski, V., Rusevski, R., Popovska, M. and Arsov, T. 2009. Evaluation of different types of media for multiplication and rooting on apple rootstock M26. Acta Horticulturae **825:** 269-272.

Jones, O.P. 1976. Effect of phloridzin and phloroglucinol on apple shoots. Nature **262**: 392-393.

Kanwar, S.M. 1987. Apples Production, technology and economics. Tata Mc Graw Hill Publishing Company Limited, New Delhi, pp.54.

Kataoka, I. and Inove, H. 1993. Micropropagation of Java apple (*Eugenia javanica* Lam.). Jap. J. Tropical Agriculture **37:** 209-213.

Kaushal, N., Modgil, M., Thakur, M. and Sharma, D.R. 2005. *In vitro* clonal multiplication of an apple rootstock by culture of shoot apices and axillary buds. Indian Journal of Experimental Biology **43:** 561-565.

Lane, W.D. 1978. Regeneration of apple plants from shoot meristem tips. Plant Science Letters **13:** 281-285.

Le, C.L. and Collet, G.F. 1991. Micropropagation of apple rootstocks III, Acclimatization of *Malus pumila* Mill. (M26, MAC9) and of *Malus domestica* Borkh. cv. Golden Delicious. Revue Suisse de Viticulture d Arboriculture et d'Horticulture **23:** 201-204.

Magyar, T.K., Dobranszki, J., Jambor, D.E., Lazanyi, J., Coalai, J. and Ferenemy A. 2002. Effects of indole-3-butyric acid levels and activated charcoal on the rooting of *in vitro* shoots of apple rootstocks. Int.J. Hort. Sci. **8:** 25-28.

Marin, M.L. and Marin, J.A. 1998. Excised rootstock roots cultured *in vitro*. Plant Cell Reports **18**: 350-355.

Miri, S.M., Livari, B.V., Khalighi, A. and Maghami, S.A.F. 2003. Effect of carbohydrate, gibberellic acid, indole butyric acid, phloroglucinol, explant orientation and culture vessels volume on optimizing *in vitro* propagation of M9 apple rootstock. Pajouhesh va Sazandegi. In: Agronomy and Horticulture **59**: 31-37.

Modgil, M. and Sharma, D.R. 2005. Micropropagation of apple. In: K.L. Chadha and R.P. Awasthi (eds.). Apple. The Malhotra Publishing House, New Delhi, pp. 112-136.

Modgil, M., Sharma, D.R. and Bhardwaj, S.V. 1999. Micropropagation of apple cv. Tydeman's Early Worcestor. Scientia Horticulturae **81**: 179-188.

Modgil, M., Sharma, T. and Thakur, M. 2009. Commercially feasible protocol for rooting and of micropropagated apple rootstocks. Acta Horticulturae **839**: 209-214.

Murashige, T. and Skoog, F. 1962. A revised medium for rapid growth and bioassays with tobacco cultures. Physiol. Plant. **15**: 473-479.

Nomura, K., Matsumoto, S., Masuda, K. and Inoue, M. 1998. Reduced glutathione promotes callus growth and shoot development in a shoot tip culture of apple rootstock M26. Plant Cell Reports **17**: 597-600.

Simmonds, J. 1983. *In vitro* propagation of Malling Merton apple rootstocks. Hort Science **15**: 597-598.

Soni, M., Thakur, M. and Modgil, M. 2011. *In vitro* multiplication of Merton793 - An apple rootstock suitable for replantation. Indian Journal of Biotechnology **10**: 362-368.

Welander, M. 1991. Micropropagation of the apple rootstock M26 – the Swedish results from an European Co-operation project, COST-87. Rapport – Institutionen – for – Tradgardsvetenskap – Sveriges – Lantbruks Universitet **58**: 25.

Zimmerman, R.H. and Broome, O.L. 1981. Phloroglucinol and *in vitro* rooting of apple cultivar cuttings. Journal of American Society for Horticulture Science **106**: 648-652.

Index